MATHEMATICS RESEARCH DEVELOPMENTS

MATHEMATICAL MODELLING WITH APPLICATIONS IN BIOSCIENCES AND ENGINEERING

MATHEMATICS RESEARCH DEVELOPMENTS

Additional books in this series can be found on Nova's website
under the Series tab.

Additional E-books in this series can be found on Nova's website
under the E-books tab.

ENGINEERING TOOLS, TECHNIQUES
AND TABLES

Additional books in this series can be found on Nova's website
under the Series tab.

Additional E-books in this series can be found on Nova's website
under the E-books tab.

MATHEMATICS RESEARCH DEVELOPMENTS

MATHEMATICAL MODELLING WITH APPLICATIONS IN BIOSCIENCES AND ENGINEERING

IOANNIS ARGYOS,
SAID HILOUT
AND
MOHAMMAD A. TABATABAI

Nova Science Publishers, Inc.
New York

Copyright © 2011 by Nova Science Publishers, Inc.

All rights reserved. No part of this book may be reproduced, stored in a retrieval system or transmitted in any form or by any means: electronic, electrostatic, magnetic, tape, mechanical photocopying, recording or otherwise without the written permission of the Publisher.

For permission to use material from this book please contact us:
Telephone 631-231-7269; Fax 631-231-8175
Web Site: http://www.novapublishers.com

NOTICE TO THE READER

The Publisher has taken reasonable care in the preparation of this book, but makes no expressed or implied warranty of any kind and assumes no responsibility for any errors or omissions. No liability is assumed for incidental or consequential damages in connection with or arising out of information contained in this book. The Publisher shall not be liable for any special, consequential, or exemplary damages resulting, in whole or in part, from the readers' use of, or reliance upon, this material. Any parts of this book based on government reports are so indicated and copyright is claimed for those parts to the extent applicable to compilations of such works.

Independent verification should be sought for any data, advice or recommendations contained in this book. In addition, no responsibility is assumed by the publisher for any injury and/or damage to persons or property arising from any methods, products, instructions, ideas or otherwise contained in this publication.

This publication is designed to provide accurate and authoritative information with regard to the subject matter covered herein. It is sold with the clear understanding that the Publisher is not engaged in rendering legal or any other professional services. If legal or any other expert assistance is required, the services of a competent person should be sought. FROM A DECLARATION OF PARTICIPANTS JOINTLY ADOPTED BY A COMMITTEE OF THE AMERICAN BAR ASSOCIATION AND A COMMITTEE OF PUBLISHERS.

Additional color graphics may be available in the e-book version of this book.

LIBRARY OF CONGRESS CATALOGING-IN-PUBLICATION DATA

Argyos, Ioannis.
  Mathematical modelling with applications in biosciences and engineering /
Ioannis Argyos and Mohammad A. Tabatabai, Said Hilout.
    p. cm.
  Includes bibliographical references and index.
  ISBN 978-1-61728-944-6 (hardcover)
  1. Nonlinear theories--Mathematical models. 2. Variational inequalities
(Mathematics) I. Tabatabai, Mohammad A. II. Hilout, Said. III. Title.
  QA427.A74 2010
  511'.8--dc22
                        2010031196

Published by Nova Science Publishers, Inc. ✦ *New York*

Contents

Preface		**vii**
PART I: EFFICIENT ITERATIVE PROCESSES		**1**
Chapter 1	Two–Step Methods of Hight Efficiency Index	**3**
Chapter 2	Numerical Methods under Slantly–Differentiability	**37**
Chapter 3	$\mathcal{H}$-Convergence	**49**
Chapter 4	Chebyshev–Secant–Type Methods	**61**
Chapter 5	Convergence Using Recurrent Functions	**73**
Chapter 6	ω–Convergence	**99**
Chapter 7	Gauss–Newton Method	**109**
Chapter 8	Convergence on K–Normed Spaces	**123**
Chapter 9	Newton's Method on Spaces with Convergence Structure	**137**
PART II. MATHEMATICAL MODELLING		**153**
Chapter 10	Newton's Method and Interior Point Techniques	**155**
Chapter 11	Finite Element Methods	**159**
Chapter 12	Convergence in Riemannian Manifolds	**165**
Chapter 13	Convergence on Lie Groups	**171**
Chapter 14	The Shadowing Lemma for Operators with Chaotic Behaviour	**181**
Chapter 15	Conditioning of Semidefinite Programs	**187**
Chapter 16	Optimal Shape Design Problems	**195**
Chapter 17	Variational Inequalities	**205**

Chapter 18	Directional Secant–Type Methods	**221**
Chapter 19	Directional Newton–Type Methods	**233**
Chapter 20	Directional Two–Step Methods	**245**
Chapter 21	τ–estimation for Nonlinear Regression Models	**265**
Chapter 22	Tabaistic Regression	**277**
Chapter 23	Hyperbolastic Growth Models and Applications	**283**
References		**291**
Glossary of Symbols		**341**
Index		**343**

The first author dedicates this book to his wife Diana, and his children Christopher, Gus, and Michael

The second author dedicates it in memory of his father Mohammed, and to Nouria, Lina and Nassim

and

The third author dedicates it to his wife Yeganeh, and his son Ali

Preface

Part 1: Efficient Iterative Processes

The field of computational sciences has seen a considerable development in mathematics, engineering sciences, and economic equilibrium theory. Researchers in this field are faced with the problem of solving a variety of equations or variational inequalities. We note that in computational sciences, the practice of numerical analysis for finding such solutions is essentially connected to variants of Newton's method. The efficient and iterative methods for finding the solutions of nonlinear equations or variational inclusions are the first goal of the present book. The second goal is the study of applications of these methods in engineering and biosciences problems whose formulations are nonlinear equations or variational inequalities. For example, dynamic systems are mathematically modelled by difference or differential equations, and their solutions represent usually the states of the systems. For the sake of simplicity, assume that a time–invariant systems is driven by the equation $x' = F(x)$, where x is the state, then the equilibrium states are determined by solving the equations $F(x) = 0$. Similar equations are used in the case of discrete systems. The unknowns of engineering equations can be:

* Functions (difference, differential, integral equations);

* Vectors (systems of linear or nonlinear algebraic equations);

* Real or complex numbers (single algebraic equations with single unknowns).

Except special cases, the most commonly used solutions methods are iterative, when starting from one or several initial approximations a sequences is constructed, which converges to a solution of the equation. Iteration methods are applied also for solving optimization problems. In such cases the iteration sequences converge to an optimal solution of the problem in hand. since all of these methods have the same recursive structure, they can be introduced and discussed in a general framework.

On the other hand, it is well–known that, a wide variety of variational problems including:

* Linear and nonlinear complementarity problems;

* Variational inequalities (for example first–order necessary conditions for nonlinear programming)

can be formulated in the form of generalized equation $0 \in F(x) + G(x)$, where F is a function, and G is a multifunction. In particular, generalized equations may characterize: Optimality or equilibrium (traffic network equilibrium, spatial price equilibrium problems, migration equilibrium problems, environmental network problems, $\cdots$) and then have several applications in engineering and economics (analysis of elastoplastic structures, Walrasian equilibrium, Nash equilibrium, financial equilibrium problems, $\cdots$).

To complicate the matter further, many of equations or generalized equations are nonlinear (i.e., F is nonlinear). However, all may be formulated in terms of operators mapping a linear space into another, the solutions being sought as points in the corresponding space. Consequently, computational methods that work in this general setting for the solution of equations apply to a large number of problems, and lead directly to the development of suitable computer programs to obtain accurate approximate solutions to equations in the appropriate space.

This monograph is intended for researchers in computational sciences, and as a reference book for an advanced numerical–functional analysis or computer science course. The goal is to introduce these powerful concepts and techniques at the earliest possible stage. The reader is assumed to have had basic courses in numerical analysis, computer programming, computational linear algebra, and an introduction to real, complex, and functional analysis.

We have divided the material into two parts. Although the monograph is of a theoretical nature, with optimization and weakening of existing hypotheses considerations each chapter contains several new theoretical results and important applications in engineering, in dynamic economic systems, in input–output systems, in the solution of nonlinear and linear differential equations, and optimization problems. The applications appear in the form of Examples or Applications or study cases or they are implied since our results improve earlier one that have already been applied in concrete problems. Sections have been written as independent of each other as possible. Hence the interested reader can go directly to a certain section and understand the material without having to go back and forth in the whole textbook to find related material.

There are four basic problems connected with iterative methods.

Problem 1. Show that the iterates are well defined. For Example, if the algorithm requires the evaluation of F at each x_n, it has to be guaranteed that the iterates remain in the domain of F. It is, in general, impossible to find the exact set of all initial data for which a given process is well defined, and we restrict ourselves to giving conditions which guarantee that an iteration sequence is well defined for certain specific initial guesses.

Problem 2. Concerns the convergence of the sequences generated by a process and the question of whether their limit points are, in fact, solutions of the equation. There are several types of such convergence results. The first, which we call a local convergence theorem, begins with the assumption that a particular solution $x^\star$ exists, and then asserts that there is a neighborhood U of $x^\star$ such that for all initial vectors in U the iterates generated by the process are well defined and converge to $x^\star$. The second type of convergence theorem, which we call semilocal, does not require knowledge of the existence of a solution, but states that, starting from initial vectors for which certain–usually stringent–conditions are satisfied, convergence to some (generally nearby) solutions $x^\star$ is guaranteed. Moreover, theorems of this type usually include computable (at least in principle) estimates for the error $x_n - x^\star$, a possibility not afforded by the local convergence theorems. Finally, the third and most elegant type of convergence result, the global theorem, asserts that starting anywhere in a linear space, or at least in a large part of it, convergence to a solution is assured.

Problem 3. Concerns the economy of the entire operations, and, in particular, the question of how fast a given sequence will converge. Here, there are two approaches, which correspond to the local and semilocal convergence theorems. As mentioned above, the analysis which leads to the semilocal type of theorem frequently produces error estimates, and these, in turn, may sometimes be reinterpreted as estimates of the rate of convergence of the sequence. Unfortunately, however, these are usually overly pessimistic. The second approach deals with the behavior of the sequence $\{x_n\}$ when n is large, and hence when x_n is near the solutions $x^\star$. This behavior may then be determined, to a first approximation, by the properties of the iteration function near $x^\star$ and leads to so–called asymptotic rates of convergence.

Problem 4. Concerns with how to best choose a method, algorithm, or software program to solve a specific type of problem and its descriptions of when a given algorithm or method succeeds or fails.

We have included a variety of new results dealing with Problems 1–4. Results by other authors (or us) are either explicitly stated or implicitly when compared with the corresponding ones of ours results.

This monograph is an outgrowth of research work undertaken by us and complements/ updates earlier works of ours focusing on in depth treatment of convergence theory for iterative methods [155]–[171], [176], [178], [181]–[233], [238]–[244]. Such a comprehensive study of optimal iterative procedures appears to be needed and should benefit not only those working in the field but also those interest in, or in need of, information about specific results or techniques. We have endeavored to make the main text as self contained as possible, to prove all results in full detail. In order to make the study useful as a reference source, we have complemented each section with a set of "Remarks" in which literature citations are given, other related results are discussed, and various possible extensions of the results of the text are indicated. For completion, the monograph ends with a comprehensive list of references and an appendix containing useful contemporary numerical algorithms. Because we believe our readers come from diverse backgrounds and have varied interests, we provide "recommended reading" throughout the textbook. Often a long monograph summarizes knowledge in a field. This monograph, however, may be viewed as a report on work in progress. We provide a foundation for a scientific field that is rapidly changing. Therefore we list numerous conjectures and open problems as well as alternative models which need to be explored.

The book is divided into two main parts: the first part presents efficient iterative methods for nonlinear equations and generalized equations (chapters 1 to 9), and the second part develops several applications of the methods studied in part 1 in biosciences and engineering (chapters 10 to 23).

Chapter 1 is devoted to a thorough study of our new idea of recurrent functions in order to provide a new semilocal convergence analysis for two–step Newton–type methods of high efficiency index. It turns out that our sufficient convergence conditions are weaker, and the error bounds are tighter than in earlier studies in many interesting cases. Throughout this chapter, we present an illustrative examples involving a nonlinear integral equation of Chandrasekhar–type, and a differential equation containing a Green's kernel, to show that the different theorems are applicable in some situations in which the other are not applicable.

Chapter 2 concerns the local convergence of the Newton–like method to a unique solution of nondifferentiable variational inclusions in a Banach space setting. Using the Lipschitz–like property of set–valued mappings and the concept of slant differentiability hypothesis on the operator involved, as was inaugurated by Chen, Nashed and Qi. The linear convergence of the Newton–like method is also established. The results of this chapter extend the applicability of Newton–like method to variational inequalities. We conclude this chapter by a conjecture on the convergence of the smoothing and the semismoothing Newton methods in the case of generalized equations.

Chapter 3 deals with the semilocal convergence analysis for Newton–like methods using ω–type conditions, in order to approximate a locally unique solution of an equation in a Banach space. Using a combination of Lipschitz/center–Lipschitz conditions, instead of only Lipschitz conditions, and our new idea of recurrent functions, we provided an analysis with the following advantages over the works: weaker sufficient convergence conditions, tighter error bounds and larger convergence domain in some interesting cases. Numerical examples and applications further validating the results are also provided in this chapter.

Chapter 4 presents a new three–step Chebyshev–Sacant–type method of high efficiency index to solve nonlinear equation in a Banach space setting. This method uses the divided difference of order one. Note also that the method presented in this chapter only requires the evaluation of a new function at each step. We provide a semilocal convergence analysis using recurrence relations. Some special cases and numerical example validating our theoretical results are also provided in this chapter.

Chapter 5 deals with a unified approach for the convergence of certain numerical algorithms, using our new idea of recurrent functions. We use a combination of Lipschitz (resp. Hölder)/center–Lipschitz (resp. center–Hölder) conditions, instead of only Lipschitz (resp. Hölder) conditions, respectively. Under the same hypotheses, and computational cost, we show that a finer convergence analysis can be provided than the popular Kantorovich analysis based on recurrent relation's approach. Our new analysis generates sufficient conditions for the convergence of numerical algorithms to a solution of a nonlinear equations as well as provides the corresponding error estimates on the distances involved. Numerical examples and comparison tables are also provided.

Chapter 6 exhibits our new idea of recurrent functions, and Zabrejko–Zincenko–type conditions to provided a semilocal convergence analysis for Newton–type methods in order to approximate a locally unique solution of an equation in a Banach space. Our analysis has the following advantages over the earlier works: weaker sufficient convergence conditions, and larger convergence domain. Note that in the case of Newton's method, these advantages are obtained under the same computational cost, since in practice the computation of the Lipschitz constant requires the computation of the center–Lipschitz constant. As in previous chapters, numerical examples are provided of this chapter further validating the different results.

Chapter 7 presents an extension of the applicability of Gauss–Newton method by replacing existing conditions by weaker ones, in order to solve systems of equations with constant rank derivatives. The famous for its simplicity and clarity Newton–Kantorovich hypothesis of Newton's method has been used for a long time as the sufficient convergence condition for solving nonlinear equations. In this chapter, we use also our new idea of recurrent functions to provide weaker hypotheses and we present special cases and numerical examples to solve equations in cases not covered before.

Chapter 8 is motivated by optimization considerations, and enlargement of applications to Banach space ordered by a closed convex cone. We use our new idea of recurrent functions to provide a new semilocal convergence result for a Newton–type method for solving a nonlinear operator equation in a Banach space ordered by a closed convex cone. Using more precise majorizing sequences than before, we show how to expand the convergence domain of Newton–type method under the same computational cost as before. A numerical examples of this chapter shows how to solve an equation in cases not covered before.

Chapter 9 presents a semilocal convergence theorem for Newton's method on spaces with a convergence structure, using the same arguments and hypotheses as the previous chapter. A general structure for the convergence analysis of Newton's method, and theorems based on monotonicity considerations are presented in previous works. In this chapter, we use our new idea of recurrent functions to provide a tighter semilocal analysis of Newton's method with weaker conditions, and under the same computational cost as before. Other approaches in literature on the Kantorovich–type semilocal convergence, and special cases are also presented in this chapter.

Part 2: Mathematical Modelling

This part is on applications of the theory developed in the previous part to problems in biosciences and engineering. First, we provide in this part a weaker Newton–Kantorovich theorem for solving equations, to analyze interior point methods and to show existence of finite element solutions of strongly nonlinear elliptic boundary value problems, respectively. The second application concerns an approach that leads to a manageable way of determining bounds for the distance of a manifold to a nearby point. We provide also an application of the local/semilocal Kantorovich–type convergence for Newton's method to solve equation on Lie groups. We use also in this part a weaker version of the celebrated Newton–Kantorovich theorem reported by us in to find solutions of discrete dynamical systems involving operators with chaotic behavior. We conclude this part with other applications to semidefinite programs, optimal shape design problems, variational inequalities, directional methods and nonlinear regression models.

Chapter 10 focuses on mathematical programming problems. We provide in this chapter a bridge between Newton–Kantorovich–type theorem and linear programming/interior point methods. We start from the point where the assumptions made can be replaced by ours which are weaker.

Chapter 11 deals with the discretization result to find finite element solutions of elliptic boundary problems using Newton–Kantorovich–type theorem. We develop also a semilocal convergence analysis for directional Newton's method in finite dimensional spaces using a combination of Lipschitz and center–Lipschitz conditions.

Chapters 12 and 13 present a finer semilocal convergence analysis of Newton's method in Riemannian manifolds and on Lie groups, respectively. Using more precise majorizing sequences than before, we obtain in these chapter a larger convergence domain, a finer error bounds on the distances involved, and a more precise information on the location of the singularity of the vector field.

Chapter 14 is motivated by finding solutions of discrete dynamical systems involving operators with chaotic behavior. It is well known that complicated behaviour of dynamical systems can easily be detected via numerical experiments. However, it is very difficult to prove mathematically in general that a given system behaves chaotically. In this chapter, We use a weaker version of the Newton–Kantorovich theorem to find such solutions. Our results are obtained by extending the application of the shadowing lemma.

Chapter 15 is an application to the conditioning of semidefinite programs. We use the same arguments and hypotheses as the previous chapter, to show how to extend the results dealing with the analyzing of the effect of small perturbations in problem data on the solution in cases not covered before.

Another application to shape optimization is studied in chapter 16. This application is described by finding the geometry of a structure which is optimal in the sense of a minimized cost function with respect to certain constraints. A Newton's mesh independence principle was very efficiently used to solve a certain class of optimal design problems. In this chapter, we show that under the same computational cost an even finer mesh independence principle can be given than the previous works.

Chapter 17 presents a development of Newton–Josephy method for variational inequalities in finite dimensional spaces. By using a combination of Lipschitz and center–Lipschitz conditions, and our new idea of recurrent functions, we provide an analysis with some advantages over the earlier works. We provide also in this chapter a numerical example and a comparison table, where we compare our error bounds with the previous works on the subject.

Chapters 18, 19 and 20 are a continuation application of part 1 of this book. We present some directional methods in both finite–dimensional space and Hilbert space. Chapters 18 and 19 concern the Secant–type and Newton–type methods, respectively. Chapter 20 presents a new directional two–step method for approximating a zero of a differentiable function defined on a convex subset of Hilbert space. For these three directional methods, two different techniques are used to generate the sufficient convergence results, as well as the corresponding error bounds. The first technique uses our new idea of recurrent functions, whereas the second uses recurrent sequences. We also compare in these three chapters the different results, and present some special cases and examples.

Chapter 21 discusses the τ–robust method for nonlinear regression as a highly efficient method with high breakdown point. Usage of such a powerful method would enhance the accuracy of parameter estimates and hypothesis testing. The least–square method is very sensitive to the presence of outliers and leverages. This sensitivity can jeopardize the accuracy of the estimated model parameters and their corresponding hypothesis testing. The usage of τ–robust method for nonlinear regression would help to rectify such problems.

Chapter 22 deals with the tabaistic family of probability distributions. The members of this family can have either unimodal or bimodal probability density function. This family can be applied to skewed or bimodal data. A major application of the unimodal member of this family is in the analysis of categorical data when covariates are present. The logistic regression and probit analysis are widely used when the Underlying distribution is symmetric. When the distribution is asymmetric, the tabaistic regression will be a better choice.

In chapter 23, an attempt is made to explain the hyperbolastic growth models and discuss their usefulness as growth models of choice for very many biomedical and engineering problems. These models in the past have shown very good results when applied to a wide range of problems including tumor growth, stem cell growth, animal or plant growth. It also can be used to analyze the time to both wound and fractured bone healings.

Part I

EFFICIENT ITERATIVE PROCESSES

Chapter 1

Two–Step Methods of Hight Efficiency Index

In this chapter, we introduce our new idea of recurrent functions to provide a new semilocal convergence analysis for two–step Newton–type methods of high efficiency index. It turns out that our sufficient convergence conditions are weaker, and the error bounds are tighter than in earlier studies in many interesting cases [157], [770]. Applications and numerical examples, involving a nonlinear integral equation of Chandrasekhar–type, and a differential equation containing a Green's kernel are also provided in this chapter.

We are concerned with the problem of approximating a locally unique solution $x^\star$ of equation

$$F(x) + G(x) = 0, \tag{1.1}$$

where F is a Fréchet–differentiable operator defined on a convex subset $\mathcal{D}$ of a Banach space X with values in a Banach space $\mathcal{Y}$, and $G : \mathcal{D} \longrightarrow \mathcal{Y}$ is a continuous operator.

Isaac Newton in 1669, inaugurated his method through the use of numerical examples to solve equations, but did not use the current iterative expression. Later, in 1690, Raphson introduced Newton's method or the also called Newton–Raphson method.

Newton's method is currently and undoubtely the most popular one–point iterative procedure for generating a sequence approximating $x^\star$. Results on local as well as semilocal convergence of Newton–type methods can be found in [151], and the references there (see, also [91], [156], [157], [770]).

One factor that is taken into account, when using one–point iterative methods is the efficiency index: $EP = p^{\frac{1}{q}}$, where p is the order of convergence of the method, and q is the number of new function information required at each step.

Recently, in the elegant study by Ezquerro, and Hernández [377] from Chebyshev's method [151], new third order multipoint iterations are constructed with efficiency index close to Newton's method, and the same region of accessibility.

Motivated by optimization considerations, and the works [136], [156], where modified Newton's method is mixed with Newton's method in order to expand the applicability of the latter, we introduce the two–step Newton–type method (TSNTM):

$$x_0 \in \mathcal{D},$$
$$y_n = x_n - A_n^{-1} \left(F(x_n) + G(x_n) \right),$$
$$z_n = x_n + \alpha \, (y_n - x_n),$$
$$x_{n+1} = x_n - A_n^{-1} \left(\beta \, (F(x_n) + G(x_n)) + \gamma \, (F(z_n) + G(z_n)) \right), \quad (n \geq 0),$$

where, $A_n := A(x_n) \in L(X, \mathcal{Y})$ the space of bounded linear operators from X to $\mathcal{Y}$, and α, β, γ are numbers chosen that sequences $\{x_n\}$, $\{x_n\}$ converge to $x^\star$.

Many iterative methods are special cases of (TSNTM). For example, if $\gamma = 0$, $\alpha = \beta = 1$, and $A(x) = F'(x)$, $(x \in \mathcal{D})$, we obtain Zincenko's method [770]. Moreover, if $G(x) = 0$, $(x \in \mathcal{D})$, we obtain Newton's method, whereas if $A(x) = [x, g(x); F]$ (g is a continuous function and $[x, y; F]$ is a divided difference of order one), we obtain the Secant method in the case $g(x) = x^+$, x^+ is the next iterate. Several other choices are also possible [91]–[770]. In particular, for

$$A(x) = F'(x), \qquad x \in \mathcal{D},$$
$$\alpha \in [0, 1], \quad \beta = \frac{\alpha^2 + \alpha - 1}{\alpha^2}, \quad \gamma = \frac{1}{\alpha^2}, \tag{1.2}$$

we obtain the method introduced in [377], denoted by (TSNM), as a special case of (TSNTM). This method was shown to be of order 3, with efficiency index $\sqrt[3]{3}$. That is the efficiency index of this method is between Newton's method $\sqrt{2}$, and Secant method $\frac{1 + \sqrt{5}}{2}$. Then, it is suggested: we can approximate $F'(x_n)$ at each step by a divided difference exactly as we do in Newton's method to obtain the Secant method. This way we save one computation, since only the evaluation of a new function is needed at each step. Instead of doing just that, we provide a semilocal convergence analysis for the more general (TSTNM), using our new idea of recurrent functions. A favorable for us comparison between (TSNTM) and Newton–type method (NTM) [377] is given in Remark 1.1.

The chapter is organized as follows: the semilocal convergence of (TSNTM) is given for $\gamma \neq 0$, and $\gamma = 0$, respectively.

Let $\alpha \geq 0$, $\gamma \geq 0$, $\mu \geq 0$, $K \geq 0$, $L \geq 0$, $M \geq 0$, $N \geq 0$, $\eta \geq 0$, and $\ell \in [0, 1)$ be given constants.

Set

$$b = \mu + N, \qquad c = 1 + \alpha \, \gamma. \tag{1.3}$$

It is convient for us to define scalar sequences $\{t_n\}$, $\{s_n\}$, $\{f_n\}$, $\{f_n^1\}$, by

$$t_0 = 0, \quad s_0 = \eta,$$

$$t_{n+1} = s_n + \frac{\alpha \, \gamma}{1 - \ell - L \, t_n} \left(\frac{\alpha \, K}{2} \, (s_n - t_n) + M \, t_n + b \right) (s_n - t_n), \tag{1.4}$$

$$t_1 = s_0 \left\{ 1 + \alpha \, \gamma \left(\frac{\alpha \, K \, s_0}{2} + b \right) \right\},$$

$$s_{n+1} = t_{n+1} + \frac{1}{1 - \ell - L \, t_{n+1}} \left(\frac{K}{2} \, (t_{n+1} - t_n)^2 + (M \, t_n + b) \, (t_{n+1} - t_n) + \frac{\alpha^2 \, \gamma \, K}{2} \, (s_n - t_n)^2 + \alpha \, \gamma \, (M \, t_n + b) \, (s_n - t_n) \right), \tag{1.5}$$

$$\begin{aligned}
f_n(w) &= \frac{\alpha^2\,\gamma\,K}{2}\,w^n\,\eta + \alpha\,\gamma\,M\,(1+2\,w\,(1+w+\cdots+w^{n-2})+ \\
&\quad w^n)\,\eta + L\,w\,(1+2\,w\,(1+w+\cdots+w^{n-2})+w^n)\,\eta \\
&\quad -(1-\ell)\,w+\alpha\,\gamma\,b,
\end{aligned} \tag{1.6}$$

$$\begin{aligned}
f_n^1(w) &= \frac{K}{2}\,(1+w)^2\,w^n\,\eta + (c+w)\,(M\,(1+2\,w\,(1+w+\cdots+ \\
&\quad w^{n-2})\,\eta+b) + \frac{\alpha^2\,\gamma\,K}{2}\,w^n\,\eta + L\,M\,w\,(1+2\,w\,(1+w+ \\
&\quad \cdots+w^{n-1})+w^{n+1})\,\eta - (1-\ell)\,w,
\end{aligned} \tag{1.7}$$

functions f_∞, g, f_∞^1, and g^1 on $[0,+\infty)$ by

$$f_\infty(w) = (L\,\eta+1-\ell)\,w^2 - (\alpha\,\gamma\,b+1-\ell-\alpha\,\gamma\,M\,\eta)\,w+\alpha\,\gamma\,b, \tag{1.8}$$

$$g(w) = L\,w^2 + \left(L+\alpha\,\gamma\,M + \frac{\alpha^2\,\gamma\,K}{2}\right)\,w+\alpha\,\gamma\,M - \frac{\alpha^2\,\gamma\,K}{2}, \tag{1.9}$$

$$f_\infty^1(w) = ((\eta+L)\,M+1-b-\ell)\,w^2 + (c\,M\,\eta-c\,b+b+\ell-1)\,w+c\,b, \tag{1.10}$$

$$\begin{aligned}
g^1(w) &= \left(\frac{K}{2}+L\,M\right)\,w^3 + \left(\frac{K}{2}+M+L\,M\right)\,w^2+ \\
&\quad \left(c+M + \frac{\alpha^2\,\gamma\,K}{2} - \frac{K}{2}\right)\,w+c\,M - \frac{K}{2} - \frac{\alpha^2\,\gamma\,K}{2}.
\end{aligned} \tag{1.11}$$

Denote by w_2, w_∞, v, w_2^1, w_∞^1, v^1 the minimal nonnegative zeros of f_2, f_∞, g, f_2^1, f_∞^1, and g^1, respectively (if they exist).

Set

$$\delta_1 = \alpha\,\gamma\,\left(\frac{\alpha\,K}{2}+b\right), \tag{1.12}$$

$$\delta_2 = \frac{1}{(1-\ell-L\,t_1)\,\eta}\,\left(\frac{K}{2}\,t_1^2+b\,t_1 + \frac{\alpha^2\,\gamma\,K}{2}\,\eta^2+\alpha\,\gamma\,b\,\eta\right), \quad \eta\neq 0, \tag{1.13}$$

$$\delta_0 = \max\,\{\delta_1,\delta_2\} \tag{1.14}$$

and

$$w_0 = \max\,\{w_\infty,\,w_\infty^1\}. \tag{1.15}$$

Hypotheses of Lemma 1.1 that follows have been left as uncluttered as possible. Note however that the verification of these hypotheses involve only computations at the initial point x_0. Stronger, but easier to verify conditions can be considered replacing all hypotheses of Lemma 1.1, except (1.17). A set of such conditions is given by:

(C_0)

$$2\,M \leq \alpha\,K, \qquad \alpha \in [0,1]$$

$$(1-\ell)\,v-\alpha\,\gamma\,b>0,$$

$$(1-\ell)\,v^1 - (c+w)\,b > 0,$$

and

$$\eta < \eta_0 = \min\{\eta_1,\eta_2,\eta_3 : \eta_1 > 0, f_2(\eta_1) = 0, \eta_2 > 0, f_2^1(\eta_2) = 0, \eta_3 > 0, L\,t_1 + \ell = 1\}.$$

Indeed, the first condition, and the intermediate value theorem (IVT) applied to functions g, g^1 defined on $[0, w]$ for sufficiently large $w > 0$, guarantee the existence of zeros v, and v^1, respectively.

The second, third condition together with (IVT), and the choices of η, η_1, η_2 guarantee the existence of w_2, w_2^1, so that (1.18), and (1.19) are satisfied. Moreover, the choice of η, and η_3 show that (1.16) is also satisfied. Hence, (1.17) together with the set of conditions $(\mathcal{C}_0)$ can certainly replace the hypotheses of Lemma 1.1.

We can show the following result on majorizing sequences for majorizing sequences for (TSNTM).

Lemma 1.1. *Assume:*

there exist minimal nonnegative zeros w_2, w_∞, v, w_2^1, w_∞^1, v^1 *of functions* f_2, f_∞, g, f_2^1, f_∞^1, *and* g^1, *respectively;*

$$L t_1 + \ell < 1, \tag{1.16}$$

$$\delta_0 \leq w_0 \leq 1, \tag{1.17}$$

$$w_2 \leq v, \tag{1.18}$$

and

$$w_2^1 \leq v^1. \tag{1.19}$$

Choose δ, *such that:*

$$\delta = \delta^\star = 2 \max\{w_2, w_2^1\}. \tag{1.20}$$

Then, scalar sequences $\{t_n\}$, $\{s_n\}$ $(n \geq 0)$ *given by* (1.4), *and* (1.5) *are increasing, bounded above by*

$$t^{\star\star} = \frac{2+\delta}{2-\delta} \, \eta, \tag{1.21}$$

and converge to their common, unique least upper bound $t^\star$ *satisfying*

$$0 \leq t^\star \leq t^{\star\star}. \tag{1.22}$$

Moreover the following estimates hold for all $n \geq 0$:

$$0 \leq t_{n+1} - s_n \leq \frac{\delta}{2} \, (s_n - t_n) \leq \left(\frac{\delta}{2}\right)^{n+1} \eta, \tag{1.23}$$

$$0 \leq s_{n+1} - t_{n+1} \leq \frac{\delta}{2} \, (s_n - t_n) \leq \left(\frac{\delta}{2}\right)^{n+1} \eta, \tag{1.24}$$

$$0 \leq t^\star - t_n \leq \frac{2+\delta}{2-\delta} \left(\frac{\delta}{2}\right)^{n} \eta, \tag{1.25}$$

and

$$0 \leq t^\star - s_n \leq \frac{3}{2-\delta} \left(\frac{\delta}{2}\right)^{n+1} \eta. \tag{1.26}$$

Chapter 1. Two–Step Methods of Hight Efficiency Index

Proof. We shall show using induction on n, that estimates (1.23), and (1.24) hold. These estimates hold (for $n = 0$), by (1.4), (1.5), (1.12), (1.13), (1.17)–(1.20).

Let us assume that estimates (1.23), and (1.24) hold for all $k \leq n$. Then, we have in turn the estimates:

$$
\begin{aligned}
s_{k+1} &\leq t_{k+1} + \frac{\delta}{2}(s_k - t_k) \\
&\leq s_k + \frac{\delta}{2}(s_k - t_k) + \frac{\delta}{2}(s_k - t_k) \\
&\leq s_k + 2\left(\frac{\delta}{2}\right)^{k+1}\eta \\
&\leq s_{k-1} + 2\left(\frac{\delta}{2}\right)^{k}\eta + 2\left(\frac{\delta}{2}\right)^{k+1}\eta \\
&\leq s_0 + 2\left\{\frac{\delta}{2} + \cdots + \left(\frac{\delta}{2}\right)^{k+1}\right\}\eta \\
&= \eta + 2\frac{\delta}{2}\left\{1 + \cdots + \left(\frac{\delta}{2}\right)^{k}\right\}\eta \\
&= \left\{1 + \frac{1 - \left(\frac{\delta}{2}\right)^{k+1}}{1 - \frac{\delta}{2}}\,\delta\right\}\eta \\
&\leq t^{\star\star} \qquad \text{by (1.21)},
\end{aligned}
\tag{1.27}
$$

and

$$
\begin{aligned}
t_{k+1} &\leq s_k + \frac{\delta}{2}(s_k - t_k) \\
&\leq \left\{1 + \frac{1 - \left(\frac{\delta}{2}\right)^{k}}{1 - \frac{\delta}{2}}\,\delta + \left(\frac{\delta}{2}\right)^{k+1}\right\}\eta \leq t^{\star\star} \qquad \text{by (1.21)}.
\end{aligned}
\tag{1.28}
$$

Estimate (1.23) certainly holds, if

$$
t_{k+1} - s_k \leq \frac{\delta}{2}(s_k - t_k)
$$

or

$$
\frac{\alpha^2\,\gamma\,K}{2}(s_k - t_k) + \alpha\,\gamma\,(M\,t_k + b) \leq \frac{\delta}{2}(1 - \ell - L\,t_k),
\tag{1.29}
$$

and

$$
L\,t_k + \ell < 1.
\tag{1.30}
$$

Estimates (1.29), and (1.30) in turn hold, if

$$\frac{\alpha^2 \gamma K}{2} \left(\frac{\delta}{2}\right)^k \eta + \alpha \gamma M \left\{ 1 + \frac{1 - \left(\frac{\delta}{2}\right)^{k-1}}{1 - \frac{\delta}{2}} \delta + \left(\frac{\delta}{2}\right)^k \right\} \eta +$$

$$L \frac{\delta}{2} \left\{ 1 + \frac{1 - \left(\frac{\delta}{2}\right)^{k-1}}{1 - \frac{\delta}{2}} \delta + \left(\frac{\delta}{2}\right)^k \right\} \eta - \frac{\delta}{2} (1 - \ell) + \alpha \gamma b \leq 0. \tag{1.31}$$

Estimate (1.31) motivates us to define functions f_k given by (1.6) for $w = \frac{\delta}{2}$, and show instead of (1.31):

$$f_k(\delta^\star) \leq 0, \quad \delta^\star = w_2, \quad (k \geq 1). \tag{1.32}$$

By letting $k \longrightarrow \infty$ in (1.31), we get

$$\alpha \gamma M \left(1 + \frac{2w}{1 - w}\right) \eta + L w \left(1 + \frac{2w}{1 - w}\right) \eta - (1 - \ell) w + \alpha \gamma b = 0$$

or

$$f_\infty(w_\infty) = 0.$$

By hypothesis, we also have $f_2(w_2) = 0$.

We need to find a relationship between two consecutive f_k:

$$\begin{aligned} f_{k+1}(w) &= f_k(w) + \frac{\alpha^2 \gamma K}{2} w^{k+1} \eta - \frac{\alpha^2 \gamma K}{2} w^k \eta + \\ &\quad \alpha \gamma M (w^k + w^{k+1}) \eta + L w (w^k + w^{k+1}) \eta \\ &= f_k(w) + g(w) w^k \eta, \end{aligned} \tag{1.33}$$

where, function g is given by (1.9).

Using (1.33) for $k = 2$, we get:

$$f_3(w_2) = f_2(w_2) + g(w_2) w_2^2 \eta = g(w_2) w_2^2 \eta \leq 0, \tag{1.34}$$

since $f_2(w_2) = 0$, and $g(w_2) \leq 0$ (by (1.18)).

We also have:

$$f_3(0) = \alpha \gamma (M \eta + b) \geq 0. \tag{1.35}$$

It follows from (1.34), (1.35), and the intermediate value theorem that there exists $w_3 \in [0, w_2]$, such that $f_3(w_3) = 0$. Denote the minimal zero of function f_3 in $[0, w_2]$ by the same symbol w_3.

Let us assume: there exists minimal $w_k \in [0, w_{k-1}]$, with $f_k(w_k) = 0$. As in (1.34), we get:

$$\begin{aligned} f_{k+1}(0) &= \alpha \gamma (M \eta + b) \geq 0, \\ f_{k+1}(w_k) &= f_k(w_k) + g(w_k) w_k^k \eta \leq 0. \end{aligned} \tag{1.36}$$

since $f_k(w_k) = 0$, and $g(w_k) \leq 0$ (by (1.18)).

Chapter 1. Two–Step Methods of Hight Efficiency Index

Hence, again we deduce that there exists minimal $w_{k+1} \in [0, w_k]$, such that $f_{k+1}(w_{k+1}) = 0$.

Sequence $\{w_k\}$ is non–increasing, bounded below by zero, and as such it converges to its unique maximum lowest bound $w^{\star\star}$ satisfying $w^{\star\star} \geq w_\infty$. It then follows by (1.20), that (1.32) holds.

Using the induction hypotheses, estimate (1.24) shall hold, if

$$0 \leq s_{n+1} - t_{n+1} \leq \frac{\delta}{2}(s_n - t_n) \tag{1.37}$$

and

$$L\, t_{n+1} + \ell < 1. \tag{1.38}$$

These estimates hold for $n = 0$ by the initial conditions.
Estimates (1.37), and (1.38) shall also hold, if

$$\frac{1}{1 - \ell - L\, t_{k+1}} \left\{ \frac{K}{2} \left((t_{k+1} - s_k) + (s_k - t_k) \right)^2 + \right.$$
$$(M\, t_k + b)\left((t_{k+1} - s_k) + (s_k - t_k) \right) + \frac{\alpha^2\, \gamma\, K}{2}(s_k - t_k)^2 + \tag{1.39}$$
$$\left. \alpha\, \gamma\, (M\, t_k + \mu)\, (s_k - t_k) + N\, \alpha\, \gamma\, (s_k - t_k) \right\} \leq \frac{\delta}{2}(s_k - t_k)$$

or

$$\frac{1}{1 - \ell - L\, t_{k+1}} \left\{ \frac{K}{2}\left(1 + \frac{\delta}{2}\right)^2 (s_k - t_k)^2 + (M\, t_k + b)\left(1 + \frac{\delta}{2}\right)(s_k - t_k) + \right.$$
$$\left. \frac{\alpha^2\, \gamma\, K}{2}(s_k - t_k)^2 + \alpha\, \gamma\, (M\, t_k + \mu)\, (s_k - t_k) + N\, \alpha\, \gamma\, (s_k - t_k) \right\} \leq \frac{\delta}{2}(s_k - t_k)$$

or

$$\frac{K}{2}\left(1 + \frac{\delta}{2}\right)^2 (s_k - t_k) + (M\, t_k + b)\left(1 + \frac{\delta}{2}\right) + \frac{\alpha^2\, \gamma\, K}{2}(s_k - t_k) +$$
$$\alpha\, \gamma\, (M\, t_k + \mu) + N\, \alpha\, \gamma + L\, \frac{\delta}{2}\, t_{k+1} - (1 - \ell)\, \frac{\delta}{2} \leq 0,$$

or

$$\frac{K}{2}\left(1 + \frac{\delta}{2}\right)^2 \left(\frac{\delta}{2}\right)^k \eta + \left(c + \frac{\delta}{2}\right)\left\{ M\left(1 + \frac{1 - \left(\frac{\delta}{2}\right)^{k-1}}{1 - \frac{\delta}{2}}\, \delta + \right.\right.$$
$$\left.\left(\frac{\delta}{2}\right)^k\right) \eta + b \right\} + \frac{\alpha^2\, \gamma\, K}{2}\left(\frac{\delta}{2}\right)^k \eta + \tag{1.40}$$
$$L\, \frac{\delta}{2}\, M \left\{ 1 + \frac{1 - \left(\frac{\delta}{2}\right)^k}{1 - \frac{\delta}{2}}\, \delta + \left(\frac{\delta}{2}\right)^{k+1} \right\} \eta - (1 - \ell)\, \frac{\delta}{2} \leq 0.$$

Let, again $w = \dfrac{\delta}{2}$, and consider functions f_k^1 given by (1.7). Then, (1.40) shall hold, if

$$f_k^1(\delta^\star) \leq 0, \quad \delta^\star = w_2^1 \qquad (k \geq 1). \tag{1.41}$$

By letting $k \longrightarrow \infty$ in (1.40), we get:

$$(c+w)\left(b + \frac{M\,w\,\eta}{1-w}\right) + \frac{L\,M\,w^2\,\eta}{1-w} - (1-\ell)\,w = 0 \tag{1.42}$$

holds if $f_\infty^1(w_\infty^1) = 0$, which is true by hypothesis.

We need to find a relationship between two consecutive f_k^1:

$$
\begin{aligned}
f_{k+1}^1(w) &= f_k^1(w) + \frac{K}{2}\,(1+w)^2\,w^{k+1}\,\eta - \frac{K}{2}\,(1+w)^2\,w^k\,\eta + \\
&\quad (c+w)\,M\,(w^k + w^{k+1})\,\eta + \frac{\alpha^2\,\gamma\,K}{2}\,(w^{k+1} - w^k)\,\eta + \\
&\quad L\,M\,w\,(w^{k+1} + w^{k+2})\,\eta \\
&= f_k^1(w) + g^1(w)\,w^k\,\eta,
\end{aligned}
\tag{1.43}
$$

where, function g^1 is given by (1.11).

Using (1.43) for $k = 2$, we get:

$$f_3(w_2^1) = f_2^1(w_2^1) + g^1(w_2^1)\,(w_2^1)^2\,\eta = g^1(w_2^1)\,(w_2^1)^2\,\eta \leq 0, \tag{1.44}$$

since $f_2^1(w_2^1) = 0$, and $g^1(w_2^1) \leq 0$ (by (1.19)).

We also have:

$$f_3^1(0) = c\,(M\,\eta + b) \geq 0. \tag{1.45}$$

Hence, there exists $w_3^1 \in [0, w_2^1]$, such that $f_3^1(w_3^1) = 0$. Denote the minimal zero of function f_3^1 in $[0, w_2^1]$ by the same symbol w_3^1.

Let us assume: there exists minimal $w_k^1 \in [0, w_{k-1}^1]$, with $f_k^1(w_k^1) = 0$. As in (1.44), we get:

$$
\begin{aligned}
f_{k+1}^1(0) &= c\,(M\,\eta + b) \geq 0, \\
f_{k+1}^1(w_k^1) &= f_k^1(w_k^1) + g^1(w_k^1)\,(w_k^1)^k\,\eta \leq 0,
\end{aligned}
\tag{1.46}
$$

which imply the existence of $w_{k+1}^1 \in [0, w_k^1]$, such that $f_{k+1}^1(w_{k+1}^1) = 0$.

Sequence $\{w_k^1\}$ is non–increasing, bounded below by zero, and as such it converges to its unique maximum lowest bound $w_{\star\star}$ satisfying $w_{\star\star} \geq w_\infty^1$. It then follows from (1.20), that (1.41) holds.

The induction for (1.24) is completed. It then follows from (1.23), and (1.24) that sequences $\{t_n\}$, $\{s_n\}$ are nondecreasing, bounded above by $t^{\star\star}$, with $t_n \leq s_n \leq t_{n+1} \leq s_{n+1} \leq t^{\star\star}$, and as such they converge to their common, unique least upper bound $t^\star \in [0, t^{\star\star}]$.

Chapter 1. Two–Step Methods of Hight Efficiency Index

We also have for $m \geq 2$:

$$
\begin{aligned}
& s_{n+m} - t_n \\
={}& (s_{n+m} - t_{n+m}) + (t_{n+m} - t_n) \\
={}& (s_{n+m} - t_{n+m}) + (t_{n+m} - s_{n+m-1}) + (s_{n+m-1} - t_n) \\
\leq{}& \left(\frac{\delta}{2}\right)^{n+m} \eta + \left(\frac{\delta}{2}\right)^{n+m} \eta + \left(\frac{\delta}{2}\right)^{n+m-1} \eta + \left(\frac{\delta}{2}\right)^{n+m-1} \eta + \cdots + \\
& \left(\frac{\delta}{2}\right)^{n+1} \eta + \left(\frac{\delta}{2}\right)^{n+1} \eta + \left(\frac{\delta}{2}\right)^{n} \eta \\
={}& 2 \left(\frac{\delta}{2}\right)^{n+1} \eta \left\{ 1 + \frac{\delta}{2} + \cdots + \left(\frac{\delta}{2}\right)^{m-2} \right\} + \left(\frac{\delta}{2}\right)^{n} \eta \\
={}& 2 \eta \left(\frac{\delta}{2}\right)^{n+1} \frac{1 - \left(\frac{\delta}{2}\right)^{m-1}}{1 - \frac{\delta}{2}} + \left(\frac{\delta}{2}\right)^{n} \eta.
\end{aligned}
\tag{1.47}
$$

By letting $m \longrightarrow \infty$ in (1.47), we obtain (1.25).
We also have:

$$
\begin{aligned}
t_{n+m} - s_n ={}& (t_{n+m} - s_{n+m-1}) + (s_{n+m-1} - t_{n+m-1}) + (t_{n+m-1} - s_n) \\
\leq{}& \left(\frac{\delta}{2}\right)^{n+m} \eta + \left(\frac{\delta}{2}\right)^{n+m-1} \eta + \left(\frac{\delta}{2}\right)^{n+m-1} \eta + \\
& \left(\frac{\delta}{2}\right)^{n+m-2} \eta + \left(\frac{\delta}{2}\right)^{n+m-2} \eta + \cdots + \\
& \left(\frac{\delta}{2}\right)^{n+1} \eta + \left(\frac{\delta}{2}\right)^{n+1} \eta \\
={}& \left(\frac{\delta}{2}\right)^{n+2} \eta \left\{ 1 + \frac{\delta}{2} + \cdots + \left(\frac{\delta}{2}\right)^{m-2} \right\} + \\
& \left(\frac{\delta}{2}\right)^{n+1} \eta \left\{ 1 + \frac{\delta}{2} + \cdots + \left(\frac{\delta}{2}\right)^{m-2} \right\} + \left(\frac{\delta}{2}\right)^{n+1} \eta \\
={}& \left(\frac{\delta}{2}\right)^{n+2} \eta \frac{1 - \left(\frac{\delta}{2}\right)^{m-1}}{1 - \frac{\delta}{2}} + \left(\frac{\delta}{2}\right)^{n+1} \eta \frac{1 - \left(\frac{\delta}{2}\right)^{m-2}}{1 - \frac{\delta}{2}} + \\
& \left(\frac{\delta}{2}\right)^{n+1} \eta.
\end{aligned}
\tag{1.48}
$$

By letting $m \longrightarrow \infty$ in (1.47), we obtain (1.26).
That completes the proof of Lemma 1.1. $\qquad\qquad \diamond$

We also need a result relating the distances involved in (TSNTM).

Lemma 1.2. *If sequences $\{x_n\}$, $\{y_n\}$ are well defined for all $n \geq 0$, and*

$$
(1 - \alpha)\, \gamma = 1 - \beta \quad (\gamma \neq 0), \quad \text{and some} \quad \beta \geq 0,
\tag{1.49}
$$

then, the following hold for all $n \geq 0$:

$$
\begin{aligned}
x_{n+1} - y_n = & -\gamma A_n^{-1} \Big\{ \alpha \int_0^1 (F'(x_n + \alpha t (y_n - x_n)) - \\
& F'(x_n)) (y_n - x_n) \, dt + \alpha (F'(x_n) - A_n) (y_n - x_n) + \\
& G(z_n) - G(x_n) \Big\},
\end{aligned}
\tag{1.50}
$$

and

$$
y_{n+1} - x_{n+1} = -A_{n+1}^{-1} B_{n+1},
\tag{1.51}
$$

where,

$$
\begin{aligned}
B_{n+1} = & \; F(x_{n+1}) + G(x_{n+1}) \\
= & \int_0^1 (F'(x_n + t (x_{n+1} - x_n)) - F'(x_n)) (x_{n+1} - x_n) \, dt + \\
& (F'(x_n) - A_n) (x_{n+1} - x_n) + G(x_{n+1}) - G(x_n) - \\
& \gamma \Big\{ \alpha \int_0^1 (F'(x_n + \alpha t (y_n - x_n)) - F'(x_n)) (y_n - x_n) \, dt + \\
& \alpha (F'(x_n) - A_n) (y_n - x_n) + G(z_n) - G(x_n) \Big\}.
\end{aligned}
\tag{1.52}
$$

Proof. By eliminating x_n from the third equation in (TSNTM), we obtain in turn:

$$
\begin{aligned}
x_{n+1} - y_n = & \; x_n - A_n^{-1} \Big\{ \beta \left(F(x_n) + G(x_n) \right) + \gamma \left(F(z_n) + G(z_n) \right) \Big\} - \\
& x_n + A_n^{-1} \left(F(x_n) + G(x_n) \right) \\
= & -A_n^{-1} \Big\{ (\beta - 1) \left(F(x_n) + G(x_n) \right) + \gamma \left(F(z_n) + G(z_n) \right) \Big\} \\
= & -\gamma A_n^{-1} \Big\{ \frac{\beta - 1}{\gamma} \left(F(x_n) + G(x_n) \right) + \left(F(z_n) + G(z_n) \right) \Big\} \\
= & \; \gamma A_n^{-1} \Big\{ (1 - \alpha) \left(F(x_n) + G(x_n) \right) - \left(F(z_n) + G(z_n) \right) \Big\}
\end{aligned}
\tag{1.53}
$$

by (1.49).

We also have:

$$
\begin{aligned}
& F(z_n) + G(z_n) \\
= & \; F(z_n) + G(x_n) + G(z_n) - G(x_n) \\
= & \; (1 - \alpha) \left(F(x_n) + G(x_n) \right) + F(z_n) - F(x_n) + \\
& +\alpha \left(F(x_n) + G(x_n) \right) + G(z_n) - G(x_n) \\
= & \; (1 - \alpha) \left(F(x_n) + G(x_n) \right) + F(z_n) - F(x_n) - \\
& \alpha A_n (y_n - x_n) \\
= & \; (1 - \alpha) \left(F(x_n) + G(x_n) \right) + \\
& \alpha \int_0^1 (F'(x_n + \alpha t (y_n - x_n)) - F'(x_n)) (y_n - x_n) \, dt + \\
& \alpha (F'(x_n) - A_n) (y_n - x_n) + G(z_n) - G(x_n).
\end{aligned}
\tag{1.54}
$$

Estimate (1.50) follows from (1.53), and (1.54).

Using (TSNTM), we have:

$$
\begin{aligned}
B_{n+1} &= F(x_{n+1}) + G(x_{n+1}) \\
&= F(x_{n+1}) + G(x_{n+1}) - A_n(y_n - x_n) - F(x_n) - G(x_n) \\
&= F(x_{n+1}) - F(x_n) - F'(x_n)(x_{n+1} - x_n) + \\
&\quad F'(x_n)(x_{n+1} - x_n) - A_n(y_n - x_n) + G(x_{n+1}) - G(x_n) \\
&= \int_0^1 \left(F'(x_n + t(x_{n+1} - x_n)) - F'(x_n)\right)(x_{n+1} - x_n)\, dt + \\
&\quad (F'(x_n) - A_n)(x_{n+1} - x_n) + A_n(x_{n+1} - x_n) - \\
&\quad A_n(y_n - x_n) + G(x_{n+1}) - G(x_n) \\
&= \int_0^1 \left(F'(x_n + t(x_{n+1} - x_n)) - F'(x_n)\right)(x_{n+1} - x_n)\, dt + \\
&\quad (F'(x_n) - A_n)(x_{n+1} - x_n) + A_n(x_{n+1} - y_n) + \\
&\quad G(x_{n+1}) - G(x_n).
\end{aligned}
\tag{1.55}
$$

Estimate (1.51) follows from (1.50), and (1.55).

That completes the proof of Lemma 1.2. $\qquad\qquad\qquad\qquad\qquad\Diamond$

We shall show the following semilocal convergence theorem for (TSNTM).

Theorem 1.1. *Let $F : \mathcal{D} \subseteq X \longrightarrow \mathcal{Y}$ be a Fréchet–differentiable operator, where X, $\mathcal{Y}$ are Banach spaces, $\mathcal{D}$ be a convex subset of X, $G : \mathcal{D} \longrightarrow \mathcal{Y}$ be a continuous operator, and let $A(x) \in L(X, \mathcal{Y})$ be an approximation of $F'(x)$. Assume that there exist a vector $x_0 \in \mathcal{D}$, a bounded inverse $A_0^{-1} := A(x_0)^{-1}$ of $A_0 := A(x_0)$, and constants $K \geq 0$, $L \geq 0$, $M \geq 0$, $N \geq 0$, $\mu \geq 0$, $\ell \in [0, 1)$, $\alpha \in [0, 1]$, $\beta \in [0, 1]$, $\gamma > 0$, $\eta \geq 0$, such that for all $x, y \in \mathcal{D}$:*

$$
\| A_0^{-1}[F(x_0) + G(x_0)] \| \leq \eta,
\tag{1.56}
$$

$$
\| A_0^{-1}[F'(x) - F'(y)] \| \leq K \, \| x - y \|,
\tag{1.57}
$$

$$
\| A_0^{-1}[F'(x) - A(x)] \| \leq M \, \| x - x_0 \| + \mu,
\tag{1.58}
$$

$$
\| A_0^{-1}[A(x) - A_0] \| \leq L \, \| x - x_0 \| + \ell,
\tag{1.59}
$$

$$
\| A_0^{-1}[G(x) - G(y)] \| \leq N \, \| x - y \|,
\tag{1.60}
$$

$$
\overline{U}(x_0, t^\star) = \{x \in X, \, \| x - x_0 \| \leq t^\star\} \subseteq \mathcal{D},
\tag{1.61}
$$

hypotheses of Lemma 1.1, *and* (1.49) *hold.*

Then, sequences $\{x_n\}$, $\{y_n\}$ $(n \geq 0)$ generated by (TSNTM) are well defined, remain in $\overline{U}(x_0, t^\star)$ for all $n \geq 0$, and converge to a solution $x^\star \in \overline{U}(x_0, t^\star)$ of equation $F(x) + G(x) = 0$. Moreover, the following estimates hold for all $n \geq 0$:

$$
\| y_n - x_n \| \leq s_n - t_n,
\tag{1.62}
$$

$$
\| x_{n+1} - y_n \| \leq t_{n+1} - s_n,
\tag{1.63}
$$

$$
\| x_{n+1} - x_n \| \leq t_{n+1} - t_n,
\tag{1.64}
$$

$$
\| y_n - x^\star \| \leq t^\star - s_n,
\tag{1.65}
$$

and

$$
\| x_n - x^\star \| \leq t^\star - t_n,
\tag{1.66}
$$

14 Ioannis K. Argyros, Saïd Hilout and Mohammad A. Tabatabai

where, sequence $\{t_n\}$, $\{s_n\}$, $(n \geq 0)$, and t^* are given in Lemma 1.1.

Furthermore, the solution x^* of equation (1.1) is unique in $\overline{U}(x_0,t^*)$ provided that:

$$\left(\frac{K}{2}+M+L\right) t^* +b+\ell < 1. \tag{1.67}$$

Proof. We shall show estimates (1.62)–(1.64) hold for all $n \geq 0$, and y_n, z_n, $x_{n+1} \in \overline{U}(x_0,t^*)$. Using (TSNTM), (1.4) for $n = 0$, and (1.56), we get:

$$\| y_0 - x_0 \| = \| A_0^{-1} \left(F(x_0) + G(x_0) \right) \| \leq \eta = s_0 - t_0, \tag{1.68}$$

which implies $y_0 \in \overline{U}(x_0,t^*)$, (1.62) holds for $n = 0$ by the definition of t^*.

We also have:

$$\begin{aligned} z_0 - x_0 = \alpha \left(y_0 - x_0 \right) &\implies \| z_0 - x_0 \| = \alpha \| y_0 - x_0 \| \leq \alpha \eta \leq \eta \\ &\implies z_0 \in \overline{U}(x_0,t^*). \end{aligned} \tag{1.69}$$

Hence, x_1 is well defined.

Using (1.3), (1.4), (1.50), (1.57), (1.58), (1.68), and (1.69), we obtain:

$$\begin{aligned} \| x_1 - y_1 \| &\leq \gamma \left(\frac{\alpha^2 K}{2} \| y_0 - x_0 \|^2 + \alpha \left(M \| x_0 - x_0 \| + \right. \right. \\ &\qquad \left. \left. \mu \right) \| y_0 - x_0 \| + N \| z_0 - x_0 \| \right) \\ &\leq \alpha \gamma \left(\frac{\alpha K}{2} (s_0 - t_0) + b \right) (s_0 - t_0) \\ &\leq t_1 - s_0, \end{aligned} \tag{1.70}$$

which show (1.63) for $n = 0$.

We also have:

$$\begin{aligned} \| x_1 - x_0 \| &= \| (x_1 - y_0) + (y_0 - x_0) \| \\ &\leq \| x_1 - y_0 \| + \| y_0 - x_0 \| \\ &\leq t_1 - s_0 + s_0 - t_0 = t_1 - t_0 \leq t^*, \end{aligned} \tag{1.71}$$

which implies (1.64) holds for $n = 0$, and $x_1 \in \overline{U}(x_0,t^*)$. Let us assume (1.62)–(1.64), and y_k, z_k, $x_{k+1} \in \overline{U}(x_0,t^*)$ hold for all $k \leq n - 1$.

Let $u \in \overline{U}(x_0,t^*)$. Then, using (1.30), and (1.59), we get:

$$\begin{aligned} \| A_0^{-1} \left[A(u) - A_0 \right] \| &\leq L \| u - x_0 \| + \ell \\ &\leq L t^* + \ell < 1. \end{aligned} \tag{1.72}$$

It follows from (1.72), and the Banach lemma on invertible operators [151], [477] that $A(u)^{-1}$ exists, with

$$\| A(u)^{-1} A_0 \| \leq (1 - \ell - L \| u - x_0 \|)^{-1}. \tag{1.73}$$

In particular, for $u = x_k$, we have:

$$\begin{aligned} \| x_k - x_0 \| &\le \sum_{i=1}^{k} \| x_i - x_{i-1} \| \\ &\le \sum_{i=1}^{k} (t_i - t_{i-1}) = t_k - t_0 \le t^\star, \end{aligned} \tag{1.74}$$

and, similarily for $u = x_{k+1}$,

$$\| x_{k+1} - x_0 \| \le t_{k+1} - t_0 \le t^\star, \tag{1.75}$$

Hence, using (1.73), we have:

$$\| A_k^{-1} A_0 \| \le (1 - \ell - L\, t_k)^{-1}, \tag{1.76}$$

and

$$\| A_{k+1}^{-1} A_0 \| \le (1 - \ell - L\, t_{k+1})^{-1}. \tag{1.77}$$

Using (1.4), (1.50), (1.57), 1.58), (1.60), (1.74), (1.76), and the induction hypotheses, we get in turn:

$$\begin{aligned} &\| x_{k+1} - y_k \| \\ &\le \gamma \, \| A_k^{-1} A_0 \| \left(\frac{\alpha^2 K}{2} \| y_k - x_k \|^2 + \right. \\ &\quad \left. \alpha \, (M \, \| x_k - x_0 \| + \mu) \, \| y_k - x_k \| + N \, \| z_k - x_k \| \right) \\ &\le \frac{\alpha \gamma}{1 - \ell - L\, t_k} \left(\frac{\alpha K}{2} \| y_k - x_k \| + M \, \| x_k - x_0 \| + b \right) \| y_k - x_k \| \\ &\le \frac{\alpha \gamma}{1 - \ell - L\, t_k} \left(\frac{\alpha K}{2} (s_k - t_k) + M\, t_k + b \right) (s_k - t_k) = t_{k+1} - s_k, \end{aligned} \tag{1.78}$$

which shows (1.63) for all $n \ge 0$.

Moreover, using (1.5), (1.51), (1.57), (1.58), (1.60), (1.77), and the induction hypotheses, we obtain as in (1.78):

$$\begin{aligned} &\| y_{k+1} - x_{k+1} \| \\ &\le \| A_{k+1}^{-1} A_0 \| \, \| A_0^{-1} \, (F(x_{k+1}) + G(x_{k+1})) \| \\ &\le \frac{1}{1 - \ell - L\, t_{k+1}} \left(\frac{K}{2} (t_{k+1} - t_k)^2 + M\, (t_k + \mu) \, (t_{k+1} - t_k) + \right. \\ &\quad N\, (t_{k+1} - t_k) + \frac{\alpha^2 \gamma K}{2} (s_k - t_k)^2 + \alpha \gamma \, (M\, t_k + \mu) \, (s_k - t_k) + \\ &\quad \left. N \, \alpha \gamma \, (s_k - t_k) \right) \\ &= s_{k+1} - t_{k+1}, \end{aligned} \tag{1.79}$$

which shows (1.62) holds for all $n \ge 0$.

We also have:

$$\begin{aligned} \| x_{k+1} - x_k \| &\le \| x_{k+1} - y_k \| + \| y_k - x_k \| \\ &\le (t_{k+1} - s_k) + (s_k - t_k) \\ &= t_{k+1} - t_k, \end{aligned} \tag{1.80}$$

which shows (1.64) for all $n \geq 0$.

Furthemore, we have:

$$
\begin{aligned}
\| y_{k+1} - x_0 \| &\leq \| y_{k+1} - x_{k+1} \| + \| x_{k+1} - x_0 \| \\
&\leq (s_{k+1} - t_{k+1}) + (t_{k+1} - t_0) \\
&= s_{k+1} - t_0 \leq t^\star,
\end{aligned}
\tag{1.81}
$$

$$
\begin{aligned}
\| z_{k+1} - x_0 \| &= \| (1-\alpha)(x_{k+1} - x_0) + \alpha (y_{k+1} - x_0) \| \\
&\leq (1-\alpha) \| x_{k+1} - x_0 \| + \alpha \| y_{k+1} - x_0 \| \\
&\leq (1-\alpha) t_{k+1} + \alpha s_{k+1} \\
&= t_{k+1} + \alpha (s_{k+1} - t_{k+1}) \\
&\leq t_{k+1} + s_{k+1} - t_{k+1} = s_{k+1} \leq t^\star,
\end{aligned}
\tag{1.82}
$$

which implies $y_n, z_n \in \overline{U}(x_0, t^\star)$ for all $n \geq 0$, and

$$
\begin{aligned}
\| z_{k+1} - x_{k+1} \| &\leq \alpha \| y_{k+1} - x_{k+1} \| \\
&\leq \alpha (s_{k+1} - t_{k+1}).
\end{aligned}
$$

That completes the induction.

Lemma 1.1 implies that sequences $\{t_n\}$, $\{s_n\}$ are Cauchy. Hence, $\{x_n\}$, $\{y_n\}$ ($n \geq 0$) are also Cauchy sequences in a Banach space X, and as such they converge to their common limit $x^\star \in \overline{U}(x_0, t^\star)$ (since $\overline{U}(x_0, t^\star)$ is a closed set).

By letting $k \longrightarrow \infty$ in (1.79), we obtain $F(x^\star) + G(x^\star) = 0$. Estimate (1.65), and (1.66) follow from (1.62), and (1.63) by using standard majorization techniques [91], [151], [477].

Finally to show uniqueness, let $y^\star \in \overline{U}(x_0, t^\star)$, with $F(y^\star) + G(y^\star) = 0$. Then, using (TSNTM), (1.3), (1.57), (1.58), (1.60), (1.67), (1.76), and the identity:

$$
\begin{aligned}
y^\star - x_{k+1} = A_k^{-1} A_0 \Big\{ A_0^{-1} \Big(&\int_0^1 (F'(x_k + \theta (y^\star - x_k)) - F'(x_k))\, d\theta + \\
&(F'(x_k) - A_k) \Big) (y^\star - x_k) + A_0^{-1} (G(y^\star) - G(x_k)) \Big\},
\end{aligned}
\tag{1.83}
$$

we obtain:

$$
\begin{aligned}
& \| y^\star - y_k \| \\
&\leq (1 - \ell - L t_k)^{-1} \Big\{ \Big(\int_0^1 \| A_0^{-1}(F'(x_k + \theta (y^\star - x_k)) - F'(x_k)) \|\, d\theta + \\
&\qquad + \| A_0^{-1}(F'(x_k) - A_k) \| \Big) \| y^\star - x_k \| + \| A_0^{-1}(G(x_k) - G(y^\star)) \| \Big\} \\
&\leq (1 - \ell - Lt^\star)^{-1} \Big(\frac{K}{2} \| y^\star - x_k \| + M \| x_k - x_0 \| + b \Big) \| y^\star - x_k \| \\
&\leq (1 - \ell - L t^\star)^{-1} \Big(\frac{K}{2} t^\star + M t^\star + b \Big) \| y^\star - x_k \| \\
&< \| y^\star - x_k \| \quad (\text{by} \quad (1.67)),
\end{aligned}
\tag{1.84}
$$

which implies $\lim_{k \longrightarrow \infty} x_k = y^\star$. But we showed $\lim_{k \longrightarrow \infty} x_k = x^\star$. Hence, we deduce $x^\star = y^\star$.

That completes the proof of Theorem 1.1. $\diamond$

Remark 1.1. (a) Note that $t^\star$ can be replaced by $t^{\star\star}$ given by (1.21) in all hypotheses of Theorem 1.1.

(b) In order for us to compare our results with the corresponding ones in [377] for $A(x) = F'(x)$, $G(x) = 0$ $(x \in \mathcal{D})$, and α, β, γ given by (1.2), let us define m ajorizing sequences $\{\bar{t}_n\}$, $\{\bar{s}_n\}$ essentialy used in [377]:

$$
\begin{aligned}
&\bar{t}_0 = 0, \qquad \bar{s}_0 = \eta, \\
&\bar{t}_{n+1} = \bar{s}_n + \frac{K\,(\bar{s}_n - \bar{t}_n)^2}{2\,(1 - K\,\bar{t}_n)}, \qquad (n \geq 0) \\
&\bar{s}_n = \bar{t}_n + \frac{K\,((\bar{s}_{n-1} - \bar{t}_{n-1})^2 + (\bar{t}_n - \bar{t}_{n-1})^2)}{2\,(1 - K\,\bar{t}_n)} \qquad n \geq 1.
\end{aligned}
\tag{1.85}
$$

The sufficient convergence conditions given in affine invariant form is:

$$
h = K\,\eta < .3266. \tag{1.86}
$$

(c) In view of the proof of Theorem 1.1, (1.4), (1.5), we note that scalar sequences $\{t_n\}$, $\{s_n\}$ given by:

$$
\begin{aligned}
&t_0 = 0, \quad s_0 = \eta, \quad t_1 = s_0 + \frac{L}{2}\,(s_0 - t_0)^2, \\
&t_{n+1} = s_n + \frac{K\,(s_n - t_n)^2}{2\,(1 - L\,t_n)} \qquad (n \geq 1), \\
&s_1 = t_1 + \frac{K\,((s_0 - t_0)^2 + (t_1 - t_0)^2)}{2\,(1 - L\,t_1)} \\
&s_n = t_n + \frac{K\,((s_{n-1} - t_{n-1})^2 + (t_n - t_{n-1})^2)}{2\,(1 - L\,t_n)} \qquad (n \geq 2),
\end{aligned}
\tag{1.87}
$$

are also majorizing sequences for $\{x_n\}$, $\{y_n\}$.

Note that in general

$$
L \leq K, \tag{1.88}
$$

and $\dfrac{K}{L}$ can be large.

An inductive argument for $L < K$ shows:

$$
t_n \leq \bar{t}_n \quad (n \geq 1), \tag{1.89}
$$

$$
s_n \leq \bar{s}_n \quad (n \geq 1), \tag{1.90}
$$

$$
t_{n+1} - s_n \leq \bar{t}_{n+1} - \bar{s}_n \quad (n \geq 0), \tag{1.91}
$$

$$
s_{n+1} - t_{n+1} \leq \bar{s}_{n+1} - \bar{t}_{n+1} \quad (n \geq 0), \tag{1.92}
$$

and

$$
t^\star \leq \overline{t^\star} = \lim_{n \longrightarrow \infty} \bar{t}_n = \lim_{n \longrightarrow \infty} \bar{s}_n. \tag{1.93}
$$

Hence, under condition (1.86), sequences $\{t_n\}$, $\{s_n\}$ are tighter than $\{\bar{t}_n\}$, $\{\bar{s}_n\}$, and are also majorizing $\{x_n\}$, $\{y_n\}$. Moreover, the information on the location of the solution is as least as precise as in [377]. Note also that a direct comparison between our results and the ones in [377] cannot be done, since our sufficient convergenc conditions (see Lemma 1.1) differ from (1.86). However, since the information $L < K$ is not used in [377], and in view of (1.89)–(1.93), one expects to be able to find cases, where (1.86) is violated by the hypotheses of Lemma 1.1 hold. Note also that the hypotheses of Lemma 1.1 involve only computations at the initial guess x_0.

In the case $\gamma = 0$, it only makes sense to also set $\alpha = \beta = 1$ in (TSNTM). Hence, (TSNTM) becomes (NTM):

$$x_{n+1} = x_n - A_n^{-1} (F(x_n) + G(x_n)), \quad (x_0 \in \mathcal{D}), \quad (n \geq 0). \tag{1.94}$$

Lemma 1.3. *Assume:*
there exist constants $K > 0$, $M > 0$, $\mu \geq 0$, $L > 0$, $\ell \geq 0$, $\eta > 0$, such that:

$$2M < K; \tag{1.95}$$

$$(K + 2L)\, \eta < 2\,(1 - \ell - \mu); \tag{1.96}$$

Quadratic polynomial f_1 given by

$$f_1(s) = 2L\,\eta\,s^2 - \left(2\,(1 - \ell - L\,\eta) - K\,\eta\right) s + 2\,(M\,\eta + \mu),$$

has a minimal root in $(0,1)$, denoted by s_1.
For

$$\delta_0 = \frac{K\,\eta + 2\,\mu}{1 - L\,\eta - \ell}, \tag{1.97}$$

$$\delta_+ = \frac{2\,(K - 2M)}{K + \sqrt{K^2 - 8L\,(2M - K)}}, \qquad \delta_\infty = 2\,s_\infty, \tag{1.98}$$

where, s_∞ is the minimal root in $(0,1)$ of equation

$$\overline{f}_\infty(s) = (1 - \ell)\,s^2 - (1 - \ell - L\,\eta + \mu)\,s + M\,\eta + \mu = 0 \tag{1.99}$$

$$\delta_0 \leq \delta_\infty; \tag{1.100}$$

and

$$s_1 \leq \delta_+. \tag{1.101}$$

Choose:

$$\delta = 2\,s_1. \tag{1.102}$$

Then, scalar sequence $\{t_n\}$ $(n \geq 0)$ given by

$$t_0 = 0, \quad t_1 = \eta, \quad t_{n+2} = t_{n+1} + \frac{K\,(t_{n+1} - t_n) + 2\,(M\,t_n + \mu)}{2\,(1 - L\,t_{n+1} - \ell)}\,(t_{n+1} - t_n) \tag{1.103}$$

is increasing, bounded above by

$$t^{\star\star} = \frac{2\,\eta}{2 - \delta}, \tag{1.104}$$

Chapter 1. Two–Step Methods of Hight Efficiency Index 19

and converges to its unique least upper bound $t^\star \in [0, t^{\star\star}]$.
Moreover the following estimates hold for all $n \geq 1$:

$$t_{n+1} - t_n \leq \frac{\delta}{2} (t_n - t_{n-1}) \leq \left(\frac{\delta}{2}\right)^n \eta, \tag{1.105}$$

and

$$t^\star - t_n \leq \frac{2\eta}{2-\delta} \left(\frac{\delta}{2}\right)^n.$$

Proof. We shall show using induction on the integer m:

$$0 < t_{m+2} - t_{m+1} = \frac{K(t_{m+1} - t_m) + 2(M t_m + \mu)}{2(1 - L t_{m+1} - \ell)} (t_{m+1} - t_m)$$

$$\leq \frac{\delta}{2} (t_{m+1} - t_m), \tag{1.106}$$

and

$$\ell + L t_{m+1} < 1. \tag{1.107}$$

If (1.106), and (1.107) hold, we have (1.105) holds, and

$$\begin{aligned}
t_{m+2} &\leq t_{m+1} + \frac{\delta}{2} (t_{m+1} - t_m) \\
&\leq t_m + \frac{\delta}{2} (t_m - t_{m-1}) + \frac{\delta}{2} (t_{m+1} - t_m) \\
&\leq \eta + \left(\frac{\delta}{2}\right) \eta + \cdots + \left(\frac{\delta}{2}\right)^{m+1} \eta \\
&= \frac{1 - \left(\frac{\delta}{2}\right)^{m+2}}{1 - \frac{\delta}{2}} \eta \\
&< \frac{2\eta}{2-\delta} = t^{\star\star} \qquad \text{by (1.104)}.
\end{aligned} \tag{1.108}$$

It will then also follow that sequence $\{t_m\}$ is increasing, bounded above by $t^{\star\star}$, and as such it will converge to some $t^\star \in [0, t^{\star\star}]$.

Estimates (1.106) and (1.107) hold by the initial conditions for $m = 0$. Indeed (1.106) and (1.107) become:

$$\begin{aligned}
0 < t_2 - t_1 &= \frac{K(t_1 - t_0) + 2(M t_0 + \mu)}{2(1 - L t_1 - \ell)} (t_1 - t_0) \\
&= \frac{K\eta + 2\mu}{2(1 - L\eta - \ell)} (t_1 - t_0) = \frac{\delta_0}{2} (t_1 - t_0) \leq \frac{\delta}{2} (t_1 - t_0),
\end{aligned}$$

$$L\eta + \ell < 1,$$

which are true by the choice of δ_0, δ, (1.96), (1.103), and the initial conditions. Let us assume (1.105)–(1.107) hold for all $m \leq n+1$.

Estimate (1.106) can be re–written as

$$K\,(t_{m+1}-t_m)+2\,(M\,t_m+\mu)\le(1-L\,t_{m+1}-\ell)\,\delta$$

or

$$K\,(t_{m+1}-t_m)+2\,(M\,t_m+\mu)+\delta\,L\,t_{m+1}+\delta\,\ell-\ell\le0, \tag{1.109}$$

or

$$K\left(\frac{\delta}{2}\right)^m\eta+2\left(M\,\frac{1-\left(\frac{\delta}{2}\right)^m}{1-\frac{\delta}{2}}\,\eta+\mu\right)+\delta\,L\,\frac{1-\left(\frac{\delta}{2}\right)^{m+1}}{1-\frac{\delta}{2}}\,\eta+\delta\,(\ell-1)\le0. \tag{1.110}$$

Replace $\frac{\delta}{2}$ by s, and define functions f_m on $[0,+\infty)$ $(m\ge1)$:

$$\begin{aligned}
f_m(s) &= K\,s^m\,\eta+2\left(M\,(1+s+s^2+\cdots+s^{m-1})\,\eta+\mu\right)+\\
&\quad 2\,s\,L\,(1+s+\cdots+s^m)\,\eta+2\,s\,(\ell-1).
\end{aligned} \tag{1.111}$$

Estimate (1.110) certainly holds, if:

$$f_m(\delta)\le0, \qquad (m\ge1). \tag{1.112}$$

We need to find a relationship between two consecutive f_m:

$$\begin{aligned}
f_{m+1}(s)\\
&= K\,s^{m+1}\,\eta+2\left(M\,(1+s+s^2+\cdots+s^{m-1}+s^m)\,\eta+\mu\right)+\\
&\quad 2\,s\,L\,(1+s+\cdots+s^m+s^{m+1})\,\eta+2\,s\,(\ell-1)\\
&= K\,s^{m+1}\,\eta-K\,s^m\,\eta+K\,s^m\,\eta+\\
&\quad 2\left(M\,(1+s+s^2+\cdots+s^{m-1})\,\eta+\mu\right)+2\,M\,s^m\,\eta+\\
&\quad 2\,s\,L\,(1+s+\cdots+s^m)\,\eta+2\,s\,L\,s^{m+1}\,\eta+2\,s\,(\ell-1)\\
&= f_m(s)+K\,s^{m+1}\,\eta-K\,s^m\,\eta+2\,M\,s^m\,\eta+2\,s\,L\,s^{m+1}\,\eta\\
&= f_m(s)+g(s)\,s^m\,\eta,
\end{aligned} \tag{1.113}$$

where,

$$g(s)=2\,L\,s^2+K\,s+2\,M-K. \tag{1.114}$$

Note that in view of (1.95), function g has a positive zero δ_+ given by (1.98), and

$$g(s)<0 \qquad s\in(0,\delta_+). \tag{1.115}$$

By hypothesis, the function f_1 has a minimal positive zero s_1. Using (1.95), it is simple algebra to show $s_1\in[0,1)$. It then follows from (1.113) and (1.114):

$$\begin{aligned}
f_2(s_1) &= f_1(s_1)+g(s_1)\,s_1^m\,\eta\\
&= g(s_1)\,s_1^m\,\eta<0,
\end{aligned} \tag{1.116}$$

since $f_1(s_1)=0$, and $g(s_1)<0$. We also have from (1.111):

$$f_m(0)=2\,(M\,\eta+\mu)>0 \quad (m\ge1). \tag{1.117}$$

Chapter 1. Two–Step Methods of Hight Efficiency Index 21

It follows from the intermediate value theorem that there exists a minimal $s_2 \in (0, s_1)$, such that $f_2(s_2) = 0$. Let us assume: there exists $s_m \in (0, s_{m-1})$, with $f_m(s_m) = 0$. As in (1.116) we have

$$f_{m+1}(s_m) = f_m(s_m) + g(s_m)\, s_m^m\, \eta < 0. \tag{1.118}$$

It follows from the intermediate value theorem that there exists a minimal $s_{m+1} \in (0, s_m)$, such that $f_{m+1}(s_{m+1}) = 0$.

In view of (1.110):

$$f_\infty(s_\infty) = 2\left(\frac{M}{1 - s_\infty}\, \eta + \mu\right) + \frac{2\, s_\infty\, L}{1 - s_\infty}\, \eta + 2\, s_\infty\, (\ell - 1) = 0,$$

by the choice of s_∞. Note also that by (1.96), and (1.99), s_∞ exists in $(0, 1)$.

Sequence $\{s_m\}$ is non–increasing, bounded below by zero, and as such it converges to its unique maximum lowest bound $s^\star$ satisfying $s^\star \geq s_\infty$. Hence, we showed (1.112). That completes the induction for (1.106), and (1.107).

Finally, sequence $\{t_n\}$ is increasing, bounded above by $t^{\star\star}$, and as such it converges to its unique least upper bound $t^\star$.

That completes the proof of Lemma 1.3. $\quad\diamond$

Remark 1.2. Note that by applying the intermediate value theorem on f_1 for $s \in [0, 1]$, we see that (1.96), and the condition on the existence of s_1 can be replaced by condition:

$$(K + 4\, L + 2\, M)\, \eta < 2\,(1 - \ell - \mu).$$

Another set of replacement conditions is given by $\Delta \geq 0$, and

$$\max\{(4\, L + 2\, M + K)\, \eta + 2\, \mu,\, (6\, L + K)\, \eta\} < 2\,(1 - \ell),$$

where, Δ is the descriminant of polynomial f_1.

We shall provide a semilocal convergence analysis for (NTM).

Theorem 1.2. *Let* $F : \mathcal{D} \subseteq X \longrightarrow \mathcal{Y}$ *be a Fréchet–differentiable operator,* $G : \mathcal{D} \longrightarrow \mathcal{Y}$ *be a continuous operator, and let* $A(x) \in L(X, \mathcal{Y})$ *be an approximation of* $F'(x)$. *Assume that there exist an open convex subset* $\mathcal{D}$ *of* X, $x_0 \in \mathcal{D}$, *a bounded inverse* A_0^{-1} *of* A_0, *and constants* $K > 0$, $M > 0$, $\mu_0 \geq 0$, $\mu_1 \geq 0$, $L > 0$, $\ell \geq 0$, $\eta > 0$, *such that for all* $x, y \in \mathcal{D}$:

$$\begin{aligned}
\| A_0^{-1}\, (F(x_0) + G(x_0))\, \| &\leq \eta, \tag{1.119}\\
\| A_0^{-1}\, (F'(x) - F'(y))\, \| &\leq K\, \| x - y \|, \tag{1.120}\\
\| A_0^{-1}\, (F'(x) - A(x))\, \| &\leq M\, \| x - x_0 \| + \mu_0, \tag{1.121}\\
\| A_0^{-1}\, (A(x) - A_0)\, \| &\leq L\, \| x - x_0 \| + \ell, \tag{1.122}\\
\| A_0^{-1}\, (G(x) - G(y))\, \| &\leq \mu_1\, \| x - y \|, \tag{1.123}\\
\overline{U}(x_0, t^\star) = \{x \in X,\, \| x - x_0 \| &\leq t^\star\} \subseteq \mathcal{D},
\end{aligned}$$

and

the hypotheses of Lemma 1.3 *hold with* $\mu = \mu_0 + \mu_1$.

Then, sequence $\{x_n\}$ $(n \geq 0)$ generated by (NTM) is well defined, remains in $\overline{U}(x_0, t^\star)$ for all $n \geq 0$, and converges to a solution $x^\star$ of equation $F(x) + G(x) = 0$ in $\overline{U}(x_0, t^\star)$.

Moreover, the following estimates hold for all $n \geq 0$:

$$\| x_{n+1} - x_n \| \leq t_{n+1} - t_n, \tag{1.124}$$

and

$$\| x_n - x^\star \| \leq t^\star - t_n, \tag{1.125}$$

where, sequence $\{t_n\}$ $(n \geq 0)$, and $t^\star$ are given in Lemma 1.3.

Furthermore, the solution $x^\star$ of equation (1.1) is unique in $\overline{U}(x_0, t^\star)$ provided that:

$$\left(\frac{K}{2} + M + L \right) t^\star + \mu + \ell < 1.$$

Proof. We shall show using induction on $m \geq 0$:

$$\| x_{m+1} - x_m \| \leq t_{m+1} - t_m, \tag{1.126}$$

and

$$\overline{U}(x_{m+1}, t^\star - t_{m+1}) \subseteq \overline{U}(x_m, t^\star - t_m). \tag{1.127}$$

For every $z \in \overline{U}(x_1, t^\star - t_1)$,

$$\begin{aligned}
\| z - x_0 \| &\leq \| z - x_1 \| + \| x_1 - x_0 \| \\
&\leq t^\star - t_1 + t_1 = t^\star - t_0,
\end{aligned}$$

implies $z \in \overline{U}(x_0, t^\star - t_0)$. We also have

$$\| x_1 - x_0 \| = \| A_0^{-1} [F(x_0) + G(x_0)] \| \leq \eta = t_1 - t_0.$$

That is (1.126), and (1.127) hold for $m = 0$. Given they hold for $n \leq m$, then:

$$\begin{aligned}
\| x_{m+1} - x_0 \| &\leq \sum_{i=1}^{m+1} \| x_i - x_{i-1} \| \\
&\leq \sum_{i=1}^{m+1} (t_i - t_{i-1}) = t_{m+1} - t_0 = t_{m+1},
\end{aligned}$$

and

$$\begin{aligned}
\| x_m + \theta (x_{m+1} - x_m) - x_0 \| &\leq t_m + \theta (t_{m+1} - t_m) \\
&\leq t^\star,
\end{aligned}$$

for all $\theta \in (0, 1)$.

Using (1.96), (1.122), the induction hypotheses, we get:

$$\begin{aligned}
\| A_0^{-1} [A_{m+1} - A_0] \| &\leq L \| x_{m+1} - x_0 \| + \ell \\
&\leq L (t_{m+1} - t_0) + \ell \\
&\leq L t_{m+1} + \ell < 1
\end{aligned} \tag{1.128}$$

by (1.109).

Chapter 1. Two–Step Methods of Hight Efficiency Index   23

It follows from (1.128), and the Banach lemma on invertible operators that A_{m+1}^{-1} exists, and

$$\| A_{m+1}^{-1} A_0 \| \leq (1 - \ell - L\, t_{m+1})^{-1}. \tag{1.129}$$

Using (1.94), we obtain the approximation:

$$
\begin{aligned}
& x_{m+2} - x_{m+1} \\
=\ & -A_{m+1}^{-1} \left(F(x_{m+1} + G(x_{m+1})) \right) \\[2mm]
=\ & -A_{m+1}^{-1} A_0 A_0^{-1} \\
& \left(\int_0^1 [F'(x_{m+1} + \theta (x_m - x_{m+1})) - F'(x_m)] (x_{m+1} - x_m)\, d\theta + \right. \\
& \left. (F'(x_m) - A_m) (x_{m+1} - x_m) + G(x_{m+1}) - G(x_m) \right)
\end{aligned}
\tag{1.130}
$$

Using (1.120), (1.121), (1.123), (1.129), (1.130), and the induction hypotheses, we obtain in turn:

$$
\begin{aligned}
& \| x_{m+2} - x_{m+1} \| \\
\leq\ & (1 - \ell - L\, t_{m+1})^{-1} \left(\frac{K}{2} \| x_{m+1} - x_m \|^2 + \right. \\
& \left. (M \| x_m - x_0 \| + \mu_0) \| x_{m+1} - x_m \| + \mu_1 \| x_{m+1} - x_m \| \right) \\
\leq\ & (1 - \ell - L\, t_{m+1})^{-1} \left(\frac{K}{2} (t_{m+1} - t_m) + M\, t_m + \mu \right) (t_{m+1} - t_m) \\
=\ & t_{m+2} - t_{m+1},
\end{aligned}
\tag{1.131}
$$

which shows (1.126) for all $m \geq 0$.

Thus, for every $z \in \overline{U}(x_{m+2}, t^\star - t_{m+2})$, we have:

$$
\begin{aligned}
\| z - x_{m+1} \| & \leq & \| z - x_{m+2} \| + \| x_{m+2} - x_{m+1} \| \\
& \leq & t^\star - t_{m+2} + t_{m+2} - t_{m+1} = t^\star - t_{m+1},
\end{aligned}
$$

which shows (1.127) for all $m \geq 0$.

Lemma 1.3 implies that sequence $\{t_n\}$ is Cauchy. Moreover, it follows from (1.126) and (1.127) that $\{x_n\}$ $(n \geq 0)$ is also a Cauchy sequence in a Banach space X, and as such it converges to some $x^\star \in \overline{U}(x_0, t^\star)$ (since $\overline{U}(x_0, t^\star)$ is a closed set).

By letting $m \longrightarrow \infty$ in (1.131), we obtain $F(x^\star) + G(x^\star) = 0$. Furthermore estimate (1.125) is obtained from (1.124) by using standard majorization techniques [91], [151], [477]. Finally to show that $x^\star$ is the unique solution of equation (1.1) in $\overline{U}(x_0, t^\star)$, as in (1.130) and (1.131), we get in turn for $y^\star \in \overline{U}(x_0, t^\star)$, with $F(y^\star) + G(y^\star) = 0$, the estimation:

$$\| y^\star - x_{m+1} \|$$

$$\leq \| A_m^{-1} A_0 \|$$

$$\left\{ \left(\int_0^1 \| A_0^{-1} \left(F'(x_m + \theta (y^\star - x_m)) - F'(x_m) \right) \| \; d\theta \right. \right.$$

$$+ \| A_0^{-1} [F'(x_m) - A_m] \| \bigg) \; \| y^\star - x_m \| +$$

$$\left. \| A_0^{-1} [G(x_m) - G(y^\star)] \| \right\} \tag{1.132}$$

$$\leq (1 - L\, t_{m+1})^{-1} \left(\frac{K}{2} \; \| y^\star - x_m \|^2 + \right.$$

$$\left. (M \; \| x_m - x_0 \| + \mu) \; \| y^\star - x_m \| \right)$$

$$\leq (1 - L\, t_{m+1})^{-1} \left(\frac{K}{2} \, (t^\star - t_m) + M\, t_m + \mu \right) \; \| y^\star - x_m \|$$

$$\leq (1 - L\, t^\star)^{-1} \left(\frac{K}{2} \, (t^\star - t_0) + M\, t^\star + \mu \right) \; \| x^\star - x_m \|$$

$$< \; \| y^\star - x_m \|,$$

by the uniqueness hypothesis.

It follows by (1.132) that $\lim_{m \longrightarrow \infty} x_m = y^\star$. But we showed $\lim_{m \longrightarrow \infty} x_m = x^\star$. Hence, we deduce $x^\star = y^\star$.

That completes the proof of Theorem 1.2.

Application 1.1. $\gamma = 0$. Using (1.119)–(1.122), and hypothesis

$$h_K = \sigma\, \eta \leq \frac{1}{2} \, (1 - b)^2, \qquad \mu + \ell < 1 \tag{1.133}$$

where, $\sigma = \max\{K, M + L\}$, with $b = \mu + \ell$, a semilocal convergence theorem was provided in [303], [348], [525], [770].

(a) Let us compare the error bounds in this case. The majorizing sequence given in [303], [348], [525], [770], is:

$$v_0 = 0, \quad v_1 = \eta,$$
$$v_{n+2} = v_{n+1} + \frac{f(v_{n+1})}{q(v_{n+1})}, \qquad (n \geq 0), \tag{1.134}$$

where,

$$f(v) = \frac{\sigma}{2} \, v^2 - (1 - b) \, v + \eta,$$

and

$$q(v) = 1 - L\, v - \ell.$$

We now show that the error bounds obtained in Theorem 1.2 are more precise than the corresponding ones in the above references using (1.133).

Chapter 1. Two–Step Methods of Hight Efficiency Index   25

Proposition 1.1. *Under the hypotheses of Theorem* 1.2, *and condition* (1.133), *the following error bounds hold:*

$$
\begin{aligned}
t_{n+1} &\leq v_{n+1} \quad (n \geq 1), & \text{(1.135)} \\
t_{n+1} - t_n &\leq v_{n+1} - v_n \quad (n \geq 1), & \text{(1.136)} \\
t^\star - t_n &\leq v^\star - v_n \quad (n \geq 0), & \text{(1.137)}
\end{aligned}
$$

and

$$
t^\star \leq v^\star. \tag{1.138}
$$

Moreover strict inequality holds in (1.135) *and* (1.136) *if* $K < M + L$.

Proof. We use mathematical induction on m to first show (1.135) and (1.136). For $n = 0$ in (1.21) we obtain:

$$
\begin{aligned}
t_2 - \eta &= \frac{\dfrac{K}{2}\eta^2 + \mu\eta}{1 - \ell - L\eta} \leq \frac{\dfrac{\sigma}{2}\eta^2 + (M \cdot 0 + \mu)\eta}{1 - \ell - L\eta} \\[2ex]
&\leq \frac{\dfrac{\sigma}{2}\eta^2 + M(\eta - 0) + \mu(\eta - 0) - q(0)(\eta - 0) + f(0)}{q(\eta)} \\[2ex]
&\leq \frac{\dfrac{\sigma}{2}v_1^2 - (1 - \mu - \ell)v_1 + \eta - (\sigma - M - L)v_0(v_1 - v_0)}{q(v_1)} \\[2ex]
&\leq \frac{f(v_1)}{q(v_1)} = v_2 - v_1,
\end{aligned}
$$

and

$$
t_2 \leq v_2.
$$

Assume:

$$
t_{i+1} \leq v_{i+1}, \quad t_{i+1} - t_i \leq v_{i+1} - v_i. \tag{1.139}
$$

Using (1.21), (1.134), and (1.139), we obtain in turn:

$$
\begin{aligned}
&t_{i+2} - t_{i+1} \\[1ex]
&= \frac{\dfrac{K}{2}(t_{i+1} - t_i)^2 + (M t_i + \mu)(t_{i+1} - t_i)}{1 - \ell - L t_{i+1}} \\[2ex]
&\leq \frac{\dfrac{\sigma}{2}(v_{i+1} - v_i)^2 + (M v_i + \mu)(v_{i+1} - v_i)}{q(v_{i+1})} \\[2ex]
&= \frac{\dfrac{\sigma}{2}(v_{i+1} - v_i)^2 + M(v_{i+1} - v_i)v_i + \mu(v_{i+1} - v_i) - q(v_i)(v_{i+1} - v_i) + f(v_i)}{q(v_{i+1})} \\[2ex]
&= \frac{\dfrac{\sigma}{2}v_{i+1}^2 - (1 - \mu - \ell)v_{i+1} + \eta - (\sigma - M - L)v_i(v_{i+1} - v_i)}{q(v_{i+1})} \\[2ex]
&\leq \frac{f(v_{i+1})}{q(v_{i+1})} = v_{i+2} - v_{i+1},
\end{aligned}
$$

which show (1.135) and (1.136) for all $(n \geq 1)$.

Let $j \geq 0$, we can get:

$$
\begin{aligned}
t_{i+j} - t_i &\leq (t_{i+j} - t_{i+j-1}) + (t_{i+j-1} - t_{i+j-2}) + \cdots + (t_{i+1} - t_i) \\
&\leq (v_{i+j} - v_{i+j-1}) + (v_{i+j-1} - v_{i+j-2}) + \cdots + (v_{i+1} - v_i) \\
&\leq v_{i+1} - v_i.
\end{aligned}
\tag{1.140}
$$

By letting $j \to \infty$ in (1.140) we obtain (1.137).

Finally (1.137) implies (1.138) (since $t_1 = v_1 = 0$). It can easily be seen from (1.21), and (1.134), that strict inequality holds in (1.135) and (1.136) if $K < M + L$.

That completes the proof of Proposition 1.1. $\qquad \diamondsuit$

Note also that the above advantages hold even if hypotheses of Theorem 1.2 are replaced by (1.133).

(b) We can now compare our Theorem 1.2 with the corresponding one in [749] in the case of Newton's method $(A(x) = F'(x), G(x) = 0, (x \in \mathcal{D}))$:

Hypothesis (1.133) reduces to the famous for its simplicity and clarity Newton–Kantorovich hypothesis [91], [151], [477] for solving nonlinear equations

$$
h_K = K \eta \leq \frac{1}{2},
\tag{1.141}
$$

since $\sigma = K$, and $\mu_0 = \mu_1 = \ell = M = 0$.

Note that in this case, functions f_m $(m \geq 1)$ should be defined by

$$
f_m(s) = \left(K s^{m-1} + 2 L \left(1 + s + s^2 + \cdots + s^m \right) \right) \eta - 2,
$$

and

$$
f_{m+1}(s) = f_m(s) + g(s) s^{m-1} \eta.
$$

But this time, the conditions corresponding to Lemma 1.3 should be:

$$
\delta_1 = \max \left\{ \frac{\delta_0}{2}, \delta_+ \right\} \leq s_\infty = 1 - L \eta,
\tag{1.142}
$$

whereas,

$$
\delta = 2 \delta_1.
\tag{1.143}
$$

However, it is simple algebra to show that conditions (1.142)–(1.143) reduce to:

$$
h_A = \overline{L} \eta \leq \frac{1}{2},
\tag{1.144}
$$

where,

$$
\overline{L} = \frac{1}{8} \left(K + 4 L + \sqrt{K^2 + 8 K L} \right).
$$

Note also that

$$L \leq K \tag{1.145}$$

holds in general, and $\dfrac{K}{L}$ can be arbitrarily large.

In view of (1.141), (1.144) and (1.145), we get

$$h_K \leq \frac{1}{2} \implies h_A \leq \frac{1}{2}, \tag{1.146}$$

but not necessarily vice verca unless if $L = K$.

In the example that follows, we show that $\dfrac{K}{L}$ can arbitrarily large. Indeed:

Example 1.1. Let $X = \mathcal{Y} = \mathbb{R}$, $x_0 = 1$, and define scalar functions F and G by

$$F(x) = c_0 x + c_1 + c_2 \sin e^{c_3 x}, \qquad G(x) = 0, \tag{1.147}$$

where, c_i, $i = 0, 1, 2, 3$ are given parameters. Using (1.147), it can easily be seen that for c_3 large and c_2 sufficiently small, $\dfrac{K}{L}$ can be arbitrarily large.

In the next examples, we show (1.133) is violated but (1.144) holds.

Example 1.2. Let $X = \mathcal{Y} = \mathbb{R}^2$ be equipped with the max–norm, and

$$x_0 = (1,1)^T, \quad U_0 = \{x : \| x - x_0 \| \leq 1 - \beta\}, \quad \beta \in \left[0, \frac{1}{2}\right).$$

Define function F on U_0 by

$$F(x) = (\xi_1^3 - \beta, \xi_2^3 - \beta), \qquad x = (\xi_1, \xi_2)^T. \tag{1.148}$$

Using hypotheses of Theorem 1.2, we get:

$$\eta = \frac{1}{3} (1 - \beta), \quad L = 3 - \beta, \quad \text{and} \quad K = 2 (2 - \beta).$$

The Newton–Kantorovich condition (1.141) is violated, since

$$\frac{4}{3} (1 - \beta) (2 - \beta) > 1 \quad \text{for all} \quad \beta \in \left[0, \frac{1}{2}\right).$$

Hence, there is no guarantee that (NTM) converges to $x^{\star} = (\sqrt[3]{\beta}, \sqrt[3]{\beta})$, starting at x_0. However, our condition (1.144) is true for all $\beta \in I = \left[.450339002, \frac{1}{2}\right)$. Hence, the conclusions of our Theorem 1.2 can apply to solve equation (1.148) for all $\beta \in I$.

Example 1.3. Let $X = Y = C[0,1]$ be the space of real–valued continuous functions defined on the interval $[0,1]$ with norm

$$\| x \| = \max_{0 \leq s \leq 1} |x(s)|.$$

Let $\theta \in [0,1]$ be a given parameter. Consider the "Cubic" integral equation

$$u(s) = u^3(s) + \lambda u(s) \int_0^1 q(s,t)\, u(t)\, dt + y(s) - \theta. \qquad (1.149)$$

Here the kernel $q(s,t)$ is a continuous function of two variables defined on $[0,1] \times [0,1]$; the parameter λ is a real number called the "albedo" for scattering; $y(s)$ is a given continuous function defined on $[0,1]$ and $x(s)$ is the unknown function sought in $C[0,1]$. Equations of the form (1.149) arise in the kinetic theory of gasses [151], [293]. For simplicity, we choose $u_0(s) = y(s) = 1$, and $q(s,t) = \dfrac{s}{s+t}$, for all $s \in [0,1]$, and $t \in [0,1]$, with $s+t \neq 0$. If we let $\mathcal{D} = U(u_0, 1 - \theta)$, and define the operator F on $\mathcal{D}$ by

$$F(x)(s) = x^3(s) - x(s) + \lambda x(s) \int_0^1 q(s,t)\, x(t)\, dt + y(s) - \theta, \qquad (1.150)$$

for all $s \in [0,1]$, then every zero of F satisfies equation (1.149). We have the estimates:

$$\max_{0 \leq s \leq 1} \left| \int \frac{s}{s+t}\, dt \right| = \ln 2.$$

Therefore, if we set $\xi = \| F'(u_0)^{-1} \|$, then it follows from hypotheses of Theorem 1.2 that

$$\eta = \xi \left(|\lambda| \ln 2 + 1 - \theta \right),$$

$$K = 2\,\xi \left(|\lambda| \ln 2 + 3\,(2 - \theta) \right) \quad \text{and} \quad L = \xi \left(2\,|\lambda| \ln 2 + 3\,(3 - \theta) \right).$$

It follows from Theorem 1.2 that if condition (1.144) holds, then problem (1.149) has a unique solution near u_0. This assumption is weaker than the one given before using the Newton–Kantorovich hypothesis (1.141).

Note also that $L < K$ for all $\theta \in [0,1]$.

Example 1.4. Consider the following nonlinear boundary value problem [151]

$$\begin{cases} u'' = -u^3 - \gamma u^2 \\ u(0) = 0, \quad u(1) = 1. \end{cases}$$

It is well known that this problem can be formulated as the integral equation

$$u(s) = s + \int_0^1 Q(s,t)\, (u^3(t) + \gamma u^2(t))\, dt \qquad (1.151)$$

where, Q is the Green function:

$$Q(s,t) = \begin{cases} t\,(1-s), & t \leq s \\ s\,(1-t), & s < t. \end{cases}$$

We observe that

$$\max_{0 \le s \le 1} \int_0^1 |Q(s,t)| = \frac{1}{8}.$$

Let $X = \mathcal{Y} = C[0,1]$, with norm

$$\| x \| = \max_{0 \le s \le 1} |x(s)|.$$

Then problem (1.151) is in the form (1.1), where, $F : \mathcal{D} \longrightarrow \mathcal{Y}$ is defined as

$$[F(x)](s) = x(s) - s - \int_0^1 Q(s,t)\,(x^3(t) + \gamma x^2(t))\,dt,$$

and

$$G(x)(s) = 0.$$

It is easy to verify that the Fréchet derivative of F is defined in the form

$$[F'(x)v](s) = v(s) - \int_0^1 Q(s,t)\,(3\,x^2(t) + 2\,\gamma x(t))\,v(t)\,dt.$$

If we set $u_0(s) = s$, and $\mathcal{D} = U(u_0, R)$, then since $\| u_0 \| = 1$, it is easy to verify that $U(u_0, R) \subset U(0, R+1)$. It follows that $2\,\gamma < 5$, then

$$\| I - F'(u_0) \| \le \frac{3\,\| u_0 \|^2 + 2\,\gamma\,\| u_0 \|}{8} = \frac{3 + 2\,\gamma}{8},$$

$$\| F'(u_0)^{-1} \| \le \frac{1}{1 - \dfrac{3 + 2\,\gamma}{8}} = \frac{8}{5 - 2\,\gamma},$$

$$\| F(u_0) \| \le \frac{\| u_0 \|^3 + \gamma\,\| u_0 \|^2}{8} = \frac{1 + \gamma}{8},$$

$$\| F(u_0)^{-1} F(u_0) \| \le \frac{1 + \gamma}{5 - 2\,\gamma}.$$

On the other hand, for $x, y \in \mathcal{D}$, we have

$$[(F'(x) - F'(y))v](s) = -\int_0^1 Q(s,t)\,(3\,x^2(t) - 3\,y^2(t) + 2\,\gamma\,(x(t) - y(t)))\,v(t)\,dt.$$

Consequently,

$$
\begin{aligned}
\| F'(x) - F'(y) \| &\le \frac{\| x - y \|\,(2\,\gamma + 3\,(\| x \| + \| y \|))}{8} \\
&\le \frac{\| x - y \|\,(2\,\gamma + 6\,R + 6\,\| u_0 \|)}{8} \\
&= \frac{\gamma + 6\,R + 3}{4}\,\| x - y \|,
\end{aligned}
$$

$$\| F'(x) - F'(u_0) \| \leq \frac{\| x - u_0 \| \, (2\gamma + 3 \, (\| x \| + \| u_0 \|))}{8}$$

$$\leq \frac{\| x - u_0 \| \, (2\gamma + 3\,R + 6\,\| u_0 \|)}{8}$$

$$= \frac{2\gamma + 3\,R + 6}{8} \, \| x - u_0 \| .$$

Therefore, conditions of Theorem 1.2 hold with

$$\eta = \frac{1+\gamma}{5-2\gamma}, \quad K = \frac{\gamma + 6\,R + 3}{4}, \quad L = \frac{2\gamma + 3\,R + 6}{8}.$$

Note also that $L < K$.

Lemma 1.4. *Assume there exist constants $L \geq 0$, $K \geq 0$, and $\eta \geq 0$, such that:*

$$h_A = \overline{L}\,\eta \leq \frac{1}{2}, \tag{1.152}$$

where,

$$\overline{L} = \frac{1}{8} \left(K + 4\,L + \sqrt{K^2 + 8\,L\,K} \right). \tag{1.153}$$

The inequality in (1.152) is strict if $L = 0$.

Then, sequence $\{t_k\}$ $(k \geq 0)$ given by

$$t_0 = 0, \quad t_1 = \eta, \quad t_{k+1} = t_k + \frac{L_1 \, (t_k - t_{k-1})^2}{2 \, (1 - L\,t_k)} \qquad (k \geq 1), \tag{1.154}$$

is well defined, nondecreasing, bounded above by $t^{\star\star}$, and converges to its unique least upper bound $t^\star \in [0, t^{\star\star}]$, where

$$L_1 = \begin{cases} L & if \quad k = 1 \\ K & if \quad k > 1 \end{cases},$$

$$t^{\star\star} = \frac{2\,\eta}{2 - \delta}, \tag{1.155}$$

$$1 \leq \delta = \frac{4\,K}{K + \sqrt{K^2 + 8\,L\,K}} < 2 \quad for \ L \neq 0. \tag{1.156}$$

Moreover the following estimates hold:

$$L\,t^\star < 1, \tag{1.157}$$

$$0 \leq t_{k+1} - t_k \leq \frac{\delta}{2} \, (t_k - t_{k-1}) \leq \cdots \leq \left(\frac{\delta}{2} \right)^k \eta, \quad (k \geq 1), \tag{1.158}$$

$$t_{k+1} - t_k \leq \left(\frac{\delta}{2} \right)^k (2\,h_A)^{2^k - 1} \, \eta, \quad (k \geq 0), \tag{1.159}$$

$$0 \leq t^\star - t_k \leq \left(\frac{\delta}{2} \right)^k \frac{(2\,h_A)^{2^k - 1} \, \eta}{1 - (2\,h_A)^{2^k}}, \quad (2\,h_A < 1), \quad (k \geq 0). \tag{1.160}$$

Chapter 1. Two–Step Methods of Hight Efficiency Index

Proof. If $L = 0$, then (1.157) holds trivially. In this case, for $K > 0$, an induction argument shows that

$$t_{k+1} - t_k = \frac{2}{K} (2 h_A)^{2^k} \qquad (k \geq 0),$$

and therefore

$$t_{k+1} = t_1 + (t_2 - t_1) + \cdots + (t_{k+1} - t_k) = \frac{2}{K} \sum_{m=0}^{k} (2 h_A)^{2^m},$$

and

$$t^\star = \lim_{k \to \infty} t_k = \frac{2}{K} \sum_{k=0}^{\infty} (2 h_A)^{2^k}.$$

Clearly, this series converges, since $k \leq 2^k$, $2 h_A < 1$, and is bounded above by the number

$$\frac{2}{K} \sum_{k=0}^{\infty} (2 h_A)^k = \frac{4}{K (2 - K \eta)}.$$

If $K = 0$, then in view of (1.154), $0 \leq L \leq K$, we deduce: $L = 0$, and $t^\star = t_k = \eta$ $(k \geq 1)$.

In the rest of the proof, we assume that $L > 0$.

The result until estimate (1.158) follows from Lemma 1.1.

In order for us to show (1.159) we need the estimate:

$$\frac{1 - \left(\frac{\delta}{2}\right)^{k+1}}{1 - \frac{\delta}{2}} \eta \leq \frac{1}{L} \left(1 - \left(\frac{\delta}{2}\right)^{k-1} \frac{K}{4 \overline{L}}\right) \qquad (k \geq 1). \tag{1.161}$$

For $k = 1$, (1.161) becomes

$$\left(1 + \frac{\delta}{2}\right) \eta \leq \frac{4 \overline{L} - K}{4 \overline{L} L}$$

or

$$\left(1 + \frac{2 K}{K + \sqrt{K^2 + 8 L K}}\right) \eta \leq \frac{4 L - K + \sqrt{K^2 + 8 L K}}{L (4 L + K + \sqrt{K^2 + 8 L K})}$$

In view of (1.152), it suffices to show:

$$\frac{L (4 L + K + \sqrt{K^2 + 8 L K}) (3 K + \sqrt{K^2 + 8 L K})}{(K + \sqrt{K^2 + 8 L K}) (4 L - K + \sqrt{K^2 + 8 L K})} \leq 2 \overline{L},$$

which is true as equality.

Let us now assume estimate (1.161) is true for all integers smaller or equal to k. We must show (1.161) holds for k being $k + 1$:

$$\frac{1 - \left(\frac{\delta}{2}\right)^{k+2}}{1 - \frac{\delta}{2}} \eta \leq \frac{1}{L} \left(1 - \left(\frac{\delta}{2}\right)^{k} \frac{K}{4 \overline{L}}\right) \qquad (k \geq 1).$$

or

$$\left(1+\frac{\delta}{2}+\left(\frac{\delta}{2}\right)^2+\cdots+\left(\frac{\delta}{2}\right)^{k+1}\right)\eta\leq\frac{1}{L}\left(1-\left(\frac{\delta}{2}\right)^k\frac{K}{4\overline{L}}\right). \tag{1.162}$$

By the induction hypothesis to show (1.162), it suffices

$$\frac{1}{L}\left(1-\left(\frac{\delta}{2}\right)^{k-1}\frac{K}{4\overline{L}}\right)+\left(\frac{\delta}{2}\right)^{k+1}\eta\leq\frac{1}{L}\left(1-\left(\frac{\delta}{2}\right)^k\frac{K}{4\overline{L}}\right)$$

or

$$\left(\frac{\delta}{2}\right)^{k+1}\eta\leq\frac{1}{L}\left(\left(\frac{\delta}{2}\right)^{k-1}-\left(\frac{\delta}{2}\right)^k\right)\frac{K}{4\overline{L}}$$

or

$$\delta^2\,\eta\leq\frac{K\,(2-\delta)}{2\,\overline{L}L}.$$

In view of (1.152) it suffices to show

$$\frac{2\,\overline{L}L\,\delta^2}{K\,(2-\delta)}\leq2\,\overline{L},$$

which holds as equality by the choice of δ given by (1.156). That completes the induction for estimates (1.161).

We shall show (1.159) using induction on $k\geq0$: estimate (1.159) is true for $k=0$ by (1.152), (1.154), and (1.156). In order for us to show estimate (1.159) for $k=1$, since $t_2-t_1=\dfrac{K\,(t_1-t_0)^2}{2\,(1-L\,t_1)}$, it suffices:

$$\frac{K\,\eta^2}{2\,(1-L\eta)}\leq\delta\,\overline{L}\,\eta^2$$

or

$$\frac{K}{1-L\eta}\leq\frac{8\,\overline{L}\,K}{K+\sqrt{K^2+8\,L\,K}}\quad(\eta\neq0)$$

or

$$\eta\leq\frac{1}{L}\left(1-\frac{K+\sqrt{K^2+8\,L\,K}}{8\,\overline{L}}\right)\quad(L\neq0,\,K\neq0).$$

But by (1.152)

$$\eta\leq\frac{4}{K+4\,L+\sqrt{K^2+8\,L\,K}}.$$

It then suffices to show

$$\frac{4}{K+4\,L+\sqrt{K^2+8\,L\,K}}\leq\frac{1}{L}\left(1-\frac{K+\sqrt{K^2+8\,L\,K}}{8\,\overline{L}}\right)$$

or

$$\frac{K+\sqrt{K^2+8\,L\,K}}{8\,\overline{L}}\leq1-\frac{4\,L}{K+4\,L+\sqrt{K^2+8\,L\,K}}$$

or

$$\frac{K+\sqrt{K^2+8\,L\,K}}{8\,\overline{L}} \le \frac{K+\sqrt{K^2+8\,L\,K}}{K+4\,L+\sqrt{K^2+8\,L\,K}},$$

which is true by (1.153).

Let us assume (1.162) holds for all integers smaller or equal to k. We shall show (1.162) holds for k replaced by $k+1$.

Using (1.154), and the induction hypothesis, we have in turn

$$
\begin{aligned}
t_{k+2}-t_{k+1} &= \frac{K}{2\,(1-L\,t_{k+1})}\,(t_{k+1}-t_k)^2 \\[2mm]
&\le \frac{K}{2\,(1-L\,t_{k+1})}\left(\left(\frac{\delta}{2}\right)^k (2\,h_A)^{2^k-1}\eta\right)^2 \\[2mm]
&\le \frac{K}{2(1-Lt_{k+1})}\left(\left(\frac{\delta}{2}\right)^{k-1}(2h_A)^{-1}\eta\right)\left(\left(\frac{\delta}{2}\right)^{k+1}(2h_A)^{2^{k+1}-1}\eta\right) \\[2mm]
&\le \left(\frac{\delta}{2}\right)^{k+1}(2\,h_A)^{2^{k+1}-1}\eta,
\end{aligned}
$$

since,

$$\frac{K}{2\,(1-L\,t_{k+1})}\left(\left(\frac{\delta}{2}\right)^{k-1}(2\,h_A)^{-1}\eta\right) \le 1, \qquad (k \ge 1). \tag{1.163}$$

Indeed, we can show instead of (1.163):

$$t_{k+1} \le \frac{1}{L}\left(1-\left(\frac{\delta}{2}\right)^{k-1}\frac{K}{4\,\overline{L}}\right),$$

which is true, since by (1.158), and the induction hypothesis:

$$
\begin{aligned}
t_{k+1} &\le t_k + \frac{\delta}{2}\,(t_k - t_{k-1}) \\[2mm]
&\le t_1 + \frac{\delta}{2}\,(t_1 - t_0) + \cdots + \frac{\delta}{2}\,(t_k - t_{k-1}) \\[2mm]
&\le \eta + \left(\frac{\delta}{2}\right)\eta + \cdots + \left(\frac{\delta}{2}\right)^k \eta \\[2mm]
&= \frac{1-\left(\dfrac{\delta}{2}\right)^{k+1}}{1-\dfrac{\delta}{2}}\,\eta \\[2mm]
&\le \frac{1}{L}\left(1-\left(\frac{\delta}{2}\right)^{k-1}\frac{K}{4\,\overline{L}}\right).
\end{aligned}
$$

That completes the induction for estimate (1.159).

Using estimate (1.162) for $j \geq k$, we obtain in turn for $2\,h_A < 1$:

$$
\begin{aligned}
t_{j+1} - t_k &= (t_{j+1} - t_j) + (t_j - t_{j-1}) + \cdots + (t_{k+1} - t_k) \\
&\leq \left(\left(\frac{\delta}{2} \right)^j (2h_A)^{2^j - 1} + \left(\frac{\delta}{2} \right)^{j-1} (2h_A)^{2^{j-1} - 1} + \right. \\
&\quad \left. \cdots + \left(\frac{\delta}{2} \right)^k (2h_A)^{2^k - 1} \right) \eta \\
&\leq \left(1 + (2h_A)^{2^k} + \left((2h_A)^{2^k} \right)^2 + \cdots \right) \left(\frac{\delta}{2} \right)^k (2h_A)^{2^k - 1} \eta \\
&= \left(\frac{\delta}{2} \right)^k \frac{(2h_A)^{2^k - 1} \eta}{1 - (2h_A)^{2^k}}.
\end{aligned}
\tag{1.164}
$$

Estimate (1.160) follows from (1.164) by letting $j \longrightarrow \infty$.

That completes the proof of Lemma 1.4. $\qquad \diamond$

Remark 1.3. Under the Newton–Kantorovich condition (1.41), the majorizing sequence

$$
\bar{\bar{t}}_0 = 0, \quad \bar{\bar{t}}_1 = \eta, \quad \bar{\bar{t}}_{k+1} = \bar{\bar{t}}_k + \frac{K\,(\bar{\bar{t}}_k - \bar{\bar{t}}_{k-1})^2}{2\,(1 - L\bar{\bar{t}}_k)} \quad (k \geq 1)
$$

was used in [335], [347], [477], [593], [770].

The corresponding ratio (see (1.159)) is given by

$$
2\,h_K = K\,\eta.
$$

But we have

$$
h_A < h_K
$$

provided that $L < K$.

Hence, sequence $\{t_n\}$ given in Lemma 1.4 is a tighter majorizing sequence that $\{\bar{\bar{t}}_n\}$, obtained under weaker sufficient convergence conditions (see (1.44)).

Application 1.2. Let

$$
A(y_n) = F'(y_n) + [y_{n-1}, y_n; G], \quad (n \geq 0)
$$

and consider (NTM) in the form

$$
y_{n+1} = y_n - \left(F'(y_n) + [y_{n-1}, y_n; G] \right)^{-1} (F(y_n) + G(y_n)) \quad (n \geq 0).
\tag{1.165}
$$

This method has order $\dfrac{1 + \sqrt{5}}{2}$ (see [151]) (same as the method of Chord), but higher than the order of

$$
z_{n+1} = z_n - F'(z_n)^{-1} (F(z_n) + G(z_n)) \quad (n \geq 0)
\tag{1.166}
$$

considered in [303], [749], [770], and the method of Chord

$$
w_{n+1} = w_n - [w_{n-1}, w_n; G]^{-1} (F(w_n) + G(w_n)) \quad (n \geq 0),
\tag{1.167}
$$

where $[x, y; G]$ denotes the divided difference of G at the points x and y [151].

Let us provide an example for this case.

Chapter 1. Two–Step Methods of Hight Efficiency Index

Example 1.5. Let $X = Y = (\mathbb{R}^2, \|\cdot\|_\infty)$. Consider the system

$$3x^2y + y^2 - 1 + |x - 1| = 0$$
$$x^4 + xy^3 - 1 + |y| = 0.$$

Set $\|x\|_\infty = \|(x', x'')\|_\infty = \max\{|x'|, |x''|\}$, $F = (F_1, F_2)$, $G = (G_1, G_2)$. For $x = (x', x'') \in \mathbb{R}^2$ we take $F_1(x', x'') = 3(x')^2 x'' + (x'')^2 - 1$, $F_2(x', x'') = (x')^4 + x'(x'')^3 - 1$, $G_1(x', x'') = |x' - 1|$, $G_2(x', x'') = |x''|$. We shall take $[x, y; G] \in M_{2 \times 2}(\mathbb{R})$ as

$$[x, y; G]_{i,1} = \frac{G_i(y', y'') - G_i(x', y'')}{y' - x'}, \quad [x, y; G]_{i,2} = \frac{G_i(x', y'') - G_i(x', x'')}{y'' - x''}, \quad i = 1, 2,$$

provided that $y' \neq x'$ and $y'' \neq x''$. Otherwise define $[x, y; G]$ to be the zero matrix in $M_{2 \times 2}(\mathbb{R})$.

Using method (1.166) with $z_0 = (1, 0)$ we obtain

n	$z_n^{(1)}$	$z_n^{(2)}$	$\|z_n - z_{n-1}\|$
0	1	0	
1	1	0.333333333333333	3.333E–1
2	0.906550218340611	0.354002911208151	9.344E–2
3	0.885328400663412	0.338027276361322	2.122E–2
4	0.891329556832800	0.326613976593566	1.141E–2
5	0.895238815463844	0.326406852843625	3.909E–3
6	0.895154671372635	0.327730334045043	1.323E–3
7	0.894673743471137	0.327979154372032	4.809E–4
8	0.894598908977448	0.327865059348755	1.140E–4
9	0.894643228355865	0.327815039208286	5.002E–5
10	0.894659993615645	0.327819889264891	1.676E–5
11	0.894657640195329	0.327826728208560	6.838E–6
12	0.894655219565091	0.327827351826856	2.420E–6
13	0.894655074977661	0.327826643198819	7.086E–7
...			
39	0.894655373334687	0.327826521746298	5.149E–19

Using the method of chord (i.e., (1.167)) with $w_{-1} = (1, 0)$, and $w_0 = (5, 5)$, we obtain:

n	$w_n^{(1)}$	$w_n^{(2)}$	$\|w_n - w_{n-1}\|$
-1	5	5	
0	1	0	5.000E+00
1	0.989800874210782	0.012627489072365	1.262E–02
2	0.921814765493287	0.307939916152262	2.953E–01
3	0.900073765669214	0.325927010697792	2.174E–02
4	0.894939851625105	0.327725437396226	5.133E–03
5	0.894658420586013	0.327825363500783	2.814E–04
6	0.894655375077418	0.327826521051833	3.045E–04
7	0.894655373334698	0.327826521746293	1.742E–09
8	0.894655373334687	0.327826521746298	1.076E–14
9	0.894655373334687	0.327826521746298	5.421E–20

Using our method (1.165) with $y_{-1} = (1,0), y_0 = (5,5)$, we obtain

n	$y_n^{(1)}$	$y_n^{(2)}$	$\|y_n - y_{n-1}\|$
-1	5	5	
0	1	0	5
1	0.909090909090909	0.363636363636364	3.636E–01
2	0.894886945874111	0.329098638203090	3.453E–02
3	0.894655531991499	0.327827544745569	1.271E–03
4	0.894655373334793	0.327826521746906	1.022E–06
5	0.894655373334687	0.327826521746298	6.089E–13
6	0.894655373334687	0.327826521746298	2.710E–20

The solution is

$$x^\star = (.894655373334687, .327826521746298)$$

chosen from the lists of the tables displayed above.

Hence method (1.165) converges faster than (1.166) suggested in Chen and Yamamoto [303], Zabrejko and Nguen [764] in this case, and the method of chord [151].

Chapter 2

Numerical Methods under Slantly–Differentiability

In this chapter, we present a new results for the local convergence of the Newton–like method to a unique solution of nondifferentiable variational inclusions in a Banach space setting. Using the Lipschitz–like property of set–valued mappings and the concept of slant differentiability hypothesis on the operator involved, as was inaugurated by Chen, Nashed and Qi [301]. The linear convergence of the Newton–like method is also established. Our results extend the applicability of Newton–like method [189], [301] to variational inclusions.

We are concerned with the problem of approximating a locally unique solution $x^\star \in \mathcal{D}$ of the generalized equation

$$0 \in F(x) + G(x), \tag{2.1}$$

where F is a continuous mapping from an open subset $\mathcal{D}$ of a Banach space X into a Banach space $\mathcal{Y}$, and G is a set–valued map from X to the subsets of $\mathcal{Y}$ with closed graph.

A large number of problems in applied mathematics and engineering are solved by finding the solutions of generalized equation (2.1), introduced by Robinson [624]. Consider an example of application to variational inequalities (see [638]). Let K be a convex set in $\mathbb{R}^n$ and φ be a function from K to $\mathbb{R}^n$. The variational inequality problem consists in seeking k_0 in K such that

$$\text{for each } k \in K, \quad (\varphi(k_0), k - k_0) \geq 0, \tag{2.2}$$

where $(.,.)$ denotes the euclidean scalar product on $\mathbb{R}^n$ defined by $(x,y) = \sum_{i=1}^{n} x_i y_i$ for all x, y in $\mathbb{R}^n$. Let I_K denote the convex indicator function of K and ∂ the subdifferential operator. Then the problem (2.2) is equivalent to

$$0 \in \varphi(k_0) + \mathcal{H}(k_0), \tag{2.3}$$

with $\mathcal{H} = \partial I_K$ (also called the normal cone of K). The variational inequality problem (2.2) is equivalent to (2.3) which is a generalized equation in formulation (2.1).

In the particular case $G = \{0\}$, (2.1) is a nonlinear equation in the form

$$F(x) = 0. \tag{2.4}$$

For example, dynamic systems are mathematically modeled by differential or difference equations, and their solutions usually represent the state of the systems, which are determined by solving equation (2.4).

Most of the numerical approximation methods of $x^\star$ of (2.1) require the expensive computation of the Fréchet–derivative $F'(x)$ of operator F at each step, for example Newton's method:

$$0 \in F(x_n) + F'(x_n)(x_{n+1} - x_n) + G(x_{n+1}), \quad (x_0 \in \mathcal{D}), \quad (n \geq 0). \tag{2.5}$$

Alternative methods to (2.5) are:

(i) The Secant method

$$0 \in F(x_n) + [x_n, x_{n-1}; F](x_{n+1} - x_n) + G(x_{n+1})$$
$$(x_{-1}, x_0 \in \mathcal{D}), \quad (n \geq 0), \tag{2.6}$$

(ii) Steffensen's method

$$0 \in F(x_n) + [g_1(x_n), g_2(x_n); F](x_{n+1} - x_n) + G(x_{n+1})$$
$$(x_0 \in \mathcal{D}), \quad (n \geq 0), \tag{2.7}$$

where g_i $(i = 1, 2)$ are continuous functions from a neighborhood $\mathcal{D}$ of $x^\star$ into X and $[x, y; F]$ is a first order divided difference of F on the points x and y.

A comprehensive bibliography on these methods and their variants is given in [137], [190].

We are interested in a numerical method for solving generalized equation (2.1), when the involved function F is slantly differentiable at $x^\star$. We proceed by replacing the term $F'(x_n)$ in method (2.5) by $A(x_n)$, where $A(x) \in L(X, \mathcal{Y})$.

We consider the Newton–like method

$$0 \in F(x_n) + A(x_n)(x_{n+1} - x_n) + G(x_{n+1}), \quad (x_0 \in \mathcal{D}), \quad (n \geq 0). \tag{2.8}$$

for approximating $x^\star$.

In the nonlinear equation case (i.e., $G = \{0\}$ in (2.1)), the method (2.8) becomes

$$x_{n+1} = x_n - A(x_n)^{-1} F(x_n), \quad (x_0 \in \mathcal{D}), \quad (n \geq 0), \tag{2.9}$$

which was considered by Chen, Nashed and Qi [301] for slantly differentiable operator. Consequently, we can approximate the solution k_0 of variational inequality (2.2) using our algorithm (2.8), when φ is slantly differentiable. Here, we are motivated by the works in [189], [301], [302], [605]. Using the Lipschitz–like concept of set–valued mappings and slant differentiability hypothesis on the involved operator, we extend the applicability of Newton's method [301], [302] to variational inequalities. We prove that Newton–like method (2.8) converges linearly.

First, we collect a number of basic definitions and properties on slant differentiability, and recall a fixed points theorem for set–valued maps. Second, we show an existence–convergence theorem of sequence given by (2.8) and some improvments and remarks are also presented.

Chapter 2. Numerical Methods under Slantly–Differentiability   39

In order to make the chapter as self–contained as possible we reintroduce some definitions and some results on fixed point theorems for set–valued maps [155], [190], [301], [356], [638]. We let X be a Banach space equipped with the norm $\| \, . \, \|$. The distance from a point x to a set A in X is defined by $\mathrm{dist}\,(x,A) = \inf\limits_{y \in A} \| x - y \|$, with the convention $\mathrm{dist}\,(x,\emptyset) = +\infty$ (according to the general convention $\inf \emptyset = +\infty$). Given a subset C of X, we denote by $e(C,A)$ the Hausdorff–Pompeiu excess of C into A, defined by

$$e(C,A) = \sup_{x \in C} \, \mathrm{dist}\,(x,A),$$

with the conventions $e(\emptyset,A) = 0$ and $e(C,\emptyset) = +\infty$ whenever $C \neq \emptyset$. For a set–valued mapping $\Lambda : X \rightrightarrows X$, we denote by $\mathrm{gph}\,\Lambda$ the set $\{(x,y) \in X \times X, \; y \in \Lambda(x)\}$ and $\Lambda^{-1}(y)$ the set $\{x \in X, \; y \in \Lambda(x)\}$. The closed ball in X centered at x with radius r will be denoted by $\mathbb{B}_r(x)$.

We need to define the pseudo–Lipschitzian concept of set–valued mapping, introduced by Aubin [248] and also known as Lipschitz–like property [538]:

Definition 2.1. A set–valued mapping Γ is pseudo–Lipschitz around $(\bar{x},\bar{y}) \in \mathrm{gph}\,\Gamma$ with modulus M_1 if there exist constants a and b such that

$$\sup_{z \in \Gamma(y') \cap \mathbb{B}_a(\bar{y})} \mathrm{dist}\,(z,\Gamma(y'')) \leq M_1 \, \| y' - y'' \|, \quad \text{for all } y' \text{ and } y'' \text{ in } \mathbb{B}_b(\bar{x}). \tag{2.10}$$

In the term of excess, we have an equivalent definition of pseudo–Lipschitzian property replacing the inequality (2.10) by

$$e(\Gamma(y') \cap \mathbb{B}_a(\bar{y}),(y'')) \leq M_1 \, \| y' - y'' \|, \quad \text{for all } y' \text{ and } y'' \text{ in } \mathbb{B}_b(\bar{x}). \tag{2.11}$$

Pseudo–Lipschitzian property plays a crutial role in many aspects of variational analysis and applications [538], [638]. Let us note that the Lipschitz–like property of Γ is equivalent to the metric regularity of Γ^{-1}, which is a basic well–posedness property in optimization problems. Another characterization of Lipschitz–like property is presented by Mordukhovich [538] via the concept of coderivative. For some characterizations and applications of the Lipschitz–like property the reader could be referred to [248], [356], [538], [637], [638] and the references given there.

We need also the following fixed point theorem [356].

Lemma 2.1. *Let* ϕ *be a set–valued mapping from* X *into the closed subsets of* X. *We suppose that for* $\eta_0 \in X$, $r \geq 0$, *and* $0 \leq \lambda < 1$, *the following properties hold*

1. $\mathrm{dist}\,(\eta_0,\phi(\eta_0)) \leq r \, (1 - \lambda)$;

2. $e(\phi(z_1) \cap \mathbb{B}_r(\eta_0),\phi(z_2)) \leq \lambda \, \| z_1 - z_2 \|, \; \forall z_1, z_2 \in \mathbb{B}_r(\eta_0)$.

Then ϕ *has a fixed point in* $\mathbb{B}_r(\eta_0)$. *That is, there exists* $x \in \mathbb{B}_r(\eta_0)$ *such that* $x \in \phi(x)$. *If* ϕ *is single–valued mapping, then* x *is the unique fixed point of* ϕ *in* $\mathbb{B}_r(\eta_0)$.

We need the notion of slant differentiability (see [301]):

Definition 2.2. An operator $F : \mathcal{D} \subseteq X \longrightarrow \mathcal{Y}$ is slantly differentiable at $x \in \mathcal{D}$, if there exists an operator $f^\circ : \mathcal{D} \longrightarrow L(X, \mathcal{Y})$, such that the family $\{f^\circ(x+h)\}$ of bounded linear mappings is uniformly bounded in the operator norm, for h sufficiently small, and

$$\lim_{h \to 0} \frac{F(x+h) - F(x) - f^\circ(x+h)\,h}{\|\,h\,\|} = 0. \tag{2.12}$$

The function f° is called a slanting function for F at x.

Definition 2.3. An operator $F : \mathcal{D} \subseteq X \longrightarrow \mathcal{Y}$ is slantly differentiable in an open domain $\mathcal{D}_0 \subseteq \mathcal{D}$, if there exists an operator $f^\circ : \mathcal{D} \longrightarrow L(X, \mathcal{Y})$, such that f° is a slantly operator for F at every point $x \in \mathcal{D}_0$. In this case, f° is called a slanting function for F in $\mathcal{D}_0$.

Definition 2.4. Assume $f^\circ : \mathcal{D} \longrightarrow L(X, \mathcal{Y})$ is a slantly operator for F at $x \in \mathcal{D}$.
The set

$$\partial_S F(x) := \{ \lim_{x_k \to x} f^\circ(x_k) \} \tag{2.13}$$

is the slant derivative of F associated with f° at $x \in \mathcal{D}$. Here, $\lim_{x_k \to x} f^\circ(x_k)$ means the limit of $f^\circ(x_k)$ for any sequence $\{x_k\} \subset \mathcal{D}$, such that $x_k \to x$, and $\lim_{x_k \to x} f^\circ(x_k)$ exists. Note that $\partial_S F(x) \neq \emptyset$, since $f^\circ(x) \in \partial_S F(x)$.

We finally need the definitions of smoothing and semismoothing functions of F (see [301], [650]).

Definition 2.5. An operator $f : \mathcal{D} \times (0, +\infty) \longrightarrow \mathcal{Y}$ is a smoothing mapping of F, if f is continuously differentiable with respect to x, and for any $x \in \mathcal{D}$, any $\varepsilon > 0$,

$$\| F(x) - f(x, \varepsilon) \| \le \mu\,\varepsilon, \tag{2.14}$$

where μ is a positive constant.
The smoothing operator is said to satisfy the slant derivative consistency property at $\hat{x}$ (in $\mathcal{D}$), if

$$\lim_{\varepsilon \to 0^+} f_x'(x, \varepsilon) = f^\circ(x) \in L(X, \mathcal{Y}) \tag{2.15}$$

exists for x in a neighborhood of $\hat{x}$ (in $\mathcal{D}$), and f° serves as a slanting mapping for F at $\hat{x}$ (in $\mathcal{D}$).

Definition 2.6. An operator $F : \mathcal{D} \subseteq X \longrightarrow \mathcal{Y}$ is semismooth at x if there is a slantly operator f° for F in a neighborhood N_x of x such that f° and the associated slant derivative satisfy the conditions:

(i) $\lim_{t \to 0^+} f^\circ(x+t\,h)\,h$ exists for every $h \in X$,

and

$$\lim_{h \to 0} \frac{\lim_{t \to 0^+} f^\circ(x+t\,h)\,h - f^\circ(x+h)\,h}{\|\,h\,\|} = 0. \tag{2.16}$$

(ii)

$$f^\circ(x+h)\,h - V\,h = o\,(\|\,h\,\|) \qquad \text{for all } V \in \partial_S F(x+h). \tag{2.17}$$

Chapter 2. Numerical Methods under Slantly–Differentiability   41

Definition 2.7.   1. A mapping $F : \mathcal{D} \subseteq \mathbb{R}^n \longrightarrow \mathbb{R}^n$ is said to be directionally differentiable at $x \in \mathcal{D}$ along direction d if the following limit

$$F'(x;d) := \lim_{t\downarrow 0} \frac{F(x+t\,d) - F(x)}{t} \tag{2.18}$$

exists.

Note that a locally Lipschitz mapping F admits directional derivative at $x \in \mathcal{D}$ if and only if (see [650]):

$$\lim_{d\to 0} \frac{F(x+d) - F(x) - F'(x;d)}{\|d\|} = 0.$$

2. For $F : \mathcal{D} \subseteq \mathbb{R}^n \longrightarrow \mathbb{R}^n$ a locally Lipschitz continuous function, the limiting Jacobian of F at $x \in \mathcal{D}$ is defined by

$$\partial F(x) = \{\mathcal{M} \in L(\mathbb{R}^n, \mathbb{R}^n) : \exists\, u^k \in \mathcal{D},\ \lim_{k\longrightarrow\infty} F'(u^k) = \mathcal{M}\}. \tag{2.19}$$

3. Let $F : \mathcal{D} \subseteq \mathbb{R}^n \longrightarrow \mathbb{R}^n$ be a locally Lipschitz continuous function. Clarke's Jacobian of F at $x \in \mathcal{D}$ is defined by

$$\partial^\circ F(x) = \overline{\text{co}}\,\partial F(x), \tag{2.20}$$

where $\overline{\text{co}}\,\mathcal{A}$ is the closed convex envelope of $\mathcal{A} \subseteq \mathbb{R}^n$.

Before presenting our main result of convergence of Newton–like method (2.8), we give the local convergence theorem restricted to the resolution of nonlinear equation (2.4) [189, 219]:

Theorem 2.1. *Assume:*

$F : \mathcal{D} \subseteq X \longrightarrow \mathcal{Y}$ *is a slantly differentiable operator at a solution $x^\star$ of equation $F(x) =$*
0.

Let f° be a slanting operator for F at $x^\star$, $A(x) \in L(X, \mathcal{Y})$ $(x \in \mathcal{D})$ with $A(x^\star)^{-1} \in$
$L(\mathcal{Y}, X)$,

$$\| A(x^\star)^{-1}\,(A(x^\star + h) - A(x^\star)) \| \le L \,\| h \|, \tag{2.21}$$

$$\| A(x^\star)^{-1}\,(A(x^\star + h) - f^\circ(x^\star + h)) \| \le M \,\| h \| + M_0, \tag{2.22}$$

as $\| h \| \to 0$.

By the definition of slant differentiability, there exist $c > 0$ and $\delta = \delta(c) > 0$ such that for all $\| h \| \le \delta$, $\| f^\circ(x^\star + h) \| \le c$, and

$$\frac{\| F(x^\star + h) - F(x^\star) - f^\circ(x^\star + h)\,h \|}{\| h \|} \le c. \tag{2.23}$$

Assume:

$$M_0 + c < 1. \tag{2.24}$$

Set

$$r = \min\left\{ \delta,\ \frac{1}{L},\ \frac{1 - c - M_0}{M + L}\right\} \tag{2.25}$$

and

$$U(x^\star, r) = \{x \in X : \| x - x^\star \| < r\} \subseteq \mathcal{D}. \tag{2.26}$$

Then, sequence $\{x_n\}$ generated by (2.9) is well defined, remains in $U(x^\star, r)$ for all $n \geq 0$ and converges to $x^\star$, provided that $x_0 \in U(x^\star, r)$.

Moreover, the following estimates hold for all $k \geq 0$:

$$\| x_{k+1} - x^\star \| \leq q \| x_k - x^\star \|, \tag{2.27}$$

where,

$$0 \leq q = \frac{c + M_0 + M \| x_0 - x^\star \|}{1 - L \| x_0 - x^\star \|} < 1. \tag{2.28}$$

We will consider the existence and the convergence of the sequence defined by (2.8) to the solution $x^\star$ of (2.1).

The main result is as follows.

Theorem 2.2. *Assume:*

(C_1) *$F : \mathcal{D} \subseteq X \longrightarrow \mathcal{Y}$ is a slantly differentiable operator at a solution $x^\star$ of generalized equation $0 \in F(x) + G(x)$, where G is a set–valued map from X to the subsets of $\mathcal{Y}$ with closed graph.*

(C_2) *Let f° be a slanting operator for F at $x^\star$, $A(x) \in L(X, \mathcal{Y})$ $(x \in \mathcal{D})$, and there exist a non–negative constants L, M and M_0 such that:*

$$\| A(x^\star + h) - A(x^\star) \| \leq L \| h \|, \tag{2.29}$$

$$\| A(x^\star + h) - f^\circ(x^\star + h) \| \leq M \| h \| + M_0, \tag{2.30}$$

as $\| h \| \to 0$.

(C_3) *Hypothesis (2.23) holds.*

(C_4) *The set–valued map $(G + A(x^\star)(. - x^\star))^{-1}$ is pseudo–Lipschitz around $(-F(x^\star), x^\star)$ with constants M_1, a and b (these constants are given in Definition 2.1).*

(C_5)

$$C_0 = M_1 (c + M \delta + M_0) < 1, \tag{2.31}$$

where, c and δ are given by the definition of slant differentiability (see (2.23)).

Then, for every constant C and $\delta^\star$ satisfying

$$1 \geq C > C_0 \tag{2.32}$$

$$0 < \delta^\star < \delta_0^\star = \min \left\{ a; \sqrt{\frac{b}{2L}}; \frac{b}{2(c + M \delta + M_0)}; \frac{C - C_0}{C M_1 L} \right\} \tag{2.33}$$

respectively, and for every starting point x_0 in $\mathbb{B}_{\delta^\star}(x^\star)$ (with $x_0 \neq x^\star$), there exists a sequence (x_k) defined by (2.8) which is convergent to $x^\star$, and satisfies the following inequality for $k \geq 0$

$$\| x_{k+1} - x^\star \| \leq C \| x_k - x^\star \|. \tag{2.34}$$

Chapter 2. Numerical Methods under Slantly–Differentiability   43

Remark 2.1. We need to introduce some notations [189], [190]. First, define the set–valued maps $\Lambda : X \rightrightarrows \mathcal{Y}$ and $\Theta_k : X \rightrightarrows X$ by

$$\Lambda(x) = F(x^\star) + A(x^\star)(x - x^\star) + G(x), \qquad \Theta_k(x) = \Lambda^{-1}(\Xi_k(x)), \quad k \geq 0 \qquad (2.35)$$

where Ξ_k is a mapping from X to $\mathcal{Y}$ defined by

$$\Xi_k(x) = F(x^\star) - F(x_k) + A(x^\star)(x - x^\star) - A(x_k)(x - x_k), \qquad k \geq 0. \qquad (2.36)$$

The proof of Theorem 2.2 is given by induction on k. We first state a result involving the starting point x_0. Let us note that the point x_1 is a fixed point of Θ_0 if and only if $0 \in F(x_1) + A(x_0)(x_1 - x_0) + G(x_1)$. Once x_k is computed, we show that the function Θ_k has a fixed point x_{k+1} in X. This process is useful to prove the existence of a sequence (x_k) satisfying (2.8).

Proposition 2.1. *Under the assumptions of Theorem 2.2, and for every distinct starting point x_0 in $\mathbb{B}_{\delta^\star}(x^\star)$ (with $x_0 \neq x^\star$), the set–valued map Θ_0 has a fixed point x_1 in $\mathbb{B}_{\delta^\star}(x^\star)$ satisfying*

$$\| x_1 - x^\star \| \leq C \| x_0 - x^\star \|, \qquad (2.37)$$

where C and $\delta^\star$ are given by Theorem 2.2.

Proof.
Step 1. We prove that the first assumption in Lemma 2.1 is satisfied.
   By the pseudo–Lipschitzness hypothesis, we have

$$e(\Lambda^{-1}(y') \cap \mathbb{B}_a(x^\star), \Lambda^{-1}(y'')) \leq M_1 \| y' - y'' \|, \ \forall y', y'' \in \mathbb{B}_b(0). \qquad (2.38)$$

By (2.36), we have:

$$
\begin{aligned}
\| \Xi_0(x^\star) \| &= \| F(x^\star) - F(x_0) - A(x_0)(x^\star - x_0) \| \\
&= \| F(x_0) - F(x^\star) - f^\circ(x_0)(x_0 - x^\star) + \\
&\quad f^\circ(x_0)(x_0 - x^\star) - A(x_0)(x_0 - x^\star) \| \\
&\leq \| F(x_0) - F(x^\star) - f^\circ(x_0)(x_0 - x^\star) \| + \\
&\quad \| f^\circ(x_0)(x_0 - x^\star) - A(x_0)(x_0 - x^\star) \| \\
&\leq \| F(x_0) - F(x^\star) - f^\circ(x_0)(x_0 - x^\star) \| + \\
&\quad \| f^\circ(x_0) - A(x_0) \| \| x_0 - x^\star \| .
\end{aligned}
\qquad (2.39)
$$

Using (3.3) and (2.30), (2.39) becomes

$$
\begin{aligned}
\| \Xi_0(x^\star) \| &\leq c \| x_0 - x^\star \| + (M \| x_0 - x^\star \| + M_0) \| x_0 - x^\star \| \\
&\leq (c + M\delta + M_0) \| x_0 - x^\star \| .
\end{aligned}
\qquad (2.40)
$$

By (2.33) and (2.40), we obtain $\Xi_0(x^\star) \in \mathbb{B}_b(0)$.
Hence from (2.38) one gets

$$
\begin{aligned}
e\left(\Lambda^{-1}(0) \cap \mathbb{B}_{\delta^\star}(x^\star), \Theta_0(x^\star)\right) &= e\left(\Lambda^{-1}(0) \cap \mathbb{B}_{\delta^\star}(x^\star), \Lambda^{-1}[\Xi_0(x^\star)]\right) \\
&\leq M_1 (c + M\delta + M_0) \| x_0 - x^\star \| \\
&= C_0 \| x_0 - x^\star \| .
\end{aligned}
\qquad (2.41)
$$

According to the definition of excess e, and using (2.41), we have

$$\begin{aligned}
\text{dist}\,(x^\star, \Theta_0(x^\star)) &\leq e\left(\Lambda^{-1}(0) \cap \mathbb{B}_{\delta^\star}(x^\star), \Theta_0(x^\star)\right) \\
&\leq C_0 \parallel x_0 - x^\star \parallel .
\end{aligned} \tag{2.42}$$

Since $C > C_0$, there exists $\lambda \in [0, 1[$ such that

$$C\,(1 - \lambda) \geq C_0,$$

and

$$\text{dist}\,(x^\star, \Theta_0(x^\star)) \leq C\,(1 - \lambda) \parallel x_0 - x^\star \parallel . \tag{2.43}$$

We can choose $\lambda = \dfrac{C - C_0}{C}$. Identifying η_0, ϕ and r in Lemma 2.1 by

$$x^\star, \quad \Theta_0 \quad \text{and} \quad r_0 = C \parallel x_0 - x^\star \parallel$$

respectively, we can deduce from the inequality (2.43) that the first assumption in Lemma 2.1 is satisfied.

Step 2. We prove now that the second assumption of Lemma 2.1 is verified.

Using (2.32) and (2.33), we have $r_0 \leq \delta^\star \leq a$. Moreover for $x \in \mathbb{B}_{\delta^\star}(x^\star)$, we get in turn

$$\begin{aligned}
&\parallel \Xi_0(x) \parallel \\
=\; & \parallel F(x^\star) - F(x_0) + A(x^\star)\,(x - x^\star) - A(x_0)\,(x - x_0) \parallel \\
\\
\leq\; & \parallel F(x_0) - F(x^\star) - f^\circ(x_0)(x_0 - x^\star) \parallel + \\
& \parallel f^\circ(x_0)(x_0 - x^\star) - A(x^\star)(x - x^\star) + A(x_0)(x - x^\star + x^\star - x_0) \parallel \\
\\
\leq\; & \parallel F(x_0) - F(x^\star) - f^\circ(x_0)(x_0 - x^\star) \parallel + \\
& \parallel f^\circ(x_0) - A(x_0) \parallel \parallel x_0 - x^\star \parallel + \parallel A(x_0) - A(x^\star) \parallel \parallel x - x^\star \parallel .
\end{aligned} \tag{2.44}$$

Using (2.23), (2.29), and (2.30), we obtain

$$\begin{aligned}
\parallel \Xi_0(x) \parallel &\leq c \parallel x_0 - x^\star \parallel + (M\,\delta + M_0) \parallel x_0 - x^\star \parallel + \\
& \quad L \parallel x_0 - x^\star \parallel \parallel x - x^\star \parallel \\
&\leq c\,\delta^\star + (M\,\delta + M_0)\,\delta^\star + L\,\delta^{\star 2}.
\end{aligned} \tag{2.45}$$

By (2.33), we deduce that for all $x \in \mathbb{B}_{\delta^\star}(x^\star)$, we have $\Xi_0(x) \in \mathbb{B}_b(0)$. Then, it follows that for all $x', x'' \in \mathbb{B}_{r_0}(x^\star)$, we have

$$e(\Theta_0(x') \cap \mathbb{B}_{r_0}(x^\star), \Theta_0(x'')) \leq e(\Theta_0(x') \cap \mathbb{B}_{\delta^\star}(x^\star), \Theta_0(x'')),$$

which yields by (2.38), and the pseudo–Lipschitzness hypothesis to:

$$\begin{aligned}
& e(\Theta_0(x') \cap \mathbb{B}_{r_0}(x^\star), \Theta_0(x'')) \\
\leq\; & M_1 \parallel \Xi_0(x') - \Xi_0(x'') \parallel \\
=\; & M_1 \parallel A(x^\star)(x' - x'') - A(x_0)\,(x' - x'') \parallel \\
\leq\; & M_1 \parallel A(x^\star) - A(x_0) \parallel \parallel x'' - x' \parallel \\
\leq\; & M_1\,L \parallel x_0 - x^\star \parallel \parallel x'' - x' \parallel \\
\leq\; & M_1\,L\,\delta^\star \parallel x'' - x' \parallel \leq \lambda \parallel x'' - x' \parallel .
\end{aligned} \tag{2.46}$$

Chapter 2. Numerical Methods under Slantly–Differentiability   45

The second condition of Lemma 2.1 is satisfied. By Lemma 2.1, we can deduce the existence of a fixed point $x_1 \in \mathbb{B}_{r_0}(x^\star)$ for the map Θ_0. The proof of Proposition 2.1 is complete. $\diamondsuit$

Proof of Theorem 2.2. Keep $\eta_0 = x^\star$, and for $k \geq 1$, $r := r_k = C \parallel x^\star - x_k \parallel$. By Remark 2.1, the application of Proposition 2.1 to the map Θ_k gives the desired result. $\diamondsuit$

Remark 2.2. We can enlarge the radius of convergence in Theorem 2.2 even further as follows: using inequalities (2.40), (2.45), and $(c + M\delta + M_0)\eta - b \leq (c + M\delta + M_0)\eta + L\eta^2 - b$, we can improve $\delta^\star$ given by (2.33) by considering the constant $\delta^{\star\prime}$:

$$\delta^{\star\prime} < \delta^{\star\prime}_0 = \min\left\{a; \delta_1\right\}$$

where,

$$\delta_1 = \max\left\{\eta > 0 : (c + M\delta + M_0)\eta + L\eta^2 - b < 0\right\}.$$

In view of Theorem 2.2, we obtain an extension of [189, Theorem 1.3.2, page 56]:

Proposition 2.2. *Assume:*

(i) *$F : \mathcal{D} \subseteq X \longrightarrow \mathcal{Y}$ be slantly differentiable operator at a solution $x^\star$ of generalized equation $0 \in F(x) + G(x)$, where G is a set–valued map from X to the subsets of $\mathcal{Y}$ with closed graph.*

(ii) *Let f° be a slanting operator for F at $x^\star$, $A(x) \in L(X, \mathcal{Y})$ $(x \in \mathcal{D})$, and there exist a non–negative constants L, ρ_1, and ρ_2, such that for any x in a neighborhood of $x^\star$:*

$$\begin{aligned} \parallel A(x^\star + h) - A(x^\star) \parallel &\leq L \parallel h \parallel, \\ \parallel A(x^\star + h) - f^\circ(x^\star + h) \parallel &\longrightarrow 0 \end{aligned} \tag{2.47}$$

as $\parallel h \parallel \to 0$,

$$\parallel F(x) - F(x^\star) - f^\circ(x)(x - x^\star) \parallel \leq \rho_1 \parallel x - x^\star \parallel, \tag{2.48}$$

and

$$\parallel A(x) - f^\circ(x) \parallel \leq \rho_2. \tag{2.49}$$

(iii) *The set–valued map $(G + A(x^\star)(. - x^\star))^{-1}$ is pseudo–Lipschitz around $(-F(x^\star), x^\star)$ with constants M_1, a and b.*

(iv)

$$\overline{C}_0 = M_1(\rho_1 + \rho_2) < 1. \tag{2.50}$$

Then, for every constant $\overline{C}$ and $\overline{\delta^\star}$ such that

$$1 \geq \overline{C} > \overline{C}_0 \tag{2.51}$$

$$0 < \overline{\delta^\star} < \overline{\delta^\star_0} = \min\left\{a; \sqrt{\frac{b}{2L}}; \frac{b}{2(\rho_1 + \rho_2)}; \frac{\overline{C} - \overline{C}_0}{\overline{C} M_1 L}\right\} \tag{2.52}$$

respectively, and for every starting point x_0 in $\mathbb{B}_{\overline{\delta^\star}}(x^\star)$ (with $x_0 \neq x^\star$), the conclusion of Theorem 2.2 holds with the constant C replaced by $\overline{C}$ in (2.34).

Idea of the proof. Proposition 2.2 can be proved in the same way as Theorem 2.2. We prove that the two assumption in Lemma 2.1 are satisfied. Using assumptions (i)–(iv), simply replace estimates (2.39) and (2.44)–(2.45) in the proof of Theorem 2.2 by

$$\| \Xi_0(x^\star) \| \leq (\rho_1 + \rho_2) \| x_0 - x^\star \|$$

and

$$\begin{aligned} \| \Xi_0(x) \| &\leq \| F(x_0) - F(x^\star) - f^\circ(x_0)(x_0 - x^\star) \| + \\ & \quad \| f^\circ(x_0) - A(x_0) \| \, \| x_0 - x^\star \| + \| A(x_0) - A(x^\star) \| \, \| x - x^\star \| \\ &\leq (\rho_1 + \rho_2) \| x_0 - x^\star \| + L \| x - x^\star \| \, \| x_0 - x^\star \| \\ &\leq (\rho_1 + \rho_2) \overline{\delta^\star} + L \overline{\delta^\star}^2 \end{aligned}$$

respectively. $\diamond$

In view of Theorem 2.2, and Definitions 2.4–2.7, we obtain the special cases:

Proposition 2.3. *Assume:*

(a) *$F : \mathcal{D} \subseteq X \longrightarrow \mathcal{Y}$ be slantly differentiable operator at a solution $x^\star$ of generalized equation $0 \in F(x) + G(x)$, where G is a set–valued map from X to the subsets of $\mathcal{Y}$ with closed graph. Let f° be a slanting operator for F.*

(b) *The set–valued map $(G + A(x^\star)(.-x^\star))^{-1}$ is pseudo–Lipschitz around $(-F(x^\star), x^\star)$.*

Then Newton–like method:

$$0 \in F(x_n) + f^\circ(x_n)(x_{n+1} - x_n) + G(x_{n+1}), \quad (x_0 \in \mathcal{D}), \quad (n \geq 0)$$

converges to $x^\star$, with $A = f^\circ$ in condition (b).

Proof. The proof of Proposition 2.3 is the same one of the proof of Theorem 2.2. It is enough to make some modification by replacing the operator A by f° in expressions (2.35) and (2.36).

Conditions (C_1), (C_3) and (C_4) of Theorem 2.2 are satisfied. The condition (C_2) of Theorem 2.2 is also satisfied with $M = M_0 = 0$. Then we can apply Theorem 2.2 by replacing (C_5) by additional condition $M_1 c < 1$. $\diamond$

Open problem. We assume that both conditions (a) and (b) of Proposition 2.3 are verified. Our conjecture is given by the following two items:

(A) If $f : \mathcal{D} \times (0, +\infty) \longrightarrow \mathcal{Y}$ is smoothing function, which satisfies the slant consistency property (2.15) in a neighborhood of $x^\star$, then the smoothing Newton method:

$$0 \in F(x_n) + f'_x(x_n, \varepsilon_n)(x_{n+1} - x_n) + G(x_{n+1}), \quad (x_0 \in \mathcal{D}), \quad (n \geq 0)$$

converges to $x^\star$, with $A = f'_x(., \varepsilon)$ ($\varepsilon > 0$ is a fixed parameter) in condition (b) of Proposition 2.3 and $\varepsilon_n > 0$ is an appropriate real sequence converging to 0 [298], [302].

(B) If F is semismooth at x^*, then the semismoothing Newton method:

$$0 \in F(x_n) + V(x_n)(x_{n+1} - x_n) + G(x_{n+1}), \quad (x_0 \in \mathcal{D}), \quad (n \geq 0)$$

converges to x^*, with $A = V$ in condition (b) of Proposition 2.3 and $V(x) \in \partial_S F(x)$ for $x \in \mathcal{D}$.

Here are some points to solve this problem, but are insufficient. For (A) we replace the mappings in (2.35) and (2.36) by

$$\Lambda(x) = F(x^*) + f'_x(x^*, \varepsilon)(x - x^*) + G(x), \quad \Theta_k(x) = \Lambda^{-1}(\Xi_k(x)),$$

and

$$\Xi_k(x) = F(x^*) - F(x_k) + f'_x(x^*, \varepsilon)(x - x^*) - f'_x(x_k, \varepsilon_k)(x - x_k), \qquad k \geq 0.$$

respectively. We suppose that ε and sequence ε_k $(k \geq 0)$ satisfy $\varepsilon \leq \| x_k - x^* \|$ and $\varepsilon_k \leq \| x_k - x^* \|$ for all $k \geq 0$.

For (B) we replace the mappings in (2.35) and (2.36) by

$$\Lambda(x) = F(x^*) + V(x^*)(x - x^*) + G(x), \quad \Theta_k(x) = \Lambda^{-1}(\Xi_k(x)),$$

and

$$\Xi_k(x) = F(x^*) - F(x_k) + V(x^*)(x - x^*) - V(x_k)(x - x_k), \qquad k \geq 0.$$

Chapter 3

$\mathcal{H}$–Convergence

In this chapter, we provide a semilocal convergence analysis for Newton–like methods using the ω–versions of the famous Newton–Kantorovich theorem [136], [151], [477]. In the special case of Newton's method, our results have the following advantages over the corresponding ones [375], [602] under the same information and computational cost: finer error estimates on the distances involved; at least as precise information on the location of the solution, and weaker sufficient convergence conditions.

We are concerned with the problem of approximating a locally unique solution $x^\star$ of equation

$$F(x) = 0, \tag{3.1}$$

where F is a Fréchet–differentiable operator defined on a subset $\mathcal{D}$ of a Banach space X, with values in a Banach space $\mathcal{Y}$.

We use the Newton–like method:

$$x_{n+1} = x_n - A(x_n)^{-1} F(x_n), \quad (n \geq 0), \tag{3.2}$$

to generate a sequence $\{x_n\}$ approximating $x^\star$. Here, $A(x) \in L(X, \mathcal{Y}), (x \in \mathcal{D})$ the space of bounded linear operators from X to $\mathcal{Y}$. $A(x)$ is an approximation to the Fréchet–derivative $F'(x)$ of operator F [151].

If we set

$$A(x) = F'(x), \quad (x \in \mathcal{D}), \tag{3.3}$$

we obtain the Newton–Kantorovich method;

$$A(x_n) = [x_{n-1}, x_n; F], \tag{3.4}$$

we obtain the Secant method;

$$A(x) = [x, g(x); F], \\ g : X \longrightarrow X \quad \text{is a continuous operator}, \tag{3.5}$$

we obtain Steffensen's method. Other choices of operator A can be found in [136], [191]. A current survey on local as well as semilocal converegence theorems for Newton–like methods (3.2) under various Lipschitz–type assumptions can be found in [151], [303], [457], and the references there (see also [568], [764]).

Here, in particular we are motivated by optimization considerations, and the elegant works by Ezquerro, Hernández in [375], and Proinov in [602]. They proved semilocal convergence results for the special case $A(x) = F'(x)$, $(x \in \mathcal{D})$ by using the affine invariant condition

$$\| F'(x_0)^{-1} (F'(x) - F'(y)) \| \leq \omega(\| x - y \|), \quad \text{for all } x, y \in \mathcal{D} \tag{3.6}$$

where, ω is a non–decreasing, non–negative function on $[0, \infty)$.

Moreover, they considered a function h on $[0, 1]$ such that:

$$\omega(st) \leq h(s)\,\omega(t) \quad for \ all \ s \in [0, 1], \quad and \ t \in [0, +\infty). \tag{3.7}$$

This condition has been succesfully used to sharpen the error bounds obtained for particular expressions [375] (see also [602, Section 7]). Note that such a function h always exists. Indeed, if ω is a nonzero function on $\mathcal{J} = [0, +\infty)$, then one can define function $h : [0, 1] \longrightarrow \mathbb{R}$ by

$$h(s) = \sup \left\{ \frac{\omega(st)}{\omega(t)} : t \in [0, \infty), \quad with \ \omega(t) > 0 \right\}. \tag{3.8}$$

Clearly, function h so defined satisfies (3.7), and has the following properties [602]:

- $h(0) = 0$, $h(1) = 1$ provided that $\omega(0) = 0$;

- h is nondecreasing on $[0, 1]$ provided that ω is nondecreasing on $\mathcal{J}$;

- h is continuous on $[0, 1]$ provided that ω is nondecreasing on $\mathcal{J}$;

- h is identical to 1 on $[0, 1]$ if ω is non–decreasing on $\mathcal{J}$ and $\omega(0) > 0$.

Several choices of function h can be found in [602].

The chapter is organized as follows: first, we provide a semilocal convergence theorem for Newton–like method (3.2), whereas in the second part, we provide an extension of this result to solve more general equations than (3.1). Finally, we consider special cases and applications. In particular, in order for us to compare our results with the corresponding ones in [375], [602], we set $A(x) = F'(x)$, $(x \in \mathcal{D})$ to show that our results have the following advantages under the same information, and computational cost:

- Finer error bounds on the distances $\| x_{n+1} - x_n \|$, $\| x_n - x^\star \|$, $(n \geq 0)$;

- At least as precise information on the location of the solution;

- Weaker sufficient conditions convergence.

Theorem 3.1. *Let $F : \mathcal{D} \subseteq X \longrightarrow \mathcal{Y}$ be a Fréchet–differentiable operator, and let $A(x) \in L(X, \mathcal{Y})$ be an approximation of $F'(x)$. Assume that there exist an open convex subset $\mathcal{D}_0$ of $\mathcal{D}$, a vector $x_0 \in \mathcal{D}_0$, a bounded inverse Γ of A $(= A(x_0))$, continuous, non-decreasing, non–negative functions ω, ω_0, ω_1, ω_2 on $[0, +\infty)$, h, h_0, h_1 on $[0, 1]$, and non–negative constants*

Chapter 3. $\mathcal{H}$–Convergence

η, ℓ_0, ℓ_1, ℓ_2, such that, for all $x, y \in \mathcal{D}_0$, $t \in [0, 1]$, $s \in [0, \infty)$, the following conditions hold:

$$\| \Gamma F(x_0) \| \leq \eta, \tag{3.9}$$

$$\| \Gamma [F'(x) - F'(y)] \| \leq \omega(\| x - y \|), \tag{3.10}$$

$$\omega(ts) \leq h(t) \omega(s), \tag{3.11}$$

$$\| \Gamma [A(x) - A(x_0)] \| \leq \omega_0(\| x - x_0 \|) + \ell_0, \tag{3.12}$$

$$\omega_0(ts) \leq h_0(t) \omega_0(s), \tag{3.13}$$

$$\| \Gamma [F'(x) - F(x_0)] \| \leq \omega_1(\| x - x_0 \|) + \ell_1, \tag{3.14}$$

$$\omega_1(ts) \leq h_1(t) \omega_1(s), \tag{3.15}$$

$$\| \Gamma [F'(x) - A(x)] \| \leq \omega_2(\| x - x_0 \|) + \ell_2, \tag{3.16}$$

$$\ell_0 + \ell_2 < 1, \tag{3.17}$$

$$\ell_1 < 1. \tag{3.18}$$

Set:

$$H = \int_0^1 h(t)\, dt, \quad H_1 = \int_0^1 h_1(t)\, dt$$

$$c_0 = \frac{H_1\, \omega_1(\eta) + \omega_2(0) + \ell_2}{1 - \ell_0 - \omega_0(\eta)}, \quad c = c(r) = \frac{H\, \omega(\eta) + \omega_2(r) + \ell_2}{1 - \ell_0 - \omega_0(r)}.$$

We also assume that the scalar equation

$$\left(1 + c_0\, (1 + c) + \frac{c^3}{1 - c} \right) \eta = r \tag{3.19}$$

has a minimum zero r_0, such that:

$$r_0 > \eta, \tag{3.20}$$

$$c(r_0) < 1, \tag{3.21}$$

$$\omega_0(r_0) < 1 - \ell_0, \tag{3.22}$$

and

$$\overline{U}(x_0, r_0) = \{ x \in X : \| x - x_0 \| \leq r_0 \} \subseteq \mathcal{D}_0. \tag{3.23}$$

Then, sequences $\{x_n\}$ ($n \geq 0$) generated by Newton–like method (3.2) *is well defined, remains in $\overline{U}(x_0, r_0)$ for all $n \geq 0$, and converges to a solution $x^\star \in \overline{U}(x_0, r_0)$ of equation $F(x) = 0$.*

Moreover, the following estimates hold for all $n \geq 0$:

$$\| x_{n+1} - x_n \| \leq t_{n+1} - t_n, \tag{3.24}$$

and

$$\| x_n - x^\star \| \leq t^\star - t_n, \tag{3.25}$$

where, scalar sequence $\{t_n\}$, ($n \geq 0$), and $t^\star$ are given by:

$$t_0 = 0, \quad t_1 = \eta \quad t_2 = t_1 + c_0\, \eta$$

$$t_{n+1} = t_n + \frac{H\, \omega(t_n - t_{n-1}) + \omega_2(t_{n-1}) + \ell_2}{1 - \ell_0 - \omega_0(t_n)}\, (t_n - t_{n-1}), \quad (n \geq 2), \tag{3.26}$$

and

$$t^\star = \lim_{n \to \infty} t_n \leq r_0. \tag{3.27}$$

Furthermore, the solution $x^\star$ of equation (3.1) is unique in

$$\mathcal{D}_1 = U(x_0, r_1) \cap \mathcal{D}_0 \tag{3.28}$$

where, r_1 is the positive root of equation

$$2\,\omega_1(r_0 + r_1) + \int_{1/2}^{1} h_1(t)\,dt + \ell_1 = 1. \tag{3.29}$$

Proof. By hypotheses (3.19)–(3.22), and definition (3.26), we have $t_0 \leq t_1 \leq t_2 \leq r_0$.

Let us assume $t_{k-1} \leq t_k \leq r_0$, for all $k \leq \eta$. Then, by the definition of ω, h functions, (3.26), and the induction hypotheses, we obtain $t_k \leq t_{k+1}$. We also have:

$$
\begin{aligned}
t_{k+1} \\
\leq \quad & t_k + c\,(t_k - t_{k-1}) \\
\leq \quad & t_{k-1} + c\,(t_{k-1} - t_{k-2}) + c\,(t_k - t_{k-1}) \\
\leq \quad & t_2 + c\,(t_2 - t_1) + \cdots + c\,(t_{k-1} - t_{k-2}) + c\,(t_k - t_{k-1}) \\
\leq \quad & t_1 + c_0\,(t_1 - t_0) + c_0\,c\,(t_1 - t_0) + \cdots + c^{k-1}\,(t_1 - t_0) + c^k\,(t_1 - t_0) \\
= \quad & (1 + c_0 + c_0\,c)\,\eta + (c^3 + c^4 + \cdots + c^k)\,\eta \\
= \quad & \left(1 + (c_0 + 1)\,c + \frac{1 - c^k}{1 - c}\,c^3\right)\eta \\
\leq \quad & \left(1 + (c_0 + 1)\,c + \frac{c^3}{1 - c}\right)\eta = r_0,
\end{aligned}
\tag{3.30}
$$

which implies $t_{k+1} \leq r_0$.

Hence, sequence $\{t_k\}$ is non–decreasing, bounded above, and as such it converges to its unique least upper bound, so that $t^\star \in [0, r_0]$.

We shall show for all $k \geq 0$:

$$\| x_{k+1} - x_k \| \leq t_{k+1} - t_k, \tag{3.31}$$

and

$$\overline{U}(x_{k+1}, t^\star - t_{k+1}) \subseteq \overline{U}(x_k, t^\star - t_k). \tag{3.32}$$

Let $z_0 \in \overline{U}(x_1, t^\star - t_1)$. Then, we have:

$$\| z - x_0 \| \leq \| z - x_1 \| + \| x_1 - x_0 \| \leq t^\star - t_1 + t_1 = t^\star - t_0,$$

which implies $z \in \overline{U}(x_0, t^\star - t_0)$. Since also

$$\| x_1 - x_0 \| = \| \Gamma F(x_0) \| \leq \eta = t_1 - t_0,$$

estimate (3.31), and (3.32) hold for $k = 0$.

Given estimates (3.31), (3.32) hold for $n = 1, \cdots, k$, we get:

$$
\begin{aligned}
\| x_{k+1} - x_0 \| \quad \leq \quad & \sum_{i=1}^{k+1} \| x_i - x_{i-1} \| \\
\leq \quad & \sum_{i=1}^{k+1} (t_i - t_{i-1}) = t_{k+1} \leq r_0,
\end{aligned}
\tag{3.33}
$$

and

$$\| x_k + \theta \, (x_{k+1} - x_k) - x_0 \| \le t_k + \theta \, (t_{k+1} - t_k) \le t^\star \tag{3.34}$$

for all $\theta \in [0,1]$.

Using (3.12), (3.22), and (3.33), we obtain:

$$\begin{aligned} \| \Gamma \, [A(x_k) - A(x_0)] \| &\le \omega_0(\| x_k - x_0 \|) + \ell_0 \\ &\le \omega_0(t_k) + \ell_0 \\ &\le \omega_0(r_0) + \ell_0 < 1. \end{aligned} \tag{3.35}$$

It follows from (3.35), and the Banach lemma on invertible operators that $A(x_k)^{-1}$ exists, and

$$\| A(x_k)^{-1} A(x_0) \| \le (1 - \ell_0 - \omega_0(t_k))^{-1} \le (1 - \ell_0 - \omega_0(r_0))^{-1}. \tag{3.36}$$

In view of (3.2), we obtain the approximation

$$\begin{aligned} x_{k+1} - x_k \;=\; &-A(x_k)^{-1} A(x_0) \left(\Gamma \int_0^1 (F'(x_k + \theta \, (x_{k-1} - x_k)) - \right. \\ &F'(x_{k-1})) \, (x_k - x_{k-1}) \, d\theta + \\ &\left. \Gamma (F'(x_{k-1}) - A(x_{k-1})) \, (x_k - x_{k-1}) \right). \end{aligned} \tag{3.37}$$

By (3.37) for $k = 1$, (3.11), (3.14)–(3.16), (3.26), (3.36), and the induction hypotheses, we obtain:

$$\begin{aligned} \| x_2 - x_1 \| & \\ \le\; & \frac{1}{1 - \ell_0 - \omega_0(t_1)} \left(\int_0^1 \omega_1((1 - \theta) \, \| x_1 - x_0 \|) \, \| x_1 - x_0 \| \, d\theta + \right. \\ & \left. (\omega_2(\| x_1 - x_0 \| + \ell_2) \, \| x_1 - x_0 \| \right) \\ \le\; & \frac{H_1 \omega_1(\eta) + \omega_2(\eta) + \ell_2}{1 - \ell_0 - \omega_0(\eta)} \, \eta = t_2 - t_1. \end{aligned} \tag{3.38}$$

Moreover, using (3.10), (3.11), (3.16), (3.26), (3.36), (3.37), and the induction hypotheses, we get in turn for $k \ge 2$:

$$\begin{aligned} \| x_{k+1} - x_k \| \;\le\; & \frac{1}{1 - \ell_0 - \omega_0(t_k)} \left(\int_0^1 \omega((1 - \theta) \, \| x_k - x_{k-1} \|) \, d\theta + \right. \\ & \left. \omega_2(\| x_k - x_0 \| + \ell_2) \right) \| x_k - x_{k-1} \| \\ \le\; & \frac{H \omega(t_k - t_{k-1}) + \omega_2(t_k) + \ell_2}{1 - \ell_0 - \omega_0(t_k)} \, (t_k - t_{k-1}) = t_{k+1} - t_k, \end{aligned} \tag{3.39}$$

which together with (3.38), shows (3.31) for all $k \ge 0$.

Then, for every $z \in \overline{U}(x_{k+1}, t^\star - t_{k+1})$, we have:

$$\begin{aligned} \| z - x_k \| &\le\; \| z - x_{k+1} \| + \| x_{k+1} - x_k \| \\ &\le\; t^\star - t_{k+1} + t_{k+1} - t_k = t^\star - t_k, \end{aligned} \tag{3.40}$$

which implies $z \in \overline{U}(x_k, t^\star - t_k)$.

That is (3.32) holds for all $k \geq 0$. The induction for (3.31), and (3.32) is now completed.

In view of (3.31), and (3.32), sequence $\{x_n\}$ is Cauchy in a Banach space X, and as such it converges to some $x^\star \in \overline{U}(x_0, r_0)$ (since $\overline{U}(x_0, r_0)$ is a closed set).

We shall show $x^\star$ is a solution of equation (3.1). We can write:

$$\| \Gamma F(x_k) \| \leq \| \Gamma A(x_k) \| \, \| A(x_k)^{-1} F(x_k) \|. \tag{3.41}$$

But, we also have:

$$\| A(x_k)^{-1} F(x_k) \| \longrightarrow 0 \text{ as } k \longrightarrow \infty, \tag{3.42}$$

and

$$\begin{aligned} \| \Gamma A(x_k) \| &\leq \| \Gamma (A(x_k) - A(x_0)) \| \, \| \Gamma A(x_0) \| \\ &\leq \omega_0(\| x_k - x_0 \|) + 1 \\ &\leq \omega_0(r_0) + 1 = B. \end{aligned} \tag{3.43}$$

In view of (3.41)–(3.43), we get by letting $k \longrightarrow \infty$ that $F(x^\star) = 0$.

Estimate (3.25) follows from (3.24) by using standard majorization techniques [136], [151], [477].

Finally, to show uniqueness of $x^\star$ in $\mathcal{D}_1$, let us assume $y^\star$ is a solution in $\mathcal{D}_1$.

We need the estimate:

$$\begin{aligned} &\int_0^1 \| \Gamma(F'(x^\star + \theta(y^\star - x^\star)) - F'(x_0)) \| \\ &\leq \int_0^1 (\omega_1(\| x_0 - x^\star - \theta(y^\star - x^\star) \|) \, d\theta + \ell_1) \\ &\leq \int_0^1 (\omega_1(\| (1-\theta)(x_0 - x^\star) + \theta(x_0 - y^\star) \|) \, d\theta + \ell_1) \\ &\leq \int_0^1 (\omega_1((1-\theta) \| x_0 - x^\star \| + \theta \| x_0 - y^\star \|) \, d\theta + \ell_1) \\ &\leq \int_0^{1/2} \omega_1((1-\theta) \| x_0 - x^\star \| + \| x_0 - y^\star \|) \, d\theta + \\ &\quad \int_{1/2}^1 \omega_1(\theta(\| x_0 - x^\star \| + \| x_0 - y^\star \|)) \, d\theta + \ell_1 \\ &< \int_0^{1/2} h_1(1-\theta) \omega_1(r_0 + r_1) \, d\theta + \int_{1/2}^1 h_1(\theta) \omega_1(r_0 + r_1) \, d\theta + \ell_1 \\ &= 2 \omega_1(r_0 + r_1) \int_{1/2}^1 h_1(\theta) \, d\theta + \ell_1 = 1. \end{aligned} \tag{3.44}$$

In view of (3.44), and the Banach lemma on invertible operators,

$$\mathcal{M} = \int_0^1 \Gamma F'(x^\star + \theta(y^\star - x^\star)) \, d\theta$$

is invertible.

We then have:

$$0 = \Gamma(F(y^\star) - F(x^\star)) = \Gamma \mathcal{M} (y^\star - x^\star),$$

from which it follows

$$x^\star = y^\star.$$

That completes the proof of Theorem 3.1.

Note that if $A(x) = F'(x)$, then in view of (3.13), h_0, ω_0 can replace h_1, ω in the definition of c_0, and t_2, respectively.

Consider the equation

$$F(x) + G(x) = 0, \tag{3.45}$$

where, F is as in the introduction of this chapter, and $G : \mathcal{D} \longrightarrow \mathcal{Y}$ is a continuous operator.

We use Newton–like method:

$$x_{n+1} = x_n - A(x_n)^{-1} (F(x_n) + G(x_n)), \quad (n \geq 0), \tag{3.46}$$

to generate a sequence approximating the solution $x^\star$ of equation (3.45).

Then, working along the lines of the proof of Theorem 3.1, we can show the following semilocal result for Newton–like method (3.46):

Theorem 3.2. *Let $F : \mathcal{D} \subseteq X \longrightarrow \mathcal{Y}$ be a Fréchet–differentiable operator, $G : \mathcal{D} \longrightarrow \mathcal{Y}$ a continuous operator, and let $A(x) \in L(X, \mathcal{Y})$ be an approximation of $F'(x)$. Assume that there exist an open convex subset $\mathcal{D}_0$ of $\mathcal{D}$, $x_0 \in \mathcal{D}$, a bounded inverse Γ of A $(= A(x_0))$, continuous, non-decreasing, non–negative functions ω, ω_i, $(i = 1, 2, 3)$ on $[0, +\infty)$, h, h_0 on $[0, 1]$, and non–negative constants η, ℓ_0, ℓ_2, such that, for all $x, y \in \mathcal{D}_0$, $t \in [0, 1]$, $s \in [0, \infty)$, the following conditions hold: (3.10)–(3.13), (3.16), (3.17), (3.19)–(3.23),*

$$\| \Gamma (F(x_0) + G(x_0)) \| \leq \eta, \tag{3.47}$$

$$\| \Gamma [G(x) - G(y)] \| \leq \omega_3 (\| x - y \|) \| x - y \|, \tag{3.48}$$

where,

$$c_0 = \frac{H_1 \, \omega_1(\eta) + \omega_2(0) + \omega_3(\eta) + \ell_2}{1 - \ell_0 - \omega_0(\eta)}, \quad c = c(r) = \frac{H \, \omega(\eta) + \omega_2(r) + \omega_3(\eta) + \ell_2}{1 - \ell_0 - \omega_0(r)}.$$

Then, sequence $\{x_n\}$ $(n \geq 0)$ generated by Newton–like method (3.46) is well defined, remains in $\overline{U}(x_0, r_0)$ for all $n \geq 0$, and converges to a solution $x^\star \in \overline{U}(x_0, r_0)$ of equation $F(x) + G(x) = 0$.

Moreover, the following estimates hold for all $n \geq 0$:

$$\| x_{n+1} - x_n \| \leq t_{n+1} - t_n, \tag{3.49}$$

and

$$\| x_n - x^\star \| \leq t^\star - t_n, \tag{3.50}$$

where, scalar sequence $\{t_n\}$, $(n \geq 0)$, and $t^\star$ are given by:

$$t_0 = 0, \quad t_1 = \eta \quad t_2 = t_1 + c_0 \, \eta$$

$$t_{n+1} = t_n + \frac{H \, \omega(t_n - t_{n-1}) + \omega_2(t_{n-1}) + \ell_2 + \omega_3(t_n - t_{n-1})}{1 - \ell_0 - \omega_0(t_n)} (t_n - t_{n-1}), \tag{3.51}$$
$$(n \geq 2),$$

and

$$t^\star = \lim_{n \to \infty} t_n \leq r_0. \tag{3.52}$$

56   Ioannis K. Argyros, Saïd Hilout and Mohammad A. Tabatabai

Furthermore, $x^\star$ is the only solution of (3.45) *in* $U(x_0, r_2)$, *where* r_2 *is the unique positive root of equation*

$$f(s) = 0, \tag{3.53}$$

where,

$$f(s) = H\omega(s) + \omega_2(s) + \omega_3(s) + \omega_0(s) + \ell_0 + \ell_2 - 1, \tag{3.54}$$

and

$$r_2 \le r_0. \tag{3.55}$$

Proof. Note that the existence of r_2 is guaranteed by the intermediate value theorem, and (3.17), since $f(0) = \ell_0 + \ell_2 - 1 < 0$, and $f(s) > 0$ for sufficient large $s > 0$. The uniqueness follows from the estimate $f'(s) \ge 0$ for all $s \in [0, \infty)$. That is the graph of function f crosses the positive axis only once.

As is Theorem 3.1, we arrive at (3.50), by simply noticing that there should be extra terms of the form: $\Gamma\left(Q(x_k) - Q(x_{k-1})\right)$ (inside the braces in (3.37)), $\omega_3(\eta)$ (at the numerator in (3.38), and (3.39)), since, we are using additional estimate (3.48), and iteration (3.46) instead of (3.2).

Hence, we simply need to show the uniqueness part whose proof differs from the corresponding one in Theorem 3.1.

Let $y^\star$ be a solution in $U(x_0, r_2)$. Using (3.46), we obtain the approximation:

$$
\begin{aligned}
y^\star & - x_{k+1} \\
= \; & A(x_k)^{-1} \Gamma^{-1} \Bigg\{ \Gamma\left(\int_0^1 (F'(x_k + \theta(y^\star - x_k)) - F'(x_k))\, d\theta + \right. \\
& \left. (F'(x_k) - A(x_k)) \right) (y^\star - x_k) + \Gamma\left(G(y^\star) - G(x_k)\right) \Bigg\}.
\end{aligned}
\tag{3.56}
$$

Using (3.10), (3.11), (3.16), (3.36), (3.50), (3.52)–(3.56), we obtain in turn:

$$
\begin{aligned}
\| y^\star & - x_k \| \\
\le \; & \frac{1}{1 - \ell_0 - \omega_0(\| y^\star - x_k \|)} \times \\
& \Bigg\{ \left(\int_0^1 \| \Gamma(F'(x_k + \theta(y^\star - x_k)) - F'(x_k)) \| \, d\theta + \right. \\
& \left. \| \Gamma(F'(x_k) - A(x_k)) \| \right) \| y^\star - x_k \| + \| \Gamma(G(y^\star) - G(x_k)) \| \Bigg\} \\
\le \; & \frac{1}{1 - \ell_0 - \omega_0(\| y^\star - x_k \|)} \Bigg\{ \int_0^1 \omega(\theta \| y^\star - x_k \|) \, d\theta + \\
& \omega_2(\| x_0 - x_k \|) + \ell_2 + \omega_3(\| y^\star - x_k \|) \Bigg\} \| y^\star - x_k \| \\
< \; & \frac{1}{1 - \ell_0 - \omega_0(r_2)} \left(H\omega(r_2) + \omega_2(r_2) + \ell_2 + \omega_3(r_2)\right) \| x_k - y^\star \| \\
= \; & \| x_k - y^\star \|,
\end{aligned}
\tag{3.57}
$$

which implies $\lim\limits_{k \to \infty} x_k = y^\star$. But we have shown $\lim\limits_{k \to \infty} x_k = x^\star$. Hence, we deduce $x^\star = y^\star$.

That completes the proof of Theorem 3.2. $\diamond$

We can also provide the following local convergence result for Newton–like method (3.46).

Chapter 3. $\mathcal{H}$–Convergence

Proposition 3.1. *Assume hypotheses* (3.10)–(3.13), (3.16), (3.17), (3.48) *hold for x_0 replaced by $x^\star$, and radius of convergence r_2 is given in* (3.53).

Then, sequence $\{x_n\}$ generated by Newton–like method (3.46) *is well defined, remains in $U(x^\star, r_2)$ for all $n \geq 0$, and converges to $x^\star$, provided that $x_0 \in U(x^\star, r_2)$.*

Proof. By hypotheses $x_0 \in U(x^\star, r_2)$. Assume $x_n \in U(x^\star, r_2)$ for all $n \leq k$. We shall show $x_{k+1} \in U(x^\star, r_2)$.

As in the proof of Theorem 3.2, using approximation (3.56), we arrive at estimate (3.57) (for $x^\star$ replacing $y^\star$). That is we have:

$$\| x_{k+1} - x^\star \| < \| x_k - x^\star \| \leq r_2,$$

which implies that $x_{k+1} \in U(x^\star, r_2)$, and $\lim_{k \longrightarrow \infty} x_k = x^\star$.

That completes the proof of Proposition 3.1. $\diamondsuit$

Remark 3.1. If $G(x) = 0$ ($x \in \mathcal{D}$), and the function Ω given by (7.6) in [602], is chosen to be equal to ω, then our Proposition 3.1 essentially reduces to Theorem 7.2 in [602].

Note that more general conditions than the ones given in [602], and ours introduced in this chapter were provided in [137] to show the local (and semilocal) convergence of two–point Newton–like methods (see also [151], [189], [477]).

Let us consider the case of Newton's method. That is: $G(x) = 0$, and $A(x) = F'(x)$ for all $x \in \mathcal{D}$.

Then, we have $h_0(s) = h_1(s)$, $\omega_0(s) = \omega_1(s)$, $\omega_2(s) = 0$ for all $s \geq 0$, $\ell_2 = 0$, and $\ell_0 = \ell_1$. We can certainly also set $\ell_0 = \ell_1 = 0$.

In this case, we note:

$$\omega_0(s) \leq \omega(s), \tag{3.58}$$

$$h_0(s) \leq h(s), \tag{3.59}$$

hold for all $s \geq 0$, and $\dfrac{\omega}{\omega_0}$, $\dfrac{h}{h_0}$ can be arbitrarily large [136], [151], [190].

Comparison with a result by Ezquerro, and Hernández [375] (see also [602, Theorem 7.3]): If one reproduces these results in affine invariant form, then the corresponding to $\{t_n\}$ majorizing sequence is essentially given by:

$$v_0 = 0, \quad v_1 = \eta, \quad v_{n+1} = v_n + \frac{H\,\omega(v_n - v_{n-1}) + \omega_2(v_{n-1})}{1 - \omega(v_n)}\,(v_n - v_{n-1}), \tag{3.60}$$
$$(n \geq 1).$$

Then, if equality holds in (3.58), and (3.59), then our Theorem 3.2 reduces to the corresponding one in [602]. Otherwise it constitues an improvement under the same computational cost. In particular, we note the following advantages:

(i) Majorizing sequence $\{t_n\}$ is finer than $\{v_n\}$, since an inductive argument shows:

$$t_n < v_n \quad (n \geq 2) \tag{3.61}$$

and

$$t_{n+1} - t_n < v_{n+1} - v_n \quad (n \geq 2). \tag{3.62}$$

(ii) The information on the location of the solution is at least as precise, since:

$$t^* = \lim_{n \to \infty} t_n \leq \lim_{n \to \infty} v_n = v^*. \tag{3.63}$$

(iii) The uniqueness radius r_1 is larger than the corresponding one in [602] (since, it is derived from (3.29) for $\omega = \omega_0$, and $h_1 = h$).

It turns out our sufficient convergence conditions are weaker. Indeed, for simplicity, let us consider the case, when

$$\omega(s) = L\,s, \quad \omega_0(s) = L_0\,s, \text{ and } h_0(t) = h(t) = \frac{1}{2}. \tag{3.64}$$

Then, the iterations $\{t_n\}$, $\{v_n\}$ become:

$$t_0 = 0, \quad t_1 = \eta, \quad t_{n+1} = t_n + \frac{L\,(t_n - t_{n-1})^2}{2\,(1 - L_0\,t_n)}, \quad (n \geq 1), \tag{3.65}$$

$$v_0 = 0, \quad v_1 = \eta, \quad v_{n+1} = v_n + \frac{L\,(v_n - v_{n-1})^2}{2\,(1 - L\,v_n)}, \quad (n \geq 1). \tag{3.66}$$

Note that iteration (3.66) converges if the famous for its simplicity and clarity Newton–Kantorovich hypothesis

$$K = L\,\eta \leq \frac{1}{2} \tag{3.67}$$

holds [477].

However, iteration (3.65) converges under weaker conditions provided that $L_0 < L$. We need the following result for the convergence of majorizing sequence $\{t_n\}$.

Lemma 3.1. *Assume:*
there exist constants $L_0 \geq 0$, $L \geq 0$ with $L_0 \leq L$, and $\eta \geq 0$, such that:

$$q_0 = \bar{L}\,\eta \leq \frac{1}{2}, \tag{3.68}$$

where,

$$\bar{L} = \frac{1}{8}\left(L + 4\,L_0 + \sqrt{L^2 + 8\,L_0\,L}\right). \tag{3.69}$$

The inequality in (3.68) is strict, if $L_0 = 0$.
Then, sequence $\{t_k\}$ $(k \geq 0)$ given by

$$t_0 = 0, \quad t_1 = \eta, \quad t_{k+1} = t_k + \frac{L\,(t_k - t_{k-1})^2}{2\,(1 - L_0\,t_k)} \quad (k \geq 1), \tag{3.70}$$

is well defined, nondecreasing, bounded above by t^{**}, and converges to its unique least upper bound $t^* \in [0, t^{**}]$, where

$$t^{**} = \frac{2\,\eta}{2 - \delta}, \tag{3.71}$$

$$1 \leq \delta = \frac{4\,L}{L + \sqrt{L^2 + 8\,L_0\,L}} < 2 \quad \text{for } L_0 \neq 0. \tag{3.72}$$

Moreover, the following estimates hold:

$$L_0 t^\star \le 1, \tag{3.73}$$

$$0 \le t_{k+1} - t_k \le \frac{\delta}{2}(t_k - t_{k-1}) \le \cdots \le \left(\frac{\delta}{2}\right)^k \eta, \quad (k \ge 1), \tag{3.74}$$

$$t_{k+1} - t_k \le \left(\frac{\delta}{2}\right)^k (2q_0)^{2^k-1} \eta, \quad (k \ge 0), \tag{3.75}$$

$$0 \le t^\star - t_k \le \left(\frac{\delta}{2}\right)^k \frac{(2q_0)^{2^k-1}\eta}{1 - (2q_0)^{2^k}}, \quad (2q_0 < 1), \quad (k \ge 0). \tag{3.76}$$

Remark 3.2. If $L_0 = L$, Lemma 3.1 provides the usual error bounds appearing essentially in the Newton–Kantorovich theorem [477].

However, if $L_0 < L$, then our sufficient convergence condition (3.68) is weaker than (3.67). Finally, our ratio $2q_0$ is also smaller than $2K$.

Example 3.1. Let $X = \mathcal{Y} = \mathbb{R}^2$ be equipped with the ℓ_∞–norm [568, p. 41]. Choose $x_0 = [1,1]^T$, $\mathcal{D}_0 = \{x : \|x - x_0\| \le 1 - b\}$ for $b \in [0,1)$, and define function $F = \begin{bmatrix} F_1 \\ F_2 \end{bmatrix}$ on $\mathcal{D}_0$, where,

$$\begin{aligned} F_1(v,w) &= v^3 + \varepsilon_1 w - b \\ F_2(v,w) &= w^3 + \varepsilon_2 v - b \end{aligned} \tag{3.77}$$

for some given constants ε_1 and ε_2, such that $\varepsilon_1 \varepsilon_2 \ne 9$.

Then, the Fréchet–derivative F' of operator F is given by

$$F'(v,w) = \begin{bmatrix} 3v^2 & \varepsilon_1 \\ \varepsilon_2 & 3w^2 \end{bmatrix},$$

and

$$F'(v,w) - F'(\overline{v},\overline{w}) = \begin{bmatrix} 3(v^2 - \overline{v}^2) & 0 \\ 0 & 3(w^2 - \overline{w}^2) \end{bmatrix}.$$

If $9v^2 w^2 \ne \varepsilon_1 \varepsilon_2$, then, we obtain

$$F'(v,w)^{-1} = \frac{1}{9v^2 w^2 - \varepsilon_1 \varepsilon_2} \begin{bmatrix} 3w^2 & -\varepsilon_1 \\ -\varepsilon_2 & 3v^2 \end{bmatrix},$$

and, in particular for x_0:

$$F'(v_0,w_0)^{-1} = \frac{1}{9 - \varepsilon_1 \varepsilon_2} \begin{bmatrix} 3 & -\varepsilon_1 \\ -\varepsilon_2 & 3 \end{bmatrix}. \tag{3.78}$$

We need the estimate

$$\frac{1}{9 - \varepsilon_1 \varepsilon_2} \begin{bmatrix} 3 & -\varepsilon_1 \\ -\varepsilon_2 & 3 \end{bmatrix} \begin{bmatrix} 3(v^2 - \overline{v}^2) & 0 \\ 0 & 3(w^2 - \overline{w}^2) \end{bmatrix}$$

$$= \frac{1}{9 - \varepsilon_1 \varepsilon_2} \begin{bmatrix} 3(v^2 - \overline{v}^2) & -\varepsilon_1(w^2 - \overline{w}^2) \\ -\varepsilon_2(v^2 - \overline{v}^2) & 3(w^2 - \overline{w}^2) \end{bmatrix}. \tag{3.79}$$

Set
$$\varepsilon = \max\left\{3 + |\varepsilon_1|, \, 3 + |\varepsilon_2|\right\} \tag{3.80}$$

Using (3.78)–(3.80), we obtain the Lipschitz constants:

$$L = \frac{6\,\varepsilon\,(2-b)}{|9 - \varepsilon_1\,\varepsilon_2|}, \tag{3.81}$$

and

$$L_0 = \frac{3\,\varepsilon\,(3-b)}{|9 - \varepsilon_1\,\varepsilon_2|}. \tag{3.82}$$

Moreover, by (3.77), and (3.78), we can set

$$\eta = \frac{1}{|9 - \varepsilon_1\,\varepsilon_2|}\, \max\left\{\overline{\alpha}, \overline{\beta}\right\}, \tag{3.83}$$

where,

$$\overline{\alpha} = |3 + 2\,\varepsilon_1 - 3\,b - \varepsilon_1\,\varepsilon_2 + \varepsilon_1\,b|, \tag{3.84}$$

$$\overline{\beta} = |3 + 2\,\varepsilon_2 - 3\,b - \varepsilon_1\,\varepsilon_2 + \varepsilon_2\,b|. \tag{3.85}$$

Let us choose for example

$$b = .49, \quad \varepsilon_1 = .01, \quad and \quad \varepsilon_2 = .02. \tag{3.86}$$

Using (3.80)–(3.86), we obtain

$$L = 3.040200893, \quad L_0 = 2.526789485, \quad \eta = .175515011, \quad \delta = 1.06222409,$$

$$\overline{\alpha} = 1.5547, \quad \overline{\beta} = 1.5796, \quad \overline{L} = 2.69449131, \quad and \quad t^{\star\star} = .374321859.$$

Condition (3.67) is violated, since

$$K = .533600893 > \frac{1}{2}.$$

Hence, there is no guarantee that Newton's method starting at x_0 converges to a zero $x^\star$ of function F.

However, our condition (3.68) is satisfied, since

$$q_0 = .472916269 < \frac{1}{2}.$$

Moreover, we have:

$$\overline{U}(x_0, t^{\star\star}) \subseteq \mathcal{D}_0.$$

In view of all the above the hypotheses of Theorem 3.1 are satisfied.

That is our Theorem 3.1 guarantees the existence of a zero $x^\star$ in $\overline{U}(x_0, t^{\star\star})$ of function F, which can be obtained as the limit of sequence $\{x_n\}$.

Chapter 4

Chebyshev–Secant–Type Methods

In this chapter, we introduce a three–step Chebyshev–Secant–type method (CSTM) of high efficiency index to solve nonlinear equation in a Banach space setting. We provide a semilocal convergnece analysis for (CSTM) using recurrence relations. Numerical example validating our theoretical result is also provided in this chapter. Here, we are concerned with the problem of approximating a locally unique solution x^* of equation

$$F(x) = 0, \tag{4.1}$$

where F is a Fréchet–differentiable operator defined on a open, convex subset $\mathcal{D}$ of a Banach space X with values in a Banach space $\mathcal{Y}$.

We introduce the Chebyshev–Secant–type method (CSTM)

$$
\begin{aligned}
&x_{-1},\, x_0 \in \mathcal{D}, \\
&y_k = x_k - A_k^{-1}\, F(x_k), \quad A_k = [x_{k-1}, x_k; F], \\
&z_k = x_k + a\,(y_k - x_k) \\
&x_{k+1} = x_k - A_k^{-1}\,(b\, F(x_k) + c\, F(z_k)),
\end{aligned}
$$

where, $[x, y; F]$ is a divided difference of order one for operator F at the points $x,\, y \in \mathcal{D}$, where as a, b, c are non–negative parameters to be chosen so that sequence $\{x_k\}$ converges to x^*. (CSTM) is introduced, since we are motivated by the cubically convergent Chebyshev–Newton–type method (CNTM):

$$
\begin{aligned}
&x_0 \in \mathcal{D}, \\
&y_k = x_k - F'(x_k)^{-1}\, F(x_k), \\
&z_k = x_k + a\,(y_k - x_k) \\
&x_{k+1} = x_k - \frac{1}{a^2}\, F'(x_k)^{-1}\,((a^2 + a - 1)\, F(x_k) + F(z_k)),
\end{aligned}
$$

where $F'(x)$ $(x \in \mathcal{D})$ is the Fréchet–derivative of F [155], [477]. A semilocal convergence analysis was provided by Ezquerro, and Hernández [377]. The following statement was given in [377]: if we consider iterative methods with memory, the efficiency index

$$EI = \delta_1^{1/\delta}$$

where, δ_1 is the order of the method, and δ is the number of computations needed at each step, the efficiency index could be improved.

The Secant method (SM)

$$x_{k+1} = x_k - A_k^{-1} F(x_k), \quad x_{-1}, x_0 \in \mathcal{D}, \ (k \geq 0)$$

is undoubtedly the must known iterative method with memory in the one dimensional case for solving nonlinear equations, with $EI = (1 + \sqrt{5})/2$.

Note that Newton's method (NM)

$$x_{k+1} = x_k - F(x_k)^{-1} F(x_k), \quad x_0 \in \mathcal{D}, \ (k \geq 0)$$

has $EI = \sqrt{2}$, and (CNTM) has efficiency index $\sqrt[3]{3}$. It then follows that

$$\sqrt{2} = 1.414213562 < \sqrt[3]{3} = 1.44224957 < \frac{1 + \sqrt{5}}{2} = 1.618033989.$$

In view of the above estimate, we are motivated to introduce (CSTM). Note that (CSTM) only requires the evaluation of a new function at each step. Bosarge and Falb [268], Dennis [335], Potra [593], Argyros [89], [155], Hernández et al. [437] and others [436], [38], [749], have provided sufficient convergence conditions for the (SM) based on Lipschitz–type conditions on divided difference operator (see, also relevant works in [191], [335], [457], [498], [593], [644], [732]). Here, we provide a semilocal convergence analysis for (CSTM) using recurrence relations along the lines of the works in [377]. Numerical example is also provided in this chapter.

It is convenient for us to define functions on $[0, \gamma]$ for $\gamma = \sqrt{5} - 1$, by:

$$f(t) = \frac{1}{1 - \dfrac{t}{2} \left(1 + \dfrac{t}{2}\right)}, \tag{4.2}$$

$$g(t) = \frac{t}{8} \left(8 + 4t + t^2\right), \tag{4.3}$$

$$h(t) = g(t) f(t)^2 - 1, \tag{4.4}$$

$$h_1(t) = h(t) + 1, \tag{4.5}$$

$$A(t) = \frac{t}{2} \left(\left(1 + \frac{t}{2} f(t) \left(f(t) g(t) + \frac{t}{2} + 1\right)\right) f(t) g(t) + \frac{t}{2} + 1\right) - \\ f(t) g(t) \left(1 + f^2(t) g(t)\right), \tag{4.6}$$

and

$$H(t) = f^2(t) g(t) \left(1 + f^2(t) g(t)\right) - 1. \tag{4.7}$$

Note that γ is the only positive zero of function

$$f_1(t) = \frac{1}{f(t)}.$$

We shall provide some properties on the above functions.

Chapter 4. Chebyshev–Secant–Type Methods

Lemma 4.1. *Let functions f, f_1, g, h, h_1, and number γ, be as defined in (4.2)–(4.7). Then, the following hold:*

(a) *Function h has a minimal positive zero in $(0,\gamma)$ denoted by $\xi = .435$.*

For $t \in (0,\xi)$:

(b) *f_1 is decreasing, and $f_1(t) \in (0,1)$;*

(c) *g, h, and h_1 are increasing, and*

$$g(t), h_1(t) \in (0,1);$$

(d)
$$f_1(\theta t) \geq f_1(t),$$
$$g(\theta t) \leq \theta g(t),$$
$$h_1(\theta t) \leq \theta h_1(t)$$

for all $\theta \in [0,1]$;

(e) *$A(t) \leq 0$ for all $t \in [0,\xi_0)$, $\xi_0 = .236067977$;*

(f) *$H(t) < 0$ for all $t \in [0,\xi_1)$, $\xi_1 = .3370255674$.*

Proof.

(a) We have by (4.4): $h(0) = -1$, and for t sufficiently close to γ from the left function h becomes positive.

It follows from the intermediate value theorem that function h has a positive zero in $(0,\gamma)$. Denote by ξ the minimal positive zero of function h in $(0,\gamma)$.

(b) Function f_1 is decreasing in $(0,\xi)$, since by (4.2), and (4.7), $f_1'(t) = -\dfrac{1}{2}(1+t) < 0$, for $t \geq 0$. Noting that $f_1(0) = 1$, $f_1(\xi) > f_1(\gamma) = 0$, we conclude $f_1(t) \in (0,1)$, for $t \in (0,\xi)$.

(c) In view of (4.3), function g is increasing on $(0,\xi)$, since $g'(t) = 1 + 2t + \dfrac{3t^2}{8} > 0$, $t \geq 0$.

Moreover,
$$h'(t) = \frac{g'(t)\,f_1(t) - 2\,f_1'(t)\,g(t)}{f_1(t)^3} > 0, \quad t \in (0,\xi).$$

That is h, h_1 are increasing on $(0,\xi)$.

For $t \in (0,\xi)$, $g(t) < f_1(t)^2$, so, we get $h_1(t) \in (0,1)$. Moreover, by (b), we conclude $g(t) \in (0,1)$.

(d) This part follows immediately from the definitions of the functions f_1, g, and h_1.

(e) – (f) We use Maple to obtain these results.

64   Ioannis K. Argyros, Saïd Hilout and Mohammad A. Tabatabai

That completes the proof of Lemma 4.1. $\qquad\qquad\qquad\qquad\qquad\qquad\Diamond$

We need the Ostrowski–type approximations for (CSTM).

Lemma 4.2. *Assume sequence $\{x_k\}$ generated by (CSTM) is well defined, and $(1-a)\,c = 1-b$ holds for $a \in [0,1]$, $b \in [0,1]$, and $c \geq 0$.*
Then, the following items hold for all $k \geq 0$:

$$
\begin{aligned}
F(z_k) \;=\; & (1-a)\,F(x_k)+ \\
& a\int_0^1 \left(F'(x_k+a\,t\,(y_k-x_k))-F'(x_k)\right)(y_k-x_k)\,dt+ \\
& a\,(F'(x_k)-A_k)\,(y_k-x_k),
\end{aligned} \tag{4.8}
$$

$$
\begin{aligned}
& x_{k+1}-y_k \\
=\; & -a\,c\,A_k^{-1}\left(\int_0^1\left(F'(x_k+a\,t\,(y_k-x_k))-F'(x_k)\right)(y_k-x_k)\,dt+\right. \\
& \left.(F'(x_k)-A_k)\,(y_k-x_k)\right),
\end{aligned} \tag{4.9}
$$

and

$$
\begin{aligned}
& F(x_{k+1}) \\
=\; & \int_0^1\left(F'(x_k+t\,(x_{k+1}-x_k))-F'(x_k)\right)(x_{k+1}-x_k)\,dt+ \\
& (F'(x_k)-A_k)\,(x_{k+1}-x_k)-a\,c\left(\int_0^1\left(F'(x_k+a\,t\,(y_k-x_k))-\right.\right. \\
& \left.\left. F'(x_k)\right)(y_k-x_k)\,dt+(F'(x_k)-A_k)\,(y_k-x_k)\right).
\end{aligned} \tag{4.10}
$$

Proof. We have in turn using (CSTM)

$$
a\,F(x_k)=a\,A_k\,(x_k-y_k)\Longrightarrow 0=-a\,F(x_k)+A_k\,(x_k-z_k)\Longrightarrow
$$

$$
\begin{aligned}
F(z_k) \;=\; & F(z_k)-a\,F(x_k)+A_k\,(x_k-z_k) \\
=\; & (1-a)\,F(x_k)+F(z_k)-F(x_k)+A_k\,(x_k-z_k) \\
=\; & (1-a)\,F(x_k)+\int_0^1\left(F'(x_k+t\,(z_k-x_k))-A_k\right)(z_k-x_k)\,dt \\
=\; & (1-a)\,F(x_k)+\int_0^1\left(F'(x_k+t\,a\,(y_k-x_k))-F'(x_k)\right)(y_k-x_k)\,dt+ \\
& a\,(F'(x_k)-A_k)\,(y_k-x_k),
\end{aligned}
$$

showing (4.8).

By eliminating x_k from first and third approximation in (CSTM), we get:

$$
\begin{aligned}
x_{k+1}-y_k \;=\; & A_k^{-1}\,F(x_k)-A_k^{-1}\,b\,F(x_k)-A_k^{-1}\,c\,F(z_k) \\
=\; & A_k^{-1}\left((1-b)\,F(x_k)-c\,F(z_k)\right) \\
=\; & A_k^{-1}\left((1-a)\,F(x_k)-F(z_k)\right)c \\
=\; & -A_k^{-1}\left(-(1-a)\,F(x_k)+F(z_k)\right)c \\
=\; & -A_k\,c\left(a\int_0^1\left(F'(x_k+a\,t\,(y_k-x_k))-F'(x_k)\right)(y_k-x_k)\,dt+\right. \\
& \left. a\,(F'(x_k)-A_k)\,(y_k-x_k)\right),
\end{aligned}
$$

Chapter 4. Chebyshev–Secant–Type Methods 65

(by (4.8)), which shows (4.9).

Finally, we have:

$$
\begin{aligned}
F(x_{k+1}) &= F(x_{k+1}) - F(x_k) - A_k (y_k - x_k) \\
&= \int_0^1 (F'(x_k + t (x_{k+1} - x_k)) - F'(x_k)) (x_{k+1} - x_k) \, dt + \\
&\quad F'(x_k) (x_{k+1} - x_k) - A_k (y_k - x_k) \\
&= \int_0^1 (F'(x_k + t (x_{k+1} - x_k)) - F'(x_k)) (x_{k+1} - x_k) \, dt + \\
&\quad (F'(x_k) - A_k) (x_{k+1} - x_k) + A_k (x_{k+1} - x_k) - A_k (y_k - x_k) \\
&= \int_0^1 (F'(x_k + t (x_{k+1} - x_k)) - F'(x_k)) (x_{k+1} - x_k) \, dt + \\
&\quad (F'(x_k) - A_k) (x_{k+1} - x_k) + A_k (x_{k+1} - y_k),
\end{aligned}
$$

which shows (4.10) (by (4.9)).

That completes the proof of Lemma 4.2. ◇

We shall show the semilocal convergence analysis of (CSTM) using the conditions:

$(\mathcal{C})$:

(C_1) $F : \mathcal{D} \subseteq X \longrightarrow \mathcal{Y}$ is a Fréchet–differentiable, with divided difference denoted by $[x, y; F]$ satisfying

$$
[x, y; F] (x - y) = F(x) - F(y), \quad \text{for all } x, y \in \mathcal{D};
$$

(C_2) There exist $x_{-1}, x_0 \in \mathcal{D}$, such that $A_0 = [x_0, x_{-1}; F]$ exists, and

$$
0 < \| A_0^{-1} \| \leq \beta;
$$

(C_3)

$$
\| x_0 - x_{-1} \| \leq d;
$$

(C_4)

$$
0 < \| A_0^{-1} F(x_0) \| \leq \eta;
$$

(C_5) There exists constant $M > 0$, such that

$$
\| A_0^{-1} ([x, y; F] - F'(z)) \| \leq \frac{M}{2} \left(\| x - z \| + \| y - z \| \right) \quad \text{for all} \quad x, y, z \in \mathcal{D},
$$

(C_6) For the function

$$
p(t) = \frac{t \, f(t)}{2} \left(1 + \frac{t}{2} + f(t) \, g(t) \right),
$$

constants

$$
a_0 = \beta M \max \{ a c (a \eta + d), \eta + d \},
$$

$$
R_0 = \beta M \max \left\{ \frac{1 + p(a_0)}{1 - f(a_0) g(a_0)}, \frac{1 + \dfrac{a_0}{2}}{1 - f(a_0)^2 g(a_0)} \right\}
$$

and
$$R = R_0 \, \eta,$$
$$U(x_0, R) = \{x \in X : \| x - x_0 \| < R\} \subseteq \mathcal{D},$$

where, functions f, g, and zero ξ were defined in Lemma 4.1;

$(\mathcal{C}_7)$
$$(1 - a) \, c = 1 - b,$$

where, $a \in [0, 1]$, $b \in [0, 1]$, and $c \geq 0$;

$(\mathcal{C}_8)$
$$a_0 < \xi_0.$$

We shall show the semilocal convergence of (CSTM) using condition (C). We have: $\| y_0 - x_0 \| \leq \eta$, and $\| z_0 - x_0 \| \leq a \, \eta$, so that $x_0, y_0 \in \mathcal{D}$.

Let
$$K = \frac{M}{2}.$$

Using (4.9) for $n = 0$, and (C_1)–(C_6), we get

$$
\begin{aligned}
\| x_1 - y_0 \| \; &\leq \; a \, c \, \| A_0^{-1} \| \, K \left(2 \int_0^1 a \, t \, \| y_0 - x_0 \|^2 \, dt + \right. \\
&\qquad \left. \| x_0 - x_{-1} \| \, \| y_0 - x_0 \| \right) \\
&\leq \; \frac{a_0}{2} \, \| y_0 - x_0 \|,
\end{aligned}
$$

and

$$\| x_1 - x_0 \| \leq \| x_1 - y_0 \| + \| y_0 - x_0 \| \leq \left(1 + \frac{a_0}{2} \right) \| y_0 - x_0 \| < R.$$

That is by $x_1 \in U(x_0, R)$. Note that the value of radius R is later deduced.
Using $(\mathcal{C}_5)$, and $(\mathcal{C}_6)$, we get in turn:

$$
\begin{aligned}
\| A_0^{-1} \| & \, \| A_1 - A_0 \| \\
&= \| A_0^{-1} \| \, \| \, [x_1, x_0; F] - F'(x_0) + F'(x_0) - [x_0, x_{-1}; F] \, \| \\
&\leq \| A_0^{-1} \| \, K \left(\| x_1 - x_0 \| + \| x_0 - x_{-1} \| \right) \\
&\leq \| A_0^{-1} \| \, K \left(\left(1 + \frac{a_0}{2} \right) \| y_0 - x_0 \| + \| x_0 - x_{-1} \| \right) \\
&= \| A_0^{-1} \| \, K \left(1 + \frac{a_0}{2} \right) \| y_0 - x_0 \| + \qquad\qquad (4.11) \\
&\quad\; \| A_0^{-1} \| \, K \, \| x_0 - x_{-1} \| \\
&= \| A_0^{-1} \| \, K \left(\| y_0 - x_0 \| + \| x_0 - x_{-1} \| \right) + \\
&\quad\; \frac{a_0}{2} \, \| A_0^{-1} \| \, K \, \| y_0 - x_0 \| \\
&\leq \frac{a_0}{2} + \frac{a_0^2}{4} < 1.
\end{aligned}
$$

Chapter 4. Chebyshev–Secant–Type Methods

It follows from (4.11), and the Banach lemma on invertible operators, that A_1^{-1} exists, and

$$
\begin{aligned}
\| A_1^{-1} \| &\leq \| A_0^{-1} \| \left(1 - \| A_0^{-1} \| \, K \left(\left(1 + \frac{a_0}{2} \right) \| y_0 - x_0 \| + \right. \right. \\
&\qquad\qquad \left. \left. \| x_0 - x_{-1} \| \right) \right)^{-1} \\
&\leq \| A_0^{-1} \| \, f(a_0).
\end{aligned}
\tag{4.12}
$$

That is, y_1, and z_1 are well defined inside $\mathcal{D}$.

In view of (4.10) for $n = 0$, (CSTM), and conditions $(\mathcal{C})$, we get:

$$
\begin{aligned}
\| F(x_1) \| &\leq K \| x_1 - x_0 \|^2 + K \| x_1 - x_0 \| \, \| x_0 - x_{-1} \| + \\
&\qquad a\, c \, (a\, K \| y_0 - x_0 \|^2 + K \| x_0 - x_{-1} \| \, \| y_0 - x_0 \|) \\
&\leq K \left((\| x_1 - x_0 \| \, \| x_0 - x_{-1} \|) \, \| x_1 - x_0 \| + \right. \\
&\qquad \left. ac \, (a \| y_0 - x_0 \| + \| x_0 - x_{-1} \|) \, \| y_0 - x_0 \| \right),
\end{aligned}
$$

and

$$
\| y_1 - x_1 \| \leq \| A_1^{-1} \| \, \| F(x_1) \| \leq \| A_0^{-1} \| \, f(a_0) \, \| F(x_1) \|.
$$

But, we also have:

$$
\begin{aligned}
&\| A_0^{-1} \| \, K \left\{ (\| x_1 - x_0 \| + \| x_0 - x_{-1} \|) \left(1 + \frac{a_0}{2} \right) + \right. \\
&\quad \left. ac \, (a \| y_0 - x_0 \| + \| x_0 - x_{-1} \|) \right\} \| y_0 - x_0 \| \\
&\leq \| A_0^{-1} \| \, K \left\{ \left(1 + \frac{a_0}{2} \right) \| y_0 - x_0 \| + \| x_0 - x_{-1} \| \right\} \left(1 + \frac{a_0}{2} \right) \\
&\leq \left(\frac{a_0}{2} + \frac{a_0^2}{4} \right) \left(1 + \frac{a_0}{2} \right),
\end{aligned}
$$

so,

$$
\| A_0^{-1} \| \, \| F(x_1) \| \leq \left(\frac{a_0}{2} + \frac{a_0^2}{4} \right) \left(1 + \frac{a_0}{2} \right) + \frac{a_0}{2} = g(a_0),
$$

and

$$
\| y_1 - x_1 \| \leq f(a_0) \, g(a_0) \, \| y_0 - x_0 \|.
$$

Moreover, we get

$$
\begin{aligned}
M \| A_1^{-1} \| \, \| y_1 - x_1 \| &\leq M \| A_0^{-1} \| \, f(a_0)^2 \, g(a_0) \, \| y_0 - x_0 \| \\
&\leq a_0 \, f(a_0)^2 \, g(a_0) < 2 \, f(a_0)^2 \, g(a_0),
\end{aligned}
$$

and

$$\| x_2 - y_1 \| \le ac \, \| A_1^{-1} \| \left\{ K \left(a \, \| y_1 - x_1 \|^2 + \| x_1 - x_0 \| \, \| y_1 - x_1 \| \right) \right\}$$

$$\le ac \, \| A_1^{-1} \| \, K \, \| y_1 - x_1 \| \left(a \, \| y_1 - x_1 \| + \| x_1 - x_0 \| \right)$$

$$\le \frac{a_0}{2} f(a_0)^2 g(a_0) \, \| y_1 - x_1 \| + acK \, \| A_1^{-1} \| \, \| x_1 - x_0 \| \, \| y_1 - x_1 \|$$

$$\le \frac{a_0}{2} f(a_0)^2 g(a_0) \, \| y_1 - x_1 \| +$$

$$acK \, \| A_0^{-1} \| \, f(a_0) \left(1 + \frac{a_0}{2} \right) \| y_0 - x_0 \| \, \| y_1 - x_1 \|$$

$$\le \left(\frac{a_0}{2} f(a_0)^2 g(a_0) + \frac{a_0}{2} f(a_0) \left(1 + \frac{a_0}{2} \right) \right) \| y_1 - x_1 \|$$

$$= \frac{a_0}{2} f(a_0) \left(f(a_0) g(a_0) + 1 + \frac{a_0}{2} \right) \| y_1 - x_1 \|$$

$$= p(a_0) \, \| y_1 - x_1 \| .$$

Then,

$$\| x_2 - x_1 \| \le \| x_2 - y_1 \| + \| y_1 - x_1 \| \le (1 + p(a_0)) \, \| y_1 - x_1 \|,$$

and

$$\| x_2 - x_0 \| \le \| x_2 - x_1 \| + \| x_1 - x_0 \|$$

$$\le \left\{ \left(1 + \frac{a_0}{2} f(a_0) \left(f(a_0) g(a_0) + 1 + \frac{a_0}{2} \right) \right) f(a_0) g(a_0) + 1 + \frac{a_0}{2} \right\} \| y_0 - x_0 \|$$

$$\le \left\{ \left(1 + \frac{a_0}{2} f(a_0)^2 g(a_0) + f(a_0) \frac{a_0}{2} \left(1 + \frac{a_0}{2} \right) \right) f(a_0) g(a_0) + 1 + \frac{a_0}{2} \right\} \| y_0 - x_0 \|$$

$$\le \left\{ \left(1 + \frac{a_0}{2} + \frac{\frac{a_0}{2} \left(1 + \frac{a_0}{2} \right)}{1 - \frac{a_0}{2} \left(1 + \frac{a_0}{2} \right)} \right) f(a_0) g(a_0) + 1 + \frac{a_0}{2} \right\} \| y_0 - x_0 \|$$

$$= \left\{ \left(1 + \frac{a_0}{2} \right) \left(1 + \frac{\frac{a_0}{2}}{1 - \frac{a_0}{2} \left(1 + \frac{a_0}{2} \right)} \right) f(a_0) g(a_0) + 1 + \frac{a_0}{2} \right\} \| y_0 - x_0 \|$$

$$= \left(1 + \frac{a_0}{2} \right) \left(1 + \frac{1 - \frac{a_0^2}{4}}{1 - \frac{a_0}{2} \left(1 + \frac{a_0}{2} \right)} f(a_0) g(a_0) \right) \| y_0 - x_0 \|$$

$$\le \left(1 + \frac{a_0}{2} \right) \left(1 + \left(1 - \frac{a_0^2}{4} \right) f(a_0)^2 g(a_0) \right) \| y_0 - x_0 \|,$$

so,

$$\| x_2 - x_0 \| \le \left(1 + \frac{a_0}{2} \right) \frac{1 - r^2}{1 - r} \, \| y_0 - x_0 \| < R,$$

where,

$$r = f(a_0)^2 \, g(a_0). \tag{4.13}$$

Then, $x_2 \in U(x_0, R)$, by (C_8). Besides, $\| A_1^{-1} \| \, \| A_2 - A_1 \| < 1$, exists by the Banach lemma on invertible operators, and $\| A_2^{-1} \| \leq f(2 f(a_0)^2 g(a_0)) \, \| A_1^{-1} \|$. Indeed, we have in turn:

$$
\begin{aligned}
&\| A_1^{-1} \| \, \| A_2 - A_1 \| \\
&= \| A_1^{-1} \| \, \| \, ([x_2, x_1; F] - F'(x_1)) + (F'(x_1) - [x_1, x_0; F]) \, \| \\
&\leq \frac{\| A_1^{-1} \| \, M}{2} \left(\| x_2 - x_1 \| + \| x_1 - x_0 \| \right) \\
&\leq \frac{\| A_1^{-1} \| \, M}{2} \left((1 + p(a_0)) \, \| y_1 - x_1 \| + (1 + \frac{a_0}{2}) \, \| y_0 - x_0 \| \right) \\
&\leq \frac{\| A_1^{-1} \| \, M}{2} \left((1 + p(a_0)) \, f(a_0) \, g(a_0) + 1 + \frac{a_0}{2} \right) \, \| y_0 - x_0 \| \\
&\leq \frac{\| A_1^{-1} \| \, M}{2} f(a_0) \left((1 + p(a_0)) \, f(a_0) \, g(a_0) + \frac{a_0}{2} + 1 \right) \, \| y_0 - x_0 \| \\
&\leq \frac{a_0}{2} f(a_0) \left((1 + p(a_0)) \, f(a_0) \, g(a_0) + \frac{a_0}{2} + 1 \right) \\
&\leq f(a_0)^2 \, g(a_0) \, (1 + f(a_0)^2 \, g(a_0)) < 1
\end{aligned}
$$

(by Lemma 4.2 (e) and (f)).

Then, we can again deduce $y_2, z_2, x_3 \in U(x_0, R)$ in an analogous way.

Set, $a_1 = 2 f(a_0)^2 g(a_0)$, and define scalar sequence $\{a_k\}$ by

$$
a_{k+1} = 2 f(a_k)^2 g(a_k) \qquad (k \geq 0).
$$

Clearly, sequence $\{a_k\}$ is decreasing, and such that

$$
\frac{a_k}{2} \left(1 + \frac{a_k}{2} \right) < 1.
$$

Finally, sequence $\{x_k\}$ satisfies the following system of recurrence relations, from which we guarantee that sequence $\{x_k\}$ generated by (CSTM) is well defined. To prove them, we follow an analogous to the above method, and then the induction hypotheses.

Lemma 4.3. *Under the (C) conditions, the following items hold for all $k \geq 1$:*

(I) A_k^{-1} *exists, and*

$$
\| A_k^{-1} \| \leq f(a_{k-1}) \, \| A_{k-1}^{-1} \|,
$$

(II)

$$
\begin{aligned}
\| y_k - x_k \| &\leq f(a_{k-1}) \, g(a_{k-1}) \, \| y_{k-1} - x_{k-1} \| \\
&\leq (f(a_0) \, g(a_0))^k \, \| y_0 - x_0 \| < \eta,
\end{aligned}
$$

(III)

$$
M \, \| A_k^{-1} \| \, \| y_k - x_k \| < a_k,
$$

(IV)

$$
\| x_{k+1} - y_k \| \leq p(a_k) \, \| y_k - x_k \|,
$$

(V)

$$\| x_{k+1} - x_k \| \le (1 + p(a_k)) \, \| y_k - x_k \|,$$

(VI)

$$\| x_{k+1} - x_0 \| \le \left(1 + \frac{a_0}{2}\right) \frac{1 - r^{k+1}}{1 - r} \, \| y_0 - x_0 \| < R,$$

where, r is given by (4.13).

We can show the main semilocal convergence theorem for (CSTM).

Theorem 4.1. *Let $F : \mathcal{D} \subseteq X \longrightarrow \mathcal{Y}$ be a Fréchet–differentiable operator defined on a non–empty open, convex domain $\mathcal{D}$ of a Banach space X, with values in a Banach space $\mathcal{Y}$. Assume the (C) conditions hold.*

Then sequence $\{x_k\}$ $(k \ge -1)$ generated by $(CSTM)$ is well defined, remains in $\overline{U}(x_0, R)$ for all $k \ge 0$, and converges to a unique solution $x^\star \in \overline{U}(x_0, R)$ of equation $F(x) = 0$.

Moreover, if for

$$R \le \frac{1}{2 \, K_0}, \qquad K_0 = \frac{M_0}{2}, \tag{4.14}$$

$$U(x_0, R_1) \subseteq \mathcal{D} \tag{4.15}$$

where,

$$R_1 \in [R, \frac{1}{K_0} - R], \tag{4.16}$$

and M_0 is such that

$$\| A_0^{-1} \left([x, y; F] - F'(x_0)\right) \| \le \frac{M_0}{2} \left(\| x - x_0 \| + \| y - x_0 \|\right) \tag{4.17}$$

$$for \ all \ x, y \in \mathcal{D},$$

then, the solution $x^\star$ is unique in $\overline{U}(x_0, R_1)$.

Note that the existence of M_0 used in Theorem 4.1 follows from (C_5), and $M_0 \in [0, M]$, and $\dfrac{M}{M_0}$ can be arbitrarily large [155], [191].

Proof. In view of the discussion before Lemma 4.3, sequence $\{x_k\}$ is well defined.

Let $m \ge 1$, then we have by (II), and (V) of Lemma 4.3:

$$
\begin{aligned}
\| x_{k+m} - x_k \| \ &\le \ \sum_{i=k}^{k+m-1} \| x_{i+1} - x_i \| \\
&\le \ \sum_{i=k}^{k+m-1} \left(1 + p(a_i)\right) \| y_i - x_i \| \\
&\le \ \left(1 + p(a_k)\right) \sum_{j=k}^{k+m-1} \left(\prod_{i=0}^{j-1} f(a_i) \, g(a_i)\right) \| y_0 - x_0 \| \\
&\le \ \left(1 + p(a_0)\right) \left(f(a_0) \, g(a_0)\right)^k \frac{1 - (f(a_0) \, g(a_0))^m}{1 - f(a_0) \, g(a_0)} \, \eta.
\end{aligned}
\tag{4.18}
$$

Chapter 4. Chebyshev–Secant–Type Methods 71

If $k = 0$ in (4.18), we get

$$\| x_{k+m} - x_k \| \ \leq \ \left(1 + p(a_0) \right) \frac{1 - (f(a_0) g(a_0))^m}{1 - f(a_0) g(a_0)} \eta < R. \tag{4.19}$$

That is $x_m \in U(x_0, R)$. Similarly, $y_m, z_m \in U(x_0, R)$ for all $m \geq 0$.

In view of (4.18), sequence $\{x_k\}$ is Cauchy in a Banach space X, and as such that it converges to some $x^\star \in \overline{U}(x_0, R)$ (since $\overline{U}(x_0, R)$ is a closed set). Indeed, by letting $k \longrightarrow \infty$, we have $\| A_k^{-1} F(x_k) \| \longrightarrow 0$, and, since $\| F(x_k) \| \leq \| A_k \| \ \| A_k^{-1} F(x_k) \|$, and the sequence $\{\| A_k \|\}$ is bounded, we have $\| F(x_k) \| \longrightarrow 0$. By continuity of F, we get $F(x^\star) = 0$.

Finally, we shall show the uniqueness of the solution $x^\star$ in $U(x_0, R_1)$. Let $y^\star$ be a solution of equation $F(x) = 0$ in $U(x_0, R_1)$. We have the approximation

$$0 = A_0^{-1} (F(y^\star) - F(x^\star)) = A_0^{-1} [y^\star, x^\star; F] (y^\star - x^\star). \tag{4.20}$$

Set

$$\mathcal{M} = [y^\star, x^\star; F].$$

Using (4.17), we obtain:

$$
\begin{aligned}
\| A_0^{-1} (\mathcal{M} - F'(x_0)) \| \ &\leq \ \frac{M_0}{2} (\| y^\star - x_0 \| + \| x^\star - x_0 \|) \\
&< \ K_0 (R_1 + R) \leq 1.
\end{aligned} \tag{4.21}
$$

It follows from (4.21), and the Banach lemma on invertible operators, that linear operator $\mathcal{M}$ is invertible. It then follows from (4.20) that $x^\star = y^\star$.

That completes the proof of Theorem 4.1. $\diamond$

Remark 4.1. Condition (C_5) implies that for $x_0 = x_{-1}$, and $x = y$

$$\| F'(x_0)^{-1} (F'(x) - F'(z)) \| \leq M \| x - z \| .$$

If one defines function f by

$$f(t) = \frac{1}{1 - t \left(1 + \dfrac{t}{2} \right)},$$

Then, the conclusions of Theorem 4.4 in [377] can be obtained from Theorem 4.1 for

$$b = \frac{a^2 + a - 1}{a^2}, \qquad c = \frac{1}{a^2},$$

with the exception of the uniqueness part.

Theorem 4.1 provides a larger uniqueness ball if $M_0 < M$. To obtain the uniqueness ball of Theorem 4.4, simply set $M = M_0$ in (4.16).

We complete this chapter with a numerical example.

Example 4.1. Let $X = Y = \mathbb{R}^2$, be equipped with the max–norm. Choose:

$$x_0 = (1,1)^T, \quad \mathcal{D} = U(x_0, 1-\alpha), \quad \alpha \in \left[0, \frac{1}{2}\right).$$

Define function F on U_0 by

$$F(x) = (\theta_1^3 - \alpha, \theta_2^3 - \alpha), \qquad x = (\theta_1, \theta_2)^T. \tag{4.22}$$

The Fréchet–derivative of operator F is given by

$$F'(x) = \begin{bmatrix} 3\,\theta_1^2 & 0 \\ 0 & 3\,\theta_2^2 \end{bmatrix}, \tag{4.23}$$

and the divided difference of F is defined by

$$[y, x; F] = \int_0^1 F'(x + t\,(y-x))\, dt.$$

Then, we have:

$$M = 6\,\beta\,(2-\alpha), \qquad M_0 = 3\,\beta\,(3-\alpha), \quad \eta = (1-\alpha)\,\beta.$$

Let $\alpha = .49$. Then, we get for $a = b = .5$, and $c = 1$:

$$\beta = .333666889, \qquad M = 3.023022014,$$

$$M_0 = 2.512511674, \qquad d = .001, \qquad \eta = .170170113,$$

$$a_0 = .172656272 < \xi_0, \qquad R = .2430163,$$

and

$$R_1 = \frac{1}{K_0} - R = .552999902, \qquad 1 - \alpha = .51.$$

The hypothese of Theorem 4.1 are satisfied. Hence, equation $F(x) = 0$ has an unique solution

$$x^\star = (\sqrt[3]{.49}, \sqrt[3]{.49})^T = (.788373516, .788373516)^T,$$

which is unique in $U(x_0, R_2)$ $(R_2 = 1 - \alpha)$, and can be obtained as the limit of sequence $\{x_k\}$ starting at x_0.

Chapter 5

Convergence Using Recurrent Functions

The Kantorovich analysis [155], [189], [477], and recurrent relation's approach [380] are the most popular ways for generating sufficient conditions for the convergence of numerical algorithms to a solution of a nonlinear equations as well as providing the corresponding error estimates on the distances involved. In this chapter, we introduce the new approach of recurrent functions to show that a finer convergence analysis can be provided under the same hypotheses, and computational cost. Numerical examples are provided where our results apply, but not earlier ones.

In this chapter, we are concerned with the problem of approximating a locally unique solution $x^\star$ of equation

$$F(x) + G(x) = 0, \tag{5.1}$$

where, F is a Fréchet–differentiable operator defined on a open convex subset $\mathcal{D}$ of a Banach space X with values in X, and $G : \mathcal{D} \longrightarrow X$ is a continuous operator.

The Newton–like method

$$x_{n+1} = x_n - A_n^{-1} \left(F(x_n) + G(x_n) \right) \quad (n \geq 0), \quad (x_0 \in \mathcal{D}), \tag{5.2}$$

has been used by many authors to generate a sequence $\{x_n\}$ converging to $x^\star$ [89], [770]. Here, $A(x) \in L(X, \mathcal{Y})$ the space of bounded linear operators from X into $\mathcal{Y}$ [155]. The most popular choice for operators A, and G is $A(x) = F'(x)$, and $G(x) = 0, (x \in \mathcal{D})$, leading to the Newton–Kantorovich method

$$x_{n+1} = x_n - F'(x_n)^{-1} F(x_n), \quad (x_0 \in \mathcal{D}), \quad (n \geq 0). \tag{5.3}$$

Several other choices for A, and G are possible [89], [155], [189], [299], [155]. The Kantorovich analysis [155], [189], [477], and recurrent relation's approach [380] have been extensively used under Lipschitz–Hölder type conditions to provide a semilocal convergence for method (5.2), and (5.3) [89], [770].

We introduce the recurrent function approach to provide a finer converegence analysis under the same computational cost and hypotheses. In particular, the advantages of our approach are:

(a) Larger convergence domain;

(b) Finer error bounds on the distances $\| x_{n+1} - x_n \|, \| x_n - x^\star \|, (n \geq 0)$;

and

(c) An at least as precise information on the location of the solution $x^\star$.

Numerical examples are also provided, where our results can apply to solve equations, but the corresponding ones [380], [477] cannot. Moreover, our results have the additional advantage that they can be used to locate roots of polynomials of high degree [89].

The chapter in organized as follows: we present some results on the convergence of scalar iterations that majorize vector iterations. We also study Newton's method in the Lipschitz and Hölder cases respectively. Finally, we deal with the study of Newton–like method (5.2).

We need some results on the convergence of scalar iterations.

Lemma 5.1. *Let $\{p_n\} \in [0,1)$ be a sequence, and constants γ_0, γ_1, α, with:*

$$0 \leq \gamma_0 \leq \gamma_1, \tag{5.4}$$

and

$$\alpha \in [0,1) \tag{5.5}$$

be given.

Define scalar iteration $\{t_n\}$ $(n \geq 0)$ by

$$t_0 = \gamma_0, \quad t_1 = \gamma_1, \quad t_{n+1} = t_n + p_n (t_n - t_{n-1}). \tag{5.6}$$

Assume

$$p_n \leq \alpha \qquad (n \geq 1). \tag{5.7}$$

Then, scalar iteration $\{t_n\}$ $(n \geq 0)$, is well defined, non–decreasing, bounded above by:

$$t^{\star\star} = \frac{\gamma}{1 - \alpha}, \tag{5.8}$$

and converges to its unique least upper bound

$$t^\star \in [0, t^{\star\star}], \tag{5.9}$$

where,

$$\gamma = \gamma_1 - \gamma_0. \tag{5.10}$$

Moreover the following estimates hold for all $n \geq 1$:

$$0 \leq t_{n+1} - t_n \leq \alpha (t_n - t_{n-1}) \leq \alpha^n \gamma, \tag{5.11}$$

and

$$0 \leq t^\star - t_n \leq \frac{\gamma}{1 - \alpha} \alpha^n. \tag{5.12}$$

Chapter 5. Convergence Using Recurrent Functions 75

Proof. By hypothesis $\gamma_1 \geq \gamma_0$, and sequence $\{p_n\}$ is non–negative. It then follows by a simple inductive argument that sequence $\{t_n\}$ is non–decreasing.

Using (5.6), (5.7), and (5.10), we get

$$
\begin{aligned}
0 \;\leq\; & t_{n+1} - t_n = p_n\,(t_n - t_{n-1}) \\
\leq\; & \alpha\,(t_n - t_{n-1}) \\
\leq\; & \alpha^2\,(t_{n-1} - t_{n-2}) \leq \alpha^n\,(t_1 - t_0) = \alpha^n\,\gamma,
\end{aligned}
\tag{5.13}
$$

which shows (5.11).

Using (5.13), we get in turn:

$$
\begin{aligned}
t_{n+1} \;\leq\; & t_n + \alpha\,(t_n - t_{n-1}) \\
\leq\; & t_{n-1} + \alpha\,(t_{n-1} - t_{n-2}) + \alpha\,(t_n - t_{n-1}) \\
\leq\; & t_1 + \alpha\,(t_1 - t_0) + \cdots + \alpha\,(t_n - t_{n-1}) \\
\leq\; & \gamma + \alpha\,\gamma + \cdots + \alpha^n\,\gamma = \frac{1 - \alpha^{n+1}}{1 - \alpha}\,\gamma < \frac{\gamma}{1 - \alpha} = t^{\star\star}.
\end{aligned}
\tag{5.14}
$$

Hence, sequence $\{t_n\}$ is non–decreasing, bounded above by $t^{\star\star}$, and as such it converges to its unique least upper bound $t^\star$ satisfying (5.9).

Finally, estimate (5.12) follows from (5.11) by using standard majorization techniques [89], [477].

That completes the proof of Lemma 5.1. $\diamond$

Note that sequence $\{p_n\}$ depends on iterates t_i, $i = 0, 1, \ldots, n$ (see, e.g. (5.22)).

Lemma 5.2. *Let* $\{p_n\}$, $\{t_n\}$, γ_0, γ_1, *and* α *be as in Lemma 5.1. Assume there exists a sequence of functions* $\{f_n\}$ *($n \geq 1$) defined on* $[0, 1)$, *such that for all* $n \geq 1$:

$$
p_n - \alpha \leq f_n(\alpha),
\tag{5.15}
$$

and

$$
f_n(\alpha) \leq 0.
\tag{5.16}
$$

Then, the conclusions of Lemma 5.1 hold.

Proof. We have by (5.15):

$$
p_n \leq f_n(\alpha) + \alpha \leq \alpha, \quad (n \geq 1)
\tag{5.17}
$$

which implies (5.7). That completes the proof of Lemma 5.2. $\diamond$

Lemma 5.3. *Let* $\{p_n\}$, $\{t_n\}$, $\{f_n(\alpha)\}$, γ_0, γ_1, *and* α *be as in Lemma 5.2.*
Assume that (5.15), *and*

$$
f_n(\alpha) \leq f_1(\alpha) \leq 0
\tag{5.18}
$$

hold for all $n \geq 1$.
Then, the conclusions of Lemma 5.1 hold.

Proof. Estimates (5.15), and (5.18) imply (5.16). That completes the proof of Lemma 5.3. $\diamond$

Lemma 5.4. *Let* $\{p_n\}$, $\{t_n\}$, $\{f_n(\alpha)\}$, γ_0, γ_1, *and* α *be as in Lemma* 5.2.
Define function f_∞ *on* $[0,1)$ *by*

$$f_\infty(s) = \lim_{n\to\infty} f_n(s). \tag{5.19}$$

Assume that (5.15), *and*

$$f_n(\alpha) \le f_\infty(\alpha) \le 0 \tag{5.20}$$

hold for all $n \ge 1$.
  Then, the conclusions of Lemma 5.1 *hold.*

Proof. Estimates (5.20) implies (5.16). That completes the proof of Lemma 5.4. ◇

We shall show how to choose items introduced in the four lemmas above. Let us consider iteration $\{t_n\}$ given by:

$$t_0 = \gamma_0, \quad t_1 = \gamma_1, \quad t_{n+1} = t_n + \frac{L\,(t_n - t_{n-1})^{p+1}}{(1+p)\,(1 - L_0\,t_n^p)}, \quad (n \ge 1), \tag{5.21}$$

for some $L_0 \ge 0$, $L \ge 0$, with $L_0 \le L$, and $p \in (0,1]$.
  That is we choose:

$$p_n = \frac{L\,(t_n - t_{n-1})^p}{(1+p)\,(1 - L_0\,t_n^p)}, \quad (n \ge 1). \tag{5.22}$$

Iteration $\{t_n\}$ appears as a majorization sequence for $\{x_n\}$ in the study of the semilocal convergence of Newton's method (5.3).

We shall study the cases $p = 1$ (Lipschitz), and $p \in (0,1)$ (Hölder), separately. Similarly, we study the Newton–like method (5.2).

We need a result on locating roots of polynomials (see, also [89]).

Theorem 5.1. *Let* $a > 0$, $b > 0$, *and* $c < 0$ *be given constants. Define polynomials* f_n $(n \ge 1)$, g *on* $[0, +\infty)$ *by:*

$$f_n(s) = b\,s^n + a\,s^{n-1} + b\,(s^{n-1} + s^{n-2} + \cdots + 1) + c, \tag{5.23}$$

and

$$g(s) = b\,s^2 + a\,s - a. \tag{5.24}$$

Set

$$\alpha = \frac{2\,a}{a + \sqrt{a^2 + 4\,a\,b}}. \tag{5.25}$$

Assume:

$$\alpha \le 1 + \frac{b}{c}, \tag{5.26}$$

and

$$a + b + c < 0. \tag{5.27}$$

Then, each polynomial f_n $(n \ge 1)$ *has a unique positive root* s_n.
  Moreover, the following estimates hold for all $n \ge 1$:

$$1 + \frac{b}{c} \le s^\star \le s_{n+1} \le s_n, \tag{5.28}$$

Chapter 5. Convergence Using Recurrent Functions 77

and

$$f_n(\alpha) \leq 0, \tag{5.29}$$

where,

$$s^* = \lim_{n \longrightarrow \infty} s_n.$$

Proof. Each polynomial f_n has a unique positive root s_n ($n \geq 1$), by the Descarte's rule of signs.

Polynomial f_n can be written for $s \in [0, 1)$:

$$f_n(s) = b\, s^n + a\, s^{n-1} + b\, \frac{1 - s^n}{1 - s} + c. \tag{5.30}$$

By letting $n \longrightarrow \infty$, we get:

$$f_\infty(s) = \lim_{n \longrightarrow \infty} f_n(s) = \frac{b}{1 - s} + c. \tag{5.31}$$

Function f_∞ has a unique positive root denoted by s_∞, and given by:

$$s_\infty = 1 + \frac{b}{c} < 1. \tag{5.32}$$

Function f_∞ is also increasing, since

$$f_\infty'(s) = \frac{b}{(1 - s)^2} > 0. \tag{5.33}$$

Furthermore, we shall show estimates (5.28), and (5.29) hold.

We need the relationship between two consecutive polynomials f_n's ($n \geq 1$):

$$\begin{aligned} f_{n+1}(s) &= b\, s^{n+1} + a\, s^n + b\, (s^n + \cdots + 1) + c \\ &= b\, s^n + a\, s^{n-1} + b\, (s^{n-1} + \cdots + 1) + c + \\ &\quad a\, s^n - a\, s^{n-1} + b\, s^{n+1} \\ &= f_n(s) + s^{n-1}\, (b\, s^2 + a\, s - a) \\ &= f_n(s) + g(s)\, s^{n-1}. \end{aligned} \tag{5.34}$$

Note that α is the unique positive root of function g.

Then, using (5.34), we obtain for all $n \geq 1$:

$$f_n(\alpha) = f_{n-1}(\alpha) = \cdots = f_1(\alpha).$$

Let i be any fixed but arbitrary natural number. Then, we get:

$$f_i(\alpha) = \lim_{n \longrightarrow \infty} f_n(\alpha) = f_\infty(\alpha) \leq f_\infty(s_\infty) = 0, \tag{5.35}$$

since, function f_∞ is increasing, and $\alpha \leq s_\infty$ by hypothesis (5.26). It follows from the definition of the zeros s_n and (5.35) that

$$\alpha \leq s_n \qquad \text{for all } n \geq 1. \tag{5.36}$$

Polynomials f_n are increasing which together with (5.36) imply

$$f_n(\alpha) \le f_n(s_n) = 0.$$

In particular

$$f_\infty(\alpha) = \lim_{n \longrightarrow \infty} f_n(\alpha) \le 0.$$

Hence, estimate (5.29) holds.

We then get from (5.34), and (5.36):

$$f_{n+1}(s_{n+1}) = f_n(s_{n+1}) + g(s_{n+1}) s_{n+1}^{n-1} \eta$$

or

$$f_n(s_{n+1}) \le 0, \tag{5.37}$$

since $f_{n+1}(s_{n+1}) = 0$, and $g(s_{n+1}) s_{n+1}^{n-1} \eta \ge 0$, which imply

$$s_{n+1} \le s_n \qquad (n \ge 1).$$

Sequence $\{s_n\}$ is non–increasing, bounded below by zero, and as such it converges to $s^\star$.

We shall show $s_\infty \le s^\star$. Using (5.34), we have:

$$\begin{aligned} f_{i+1}(s_i) &= f_i(s_i) + g(s_i) s_i^{i-1} \eta \\ &= g(s_i) s_i^{i-1} \eta \ge 0, \end{aligned}$$

so,

$$f_{i+2}(s_i) = f_{i+1}(s_i) + g(s_i) s_i^i \eta \ge 0.$$

If, $f_{i+m}(s_i) \ge 0$, $m \ge 0$, then

$$f_{i+m+1}(s_i) = f_{i+m}(s_i) + g(s_i) s_i^{i+m-1} \eta \ge 0.$$

Hence, by the definition of function f_∞, we get:

$$f_\infty(s_n) \ge 0 \qquad \text{for all } n \ge 1.$$

But we also have $f_\infty(0) = b + c < 0$. That is $s_\infty \le s_n$ for all $n \ge 1$, and consequently $s_\infty \le s^\star$.

That completes the proof of Theorem 5.1. $\diamond$

Set $a = L\eta$, $b = 2L_0\eta$, and $c = -2$, in Theorem 5.1.

It is simple algebra to show that conditions (5.26), and (5.27) reduce to (5.38) in the majorizing lemma that follows:

Lemma 5.5. *Assume there exist constants $L_0 \ge 0$, $L \ge 0$, and $\eta \ge 0$, with $L_0 \le L$, such that:*

$$q_0 = \overline{L}\eta \begin{cases} \le \dfrac{1}{2} & \text{if } L_0 \ne 0 \\[2mm] \\ < \dfrac{1}{2} & \text{if } L_0 = 0 \end{cases}, \tag{5.38}$$

Chapter 5. Convergence Using Recurrent Functions

where,

$$\overline{L} = \frac{1}{8}\left(L + 4\,L_0 + \sqrt{L^2 + 8\,L_0\,L}\right). \tag{5.39}$$

Then, sequence $\{t_n\}$ $(n \geq 0)$ given by

$$t_0 = 0, \quad t_1 = \eta, \quad t_{n+1} = t_n + \frac{L\,(t_n - t_{n-1})^2}{2\,(1 - L_0\,t_n)} \quad (n \geq 1), \tag{5.40}$$

is nondecreasing, bounded above by $t^{\star\star}$, and converges to its unique least upper bound $t^\star \in [0, t^{\star\star}]$, where

$$t^{\star\star} = \frac{\eta}{1 - \alpha}, \tag{5.41}$$

$$\alpha = \frac{2\,L}{L + \sqrt{L^2 + 8\,L_0\,L}} < 1 \quad \text{for } L_0 \neq 0. \tag{5.42}$$

Moreover the following estimates hold:

$$L_0\,t^\star \leq 1, \tag{5.43}$$

$$0 \leq t_{n+1} - t_n \leq \alpha\,(t_k - t_{k-1}) \leq \cdots \leq \alpha^n\,\eta, \quad (n \geq 1), \tag{5.44}$$

$$t_{n+1} - t_n \leq \alpha^n\,(2\,q_0)^{2^n - 1}\,\eta, \quad (n \geq 0), \tag{5.45}$$

$$0 \leq t^\star - t_n \leq \alpha^n\,\frac{(2\,q_0)^{2^n - 1}\,\eta}{1 - (2\,q_0)^{2^n}}, \quad (2\,q_0 < 1), \quad (n \geq 0). \tag{5.46}$$

Proof. We shall show using induction on n that for all $n \geq 0$:

$$L\,(t_{n+1} - t_n) + 2\,\alpha\,L_0\,t_{n+1} < 2\,\alpha, \tag{5.47}$$

$$0 < t_{n+1} - t_n, \tag{5.48}$$

$$L_0\,t_{n+1} < 1, \tag{5.49}$$

and

$$0 < t_{n+1} < t^{\star\star}. \tag{5.50}$$

Estimates (5.47)–(5.50) hold true for $n = 0$ by the initial condition $t_1 = \eta$, and hypothesis (5.38). It then follows from (5.40) that

$$0 < t_2 - t_1 \leq \alpha\,(t_1 - t_0) \quad \text{and} \quad t_2 \leq \eta + \alpha\,\eta = (1 + \alpha)\,\eta < t^{\star\star}.$$

Let us assume estimates (5.47)–(5.50) hold true for all integer values k: $k \leq n$ $(k \geq 0)$. We also get

$$\begin{aligned}
t_{n+1} &\leq t_n + \alpha\,(t_n - t_{n-1}) \\
&\leq t_{n-1} + \alpha\,(t_{n-1} - t_{n-2}) + \alpha\,(t_n - t_{n-1}) \\
&\leq \eta + \alpha\,\eta + \cdots + \alpha^n\,\eta \\
&= \frac{1 - \alpha^{n+1}}{1 - \alpha}\,\eta < \frac{\eta}{1 - \alpha} = t^{\star\star}.
\end{aligned} \tag{5.51}$$

We have:

$$L\,(t_{n+1}-t_n)+2\,\alpha\,L_0\,t_{n+1}\leq L\,\alpha^n\,\eta+2\,L_0\,\alpha\,\frac{1-\alpha^{n+1}}{1-\alpha}\,\eta. \tag{5.52}$$

In view of estimate (5.52), (5.47) holds, if

$$\left\{L\,\alpha^n+2\,\alpha\,L_0\,\frac{1-\alpha^{n+1}}{1-\alpha}\right\}\eta\leq\delta. \tag{5.53}$$

Estimate (5.53) motivates us to define for $s=\dfrac{\delta}{2}$, the sequence $\{f_n\}$ of polynomials on $[0,+\infty)$ by

$$f_n(s)=\left(L\,s^{n-1}+2\,L_0\,(1+s+s^2+\cdots+s^n)\right)\eta-2. \tag{5.54}$$

In view of Theorem 5.1, the induction for (5.47)–(5.50) is completed.

Hence, sequence $\{t_n\}$ is non–decreasing, bounded above by $t^{\star\star}$, and as such that it converges to its unique least upper bound $t^\star$. The induction is completed for (5.43), and (5.44). Estimates (5.45), (5.46) have been shown in [155], [189].

That completes the proof of Lemma 5.5. $\diamondsuit$

As a first application, we show how to locate a root of a polynomial f_n ($n\geq 2$), using, say e.g. s_{n-1}, and s_∞.

Application 5.1. Let $a=b=1$, $c=-3$, and $n=2$. We obtain using (5.23)–(5.25), and (5.32):

$$f_1(s)=s-1,\qquad f_2(s)=s^2+2\,s-2,$$

$$s_1=1,\qquad s_\infty=\frac{2}{3},\qquad \alpha=.618033989.$$

Conditions (5.26), and (5.27) become:

$$.618033989<\frac{2}{3},$$

and

$$-1<0.$$

Hence, the conclusions of Theorem 5.1 hold. In particular, we know $s_2\in(s_\infty,s_1)$. Actual direct computation justifies the theoretical claim, since

$$s_2=\sqrt{3}-1=.732050808\in\left(\frac{2}{3},1\right).$$

Application 5.2. As a second application, we show how to use Theorem 5.1 to derive sufficient convergence conditions for scalar majorizing sequence (5.40) of Newton's method (5.3).

Let us consider the famous Kantorovich hypotheses for solving nonlinear equations, using Newton's method (5.3) [477]:

$(\mathcal{K})$:

Chapter 5. Convergence Using Recurrent Functions

$x_0 \in \mathcal{D}$, with $F'(x_0)^{-1} \in L(\mathcal{Y}, \mathcal{X})$,

$$\| F'(x_0)^{-1} F(x_0) \| \leq \eta,$$

$$\| F'(x_0)^{-1} (F'(x) - F'(y)) \| \leq L \| x - y \| \quad \text{for all} \quad x, y \in \mathcal{D},$$

$$q_K = L \eta \leq \frac{1}{2}, \tag{5.55}$$

$$\overline{U}(x_0, r) = \{x \in \mathcal{X} : \| x - x_0 \| \leq r\} \subseteq \mathcal{D},$$

for

$$r = \frac{1 - \sqrt{1 - 2 q_K}}{\ell}.$$

Under the $(\mathcal{K})$ hypotheses, the Newton–Kantorovich method converges quadratically (if $2 q_K < 1$) to a unique solution $x^\star \in \overline{U}(x_0, r)$ of equation $F(x) = 0$.

Moreover, scalar iteration $\{v_n\}$ ($n \geq 0$), given by

$$v_0 = 0, \quad v_1 = \eta, \quad v_{n+2} = v_{n+1} + \frac{L (v_{n+1} - v_n)^2}{2 (1 - L v_{n+1})},$$

is a majorizing sequence for $\{x_n\}$ in the sense that for all $n \geq 0$:

$$\| x_{n+1} - x_n \| \leq v_{n+1} - v_n$$

and

$$\| x_n - x^\star \| \leq r - v_n.$$

Let us consider our hypotheses:

$(\mathcal{A})$:

$x_0 \in \mathcal{D}$, with $F'(x_0)^{-1} \in L(\mathcal{Y}, \mathcal{X})$,

$$\| F'(x_0)^{-1} F(x_0) \| \leq \eta,$$

$$\| F'(x_0)^{-1} (F'(x) - F'(y)) \| \leq L \| x - y \|,$$

$$\| F'(x_0)^{-1} (F'(x) - F'(x_0)) \| \leq L_0 \| x - x_0 \| \quad \text{for all} \quad x, y \in \mathcal{D},$$

$$q_0 \leq \frac{1}{2},$$

$$\overline{U}(x_0, t^\star) \subseteq \mathcal{D} \quad (\text{or } \overline{U}(x_0, t^{\star\star}) \subseteq \mathcal{D}).$$

In [136], we showed that if $(L + L_0) \eta \leq 1$, then, $\{t_n\}$ is a majorizing sequence for $\{x_n\}$. By simply replacing this condition by weaker $2 q_0 \leq 1$, and the $(\mathcal{A})$ conditions, a weaker than the Newton–Kantorovich theorem [477] is given (see also Theorem 5.2).

Note that the center–Lipschitz condition is not an additional hypothesis, since, in practice the computation of L requires that of L_0.

The scalar iteration $\{t_n\}$ is a finer majorizing sequence for $\{x_n\}$ than $\{v_n\}$ provided that $L_0 < L$ [136], [157], [155].

Remark 5.1. In general

$$L_0 \leq L$$

holds, and $\dfrac{L}{L_0}$ can be arbitrarily large [89], [157].

Condition (5.38) coincides with the Newton–Kantorovich hypothesis (5.55),

if $L = L_0$. Otherwise (5.38) is weaker than (5.55). Moreover the ratio $2\, q_0$ is also smaller than $2\, q_K$.

That is, (5.38) can always replace (5.55) in the Newton–Kantorovich theorem [477]. Hence, the applicability of Newton's method has been extended.

Other applications, where $L_0 < L$ can be found in [155], [189].

Remark 5.2. Define scalar sequence $\{\alpha_n\}$ by

$$\alpha_0 = 0, \quad \alpha_1 = \eta, \quad \alpha_{n+2} = \alpha_n + \frac{L_1\,(\alpha_{n+1} - \alpha_n)^2}{2\,(1 - L_0\,\alpha_{n+1})} \qquad (n \geq 0), \tag{5.56}$$

where,

$$L_1 = \begin{cases} L_0, & \text{if} \quad n = 0 \\ L, & \text{if} \quad n > 0 \end{cases}.$$

It follows from (5.40), and (5.56) that $\{\alpha_n\}$ converges under the hypotheses of Lemma 5.5. Note that, under the $(\mathcal{K})$ hypotheses, $\{\alpha_n\}$ is a finer majorizing sequence for $\{x_n\}$ than $\{t_n\}$. More precisely, we have the following estimates (if $L_0 < L$):

$$\| x_{n+1} - x_n \| \leq \alpha_{n+1} - \alpha_n < t_{n+1} - t_n < v_{n+1} - v_n, \quad (n \geq 1),$$

$$\| x_n - x^\star \| \leq \alpha^\star - \alpha_n < t^\star - t_n < v^\star - v_n, \quad (n \geq 0),$$

$$\alpha_n < t_n < v_n, \quad (n \geq 2),$$

and

$$\alpha^\star \leq t^\star \leq r,$$

where,

$$\alpha^\star = \lim_{n \longrightarrow \infty} \alpha_n.$$

In Theorem 5.1, we wanted to show a result relating the roots of polynomials f_n. There is a simpler proof, which however is not involving these roots.

Lemma 5.6. *Under the hypotheses of Lemma 5.2, the conclusions of Theorem 5.1 hold.*

Proof. We shall show

$$f_n(\alpha) \leq 0 \quad \text{for all} \quad n \geq 1, \tag{5.57}$$

and

$$\frac{t_2 - t_1}{t_1 - t_0} \leq \alpha. \tag{5.58}$$

Using (5.24), (5.25), and (5.34), we get

$$f_n(\alpha) = f_{n-1}(\alpha) = \cdots = f_1(\alpha). \tag{5.59}$$

Chapter 5. Convergence Using Recurrent Functions

It follows from (5.57), and (5.59), that we only need to show

$$f_1(\alpha) \leq 0, \tag{5.60}$$

which is true by (5.30) (for $n = 1$). Estimate (5.58) also holds by (5.30).

Finaly, we have:

$$f_\infty(\alpha) = \lim_{n\to\infty} f_n(\alpha) \leq 0. \tag{5.61}$$

That completes the proof of Lemma 5.6. $\diamond$

Remark 5.3. In view of (5.21) (for $\gamma_0 = 0$, $\gamma_1 = \eta$), Lemma 5.4, Theorem 5.1, and Lemma 5.5, we have functions on $[0, 1)$ given by:

$$f_n(s) = L\, s^{n-1}\, \eta^p + L_0\,(1+p)\,(1+s+\cdots+s^n)^p\,\eta^p - (1+p), \tag{5.62}$$

$$f_{n+1}(s) = f_n(s) + g_n(s)\,\eta^p, \tag{5.63}$$

$$
\begin{aligned}
g_n(s) = {}& L\,s^n - L\,s^{n-1} + \\
& L_0\,(1+p)\,((1+s+\cdots+s^{n+1})^p - (1+s+\cdots+s^n)^p),
\end{aligned} \tag{5.64}
$$

$$g_{n+1}(s) = g_n(s) + g_n^1(s), \tag{5.65}$$

where,

$$
\begin{aligned}
g_n^1(s) = {}& (1-s)^2 L s^{n-1} + L_0\,(1+p)\,((1+s+\cdots+s^{n+2})^p + \\
& (1+s+\cdots+s^n)^p - 2\,(1+s+\cdots+s^{n+1})^p) \geq 0.
\end{aligned} \tag{5.66}
$$

Function

$$g_1(s) = L\,s - L + L_0\,(1+p)\,((1+s+s^2)^p - (1+s)^p) \tag{5.67}$$

is continuous, with

$$g_1(0) = -L, \qquad g_1(s) = L_0\,(1+p)\,(3^p - 2^p) > 0.$$

It follows by the intermediate value theorem, that g_1 has a zero $\alpha \in (0,1)$.

Using (5.62), we obtain:

$$f_\infty(s) = \lim_{n\to\infty} f_n(s) = (1+p)\left(L_0\left(\frac{\eta}{1-s}\right)^p - 1\right). \tag{5.68}$$

According to (5.63)–(5.67), we have:

$$
\begin{aligned}
f_{n+1}(\alpha) &= f_n(\alpha) + g_n(\alpha)\,\eta^p \\
&= f_n(\alpha) + g_{n-1}(\alpha) + g_{n-1}^1(\alpha) \\
&\geq f_n(\alpha) + g_{n-1}(\alpha) \geq f_n(\alpha) + g_1(\alpha) = f_n(\alpha),
\end{aligned}
$$

so, we conclude

$$f_\infty(\alpha) \geq f_n(\alpha), \qquad (n \geq 1). \tag{5.69}$$

Then, the conditions of Lemma 5.4 become

$$\frac{t_2 - t_1}{t_1 - t_0} \leq \alpha, \tag{5.70}$$

84   Ioannis K. Argyros, Saïd Hilout and Mohammad A. Tabatabai

and

$$f_\infty(\alpha) \leq 0. \tag{5.71}$$

Define constants η_1, η_2, and η_0 by:

$$\eta_1 = \left(\frac{\alpha\,(1+p)}{L + \alpha\,(1+p)\,L_0} \right)^{1/p}, \tag{5.72}$$

$$\eta_2 = (1-\alpha)\,L_0^{-1/p}, \tag{5.73}$$

and

$$\eta_0 = \min\{\eta_1, \eta_2\}. \tag{5.74}$$

If $\eta_1 \leq \eta_1$, and $\eta \leq \eta_2$, then, (5.70), and (5.71) hold, respectively.

Hence, we arrived at:

Proposition 5.1. *Let $L_0 \geq 0$, $L \geq 0$, with $L_0 \leq L$, and $p \in (0,1]$.*
If

$$\eta \leq \eta_0, \tag{5.75}$$

then, the conclusions of Lemma 5.4 hold for iteration (5.21).

We can show the following semilocal convergence result for Newton's method (5.3).

Theorem 5.2. *Let $F : \mathcal{D} \subseteq X \longrightarrow \mathcal{Y}$ be a Fréchet–differentiable operator.*
Assume:
there exist a point $x_0 \in \mathcal{D}$, and constants $\eta > 0$, $L_0 \geq 0$, $L \geq 0$, with $L_0 \leq L$, and $p \in (0,1]$, such that:

$$F'(x_0)^{-1} \in L(\mathcal{Y}, X), \tag{5.76}$$

$$\| F'(x_0)^{-1}\,F(x_0) \| \leq \eta, \tag{5.77}$$

$$\| F'(x_0)^{-1}\,(F'(x) - F'(x_0)) \| \leq L_0\,\| x - x_0 \|^p, \tag{5.78}$$

$$\| F'(x_0)^{-1}\,(F'(x) - F'(y)) \| \leq L\,\| x - y \|^p, \tag{5.79}$$

for all x, $y \in \mathcal{D}$,

$$\overline{U}(x_0, t^\star) \subseteq \mathcal{D}, \tag{5.80}$$

and hypotheses of Proposition 5.1 hold.
    Then, $\{x_n\}$ $(n \geq 0)$ generated by Newton's method (5.3) is well defined, remains in $\overline{U}(x_0, t^\star)$ for all $n \geq 0$, and converges to a unique solution $x^\star \in \overline{U}(x_0, t^\star)$ of equation $F(x) = 0$.

    Moreover, the following estimates bounds hold for all $n \geq 1$:

$$\| x_{n+1} - x_n \| \leq \frac{L\,\| x_n - x_{n-1} \|^{1+p}}{(1+p)\,(1 - L_0\,\| x_n - x_0 \|^p)} \leq t_{n+1} - t_n, \tag{5.81}$$

and

$$\| x_n - x^\star \| \leq t^\star - t_n, \tag{5.82}$$

where, iteration $\{t_n\}$ $(n \geq 0)$ is given by (5.21), and $t^\star$ is defined in Lemma 5.1.

Chapter 5. Convergence Using Recurrent Functions 85

Furthermore, if there exists $R > t^*$, such that

$$\overline{U}(x_0, R) \subseteq \mathcal{D}, \qquad (5.83)$$

and

$$L_0 \int_0^1 (\theta t^* + (1 - \theta) R)^p \, d\theta < 1, \qquad (5.84)$$

then, the solution x^* is unique in $U(x_0, R)$.

Proof. Simply replace the sufficient convergence conditions for the convergence of majorizing sequence $\{t_n\}$ given in [139, Theorem 5], by the corresponding ones in Proposition 5.1. That completes the proof of Theorem 5.2. ◇

Note that t^{**} given in closed form by (5.8) can replace t^* in the hypotheses of Theorem 5.2.

Remark 5.4. Recently, a new semilocal convergence theorem was given in [308], which improves earlier sufficient convergence conditions, but not necessarily the error bounds.

Let us define function $h : [1, +\infty) \longrightarrow \mathbb{R}$ by:

$$h(t) = \left(1 - \frac{1}{t}\right)^p \frac{1 + p}{\left((L_0 (1 + p))^{\frac{1}{1-p}} + (L t (t - 1))^{\frac{1}{1-p}}\right)^{1-p}} \qquad (5.85)$$

and point

$$c_0 = \frac{L + \sqrt{L^2 + 4 L_0 L (1 + p)^p p^{1-p}}}{2 L}. \qquad (5.86)$$

Then, the sufficient convergence condition corresponding to (5.75) is given by:

$$\eta \leq h_0(c_0)^{1/p}. \qquad (5.87)$$

Note that (5.87) was given in a non–affine invariant form [308]. However, if we apply the results to $F'(x_0)^{-1} F$, then we obtain (5.87).

A direct comparison between (5.75), and (5.87) does not seem possible. In practice, we shall test both conditions to see if any of them is satisfied.

However, we can give a favorable to us comparison between the error bounds (5.11), and

$$r_0 = 0, \quad r_1 = \eta, \quad r_{n+1} - r_n \leq r_n \left(\frac{c_0 - 1}{c_0^n - 1}\right), \qquad (n \geq 1) \qquad (5.88)$$

where, the sequence $\{r_n\}$ $(n \geq 0)$ is the same with $\{t_n\}$.

Example 5.1. Let $L_0 = 5$, $L = 7$, and $p = .5$.
Using (5.67), and (5.72)–(5.75), we get

$$\alpha = .7757109976, \quad \eta_1 = .008240494,$$

$$\eta_2 = .00897156, \quad \eta_0 = \eta_1, \quad \text{and} \quad \eta < \eta_0.$$

Hence, the conclusions of Proposition 5.1 for iteration (5.21) hold for $\eta \leq \eta_0$.

Moreover, using (5.85)–(5.87), we get:

$$c_0 = 1.431981531, \quad h(c_0) = .082528964,$$

and

$$\eta \leq h(c_0)^2 = .00681103.$$

That is in this case, our estimate (5.75) allows a wider range for η than (5.87).

We shall provide a comparison table of the error bounds for $\eta = .0068$ (see Comparison Table 1).

Comparison table 1.

n	Actual (5.21)	(5.11)	(5.88)
1	0.3671780586e-3	0.5274834784e-2	0.4748664598e-2
2	0.1090955945e-5	0.4091747352e-2	0.2058024332e-2
3	0.9631467375e-11	0.3174013421e-2	0.1116738479e-2
4	0.7506940047e-21	0.2462117117e-2	0.6747384428e-3
5	0.4560410391e-41	0.1909891325e-2	0.4306552097e-3

The table shows that our error bounds are better in this case.

Note however that our result provides an at least as good information on the location of the solution, since the uniqueness condition in [308] is:

$$L t^{\star p} < 1 \tag{5.89}$$

(see also (5.84), and note that $L_0 \leq L$, and $R \geq t^{\star}$).

Remark 5.5. As in Remark 5.3, following the proofs of Lemma 5.7, Theorem 5.1, and Lemma 5.5, we also have functions on $[0,1)$ given by:

$$f_n(s) = L s^{np-1} \eta^p + L_0 (1+p)(1+s+\cdots+s^n)^p \eta^p - (1+p), \tag{5.90}$$

$$f_{n+1}(s) = f_n(s) + g_n(s)\,\eta^p, \tag{5.91}$$

$$g_n(s) = L s^{(n+1)p-1} - L s^{np-1} + \\ L_0 (1+p)((1+s+\cdots+s^{n+1})^p - (1+s+\cdots+s^n)^p), \tag{5.92}$$

$$g_{n+1}(s) = g_n(s) + g_n^1(s), \tag{5.93}$$

where,

$$g_n^1(s) = (1-s^p)^2 L s^{np-1} + L_0(1+p)((1+s+\cdots+s^{n+2})^p + \\ (1+s+\cdots+s^n)^p - 2(1+s+\cdots+s^{n+1})^p) \geq 0. \tag{5.94}$$

In particular, function g_1 is given by

$$g_1(s) = L s^{2p-1} - L s^{p-1} + L_0 (1+p)((1+s+s^2)^p - (1+s)^p). \tag{5.95}$$

Then, with these changes, the conclusions of Proposition 5.1, and Theorem 5.2 hold by simply replacing the zero α of function g_1 given by (5.67) by the zero (also called α) of function g_1 given by (5.95).

Chapter 5. Convergence Using Recurrent Functions 87

Returning back to Example 5.1, we have

$$\alpha = .7050696619, \qquad \eta_1 = .007407686,$$

$$\eta_2 = .011797214, \qquad \eta_0 = \eta_1$$

and

$$\eta < \eta_0.$$

Hence, the conclusions of Proposition 5.1 for iteration (5.81) hold. If $\eta \leq \eta_0$, then again, we give a wider range of values η than (5.87). We shall provide a comparison table of the error bounds for $\eta = .0068$ (see Comparison Table 2).

Comparison table 2.

n	Actual (5.21)	(5.11)	(5.88)
1	0.3671780586e-3	0.4794473701e-2	0.4748664598e-2
2	0.1090955945e-5	0.3380437951e-2	0.2058024332e-2
3	0.9631467375e-11	0.2383444244e-2	0.1116738479e-2
4	0.7506940047e-21	0.1680494227e-2	0.6747384428e-3
5	0.4560410391e-41	0.1184865496e-2	0.4306552097e-3

By comparing Table 1 to Table 2, we see that the bounds in Table 2 compare even better to ones in [308].

We shall complete this chapter by providing another set of bounds for majorizing sequence $\{t_n\}$ given by (5.21).

Proposition 5.2. *Let $L_0 \geq 0$, $L \geq 0$, $\eta \geq 0$, $\lambda > 0$, and $p \in (0,1]$ be given constants. Assume:*

$$L_0\,\eta^p < 1, \tag{5.96}$$

$$\frac{L}{(1+p)\,(1-L_0\,\eta^p)} \leq \left(\frac{\lambda}{1+p}\right)^p, \tag{5.97}$$

$$\frac{\lambda\,\eta}{1+p} < 1, \tag{5.98}$$

and

$$(1+p)\,L_0\left(\frac{\lambda}{1+p}\right)^p\left\{\eta+\frac{1+p}{\lambda}\,\frac{1}{1-\left(\dfrac{\lambda\,\eta}{1+p}\right)^{1+p}}\right\}^p + \tag{5.99}$$

$$L-(1+p)\left(\frac{\lambda}{1+p}\right)^p \leq 0.$$

Then, iteration $\{t_n\}$ $(n \geq 0)$ given by (5.21) for $\gamma_0 = 0$, and $\gamma_1 = \eta$, is increasing, bounded above by

$$t^{**} = \eta + \frac{1+p}{\lambda}\,\frac{1}{1-\left(\dfrac{\lambda\,\eta}{1+p}\right)^{1+p}}, \tag{5.100}$$

and converges to its unique least upper bound $t^\star$ satisfying

$$t^\star \in [0, t^{\star\star}].\qquad(5.101)$$

Moreover the following estimates hold for all $n \geq 0$:

$$t_{n+1} - t_n \leq \frac{1+p}{\lambda} e_0^{(1+p)^n},\qquad(5.102)$$

and

$$t^\star - t_n \leq \frac{1+p}{\lambda\,(1 - e_0^{1+p})} e_0^{(1+p)^n},\qquad(5.103)$$

where,

$$e_n = \frac{\lambda}{1+p}\,(t_{n+1} - t_n).\qquad(5.104)$$

Proof. We shall show using induction on n:

$$0 \leq t_{n+2} - t_{n+1} = \frac{L\,(t_{n+1} - t_n)^{p+1}}{(1+p)\,(1 - L_0\,t_{n+1}^p)} \leq e_n^p\,(t_{n+1} - t_n)\qquad(5.105)$$

or

$$0 \leq \frac{L}{(1+p)\,(1 - L_0\,t_{n+1}^p)} \leq \left(\frac{\lambda}{1+p}\right)^p.\qquad(5.106)$$

Estimate (5.106) holds for $n = 0$ by (5.96), and (5.97). Moreover, estimate (5.102) holds for $n = 1$. Let us assume estimate (5.106) holds for all $k \leq n$. Then, estimate (5.102) also holds.

Using (5.102), and (5.104), we have in turn:

$$e_{n+1} \leq e_n^{1+p} \leq (e_{n-1}^{1+p})^{1+p} = e_{n-1}^{(1+p)^2} \leq \cdots \leq e_0^{(1+p)^{n+1}},\qquad(5.107)$$

so

$$e_n \leq e_0^{(1+p)^n},\qquad(5.108)$$

$$t_{n+1} - t_n \leq \frac{1+p}{\lambda} e_0^{(1+p)^n},\qquad(5.109)$$

and

$$\begin{aligned}
t_{n+1} &\leq t_n + \frac{1+p}{\lambda} e_0^{(1+p)^n} \\
&\leq t_{n-1} + \frac{1+p}{\lambda}\left(e_0^{(1+p)^{n-1}} + e_0^{(1+p)^n}\right) \\
&\leq \eta + \frac{1+p}{\lambda}\left\{\left(\frac{\lambda\,\eta}{1+p}\right)^{(1+p)^0} + \cdots + \left(\frac{\lambda\,\eta}{1+p}\right)^{(1+p)^n}\right\} < t^{\star\star}.
\end{aligned}\qquad(5.110)$$

Estimate (5.106) is true, if

$$(1+p)\left(\frac{\lambda}{1+p}\right)^p L_0 \left\{\eta + \frac{1+p}{\lambda}\left(\left(\frac{\lambda\,\eta}{1+p}\right)^{(1+p)^0} + \cdots + \left(\frac{\lambda\,\eta}{1+p}\right)^{(1+p)^n}\right)\right\}^p + L - (1+p)\left(\frac{\lambda}{1+p}\right)^p \leq 0\qquad(5.111)$$

Chapter 5. Convergence Using Recurrent Functions   89

or

$$(1+p)\left(\frac{\lambda}{1+p}\right)^p L_0 \left\{\eta + \frac{1+p}{\lambda}\left(1+\cdots+\left(\frac{\lambda\eta}{1+p}\right)^{n(1+p)}\right)\right\}^p +$$
$$L - (1+p)\left(\frac{\lambda}{1+p}\right)^p \le 0. \tag{5.112}$$

Estimate (5.112) motivates us to define on $[0,1)$ for $s = \left(\frac{\lambda\eta}{1+p}\right)^{1+p}$ by

$$\begin{aligned} f_n(s) &= (1+p)\left(\frac{\lambda}{1+p}\right)^p L_0 \left(\eta + \frac{1+p}{\lambda}(1+s+\cdots+s^n)\right)^p + \\ &\quad L - (1+p)\left(\frac{\lambda}{1+p}\right)^p \le 0. \end{aligned} \tag{5.113}$$

We need a relationship between two consecutive functions f_n:

$$\begin{aligned} f_{n+1}(s) &= (1+p)\left(\frac{\lambda}{1+p}\right)^p L_0 \left(\eta + \frac{1+p}{\lambda}(1+s+\cdots+s^{n+1})\right)^p + \\ &\quad L - (1+p)\left(\frac{\lambda}{1+p}\right)^p \\ &= f_n(s) + (1+p)\left(\frac{\lambda}{1+p}\right)^p L_0 \left\{\left(\eta + \frac{1+p}{\lambda}(1+s+\cdots+ \right.\right. \\ &\quad \left.\left. s^{n+1})\right)^p - \left(\eta + \frac{1+p}{\lambda}(1+s+\cdots+s^n)\right)^p\right\} \ge f_n(s). \end{aligned} \tag{5.114}$$

In view of (5.114), estimate (5.111) certainly holds if

$$f_\infty\left(\left(\frac{\lambda\eta}{1+p}\right)^{1+p}\right) \le 0, \tag{5.115}$$

where,

$$\begin{aligned} f_\infty(s) &= \lim_{n\to\infty} f_n(s) \\ &= (1+p)\left(\frac{\lambda}{1+p}\right)^p L_0 \left(\eta + \frac{1+p}{\lambda}\frac{1}{1-s}\right)^p + \\ &\quad L - (1+p)\left(\frac{\lambda}{1+p}\right)^p. \end{aligned} \tag{5.116}$$

But, estimate (5.115) holds by (5.99), and (5.116). The induction is completed.

It then follows that iteration $\{t_n\}$ is non–decreasing, bounded above by t^{**}, and as such it converges to t^*.

Let $m \ge 0$. Then, using (5.102) we have in turn:

$$\begin{aligned} &t_{n+m} - t_n \\ &\le (t_{n+m} - t_{n+m-1}) + (t_{n+m-1} - t_{n+m-2}) + \cdots + (t_{n+1} - t_n) \\ &\le \frac{1+p}{\lambda}\left(e_0^{(1+p)^{n+m-1}} + e_0^{(1+p)^{n+m-2}} + \cdots + e_0^{(1+p)^n}\right) \\ &= \frac{1+p}{\lambda} e_0^{(1+p)^n}\left(1 + e_0^{(1+p)^1} + \cdots + e_0^{(1+p)^{m-1}}\right) \\ &= \frac{1+p}{\lambda} e_0^{(1+p)^n} \frac{1 - e_0^{(1+p)^m}}{1 - e_0^{(1+p)}}. \end{aligned} \tag{5.117}$$

By letting $m \longrightarrow \infty$ in (5.117), we obtain (5.103).

That completes the proof of Proposition 5.2. $\qquad\qquad\qquad\qquad\Diamond$

Now we are concerned with Newton–like method (5.2) for approximating a locally unique solution $x^\star$ of equation (5.1). Note that if $A(x) = F'(x)$, (5.2) reduces to Zincenko's method [155], [770]. Furthemore, if $A(x) = F'(x) + [x, x_-; G]$, (5.2) reduces to Cătinaş's method [290], where $[x, x_-; G]$ is divided difference of order one, and x_- denotes the previous iterate of x.

(a) **Case $G = 0$.** If $A(x) = F'(x)$ for $x \in \mathcal{D}$, and under Kantorovich–type assumptions, Rheinboldt [615] established a convergence theorem for Newton–type method (5.2), which includes the Kantorovich theorem for the Newton method as a special case [155], [477]. A further generalization was given by Dennis in [335], Deuflhard and Heindl in [347], [348], Potra in [591], [593], and Păvăloiu in [575], [578]. Miel [524], [525] improved the error bounds for Rheinboldt [615]. Moret [539] obtained a convergence theorem as well as error bounds for (5.2) under stronger conditions than those of Rheinboldt.

(b) **Case $G \neq 0$.** If $A(x) = F'(x)$, $(x \in \mathcal{D})$, Zabrejko and Nguen [764] established a convergence theorem for the Krasnoselskii–Zincenko–type iteration [770]. Later, Yamamoto and Chen [303] extended the results in [764], [770], when $A(x)$ is not necessarily equal to $F'(x)$. More recently, Argyros [157] provided a unified convergence theory for even more general (5.2).

We need the following result on majorizing sequence for (5.2).

Lemma 5.7. *Assume:*

there exist constants $L > 0$, $M > 0$, $\mu \geq 0$, $L_0 > 0$, and $\eta > 0$, such that:

$$2\,M < L; \qquad\qquad (5.118)$$

Quadratic polynomial f_1 given by

$$f_1(s) = 2\,L_0\,\eta\,s^2 - \left(2\,(1 - L_0\,\eta) - L\,\eta\right)s + 2\,(M\,\eta + \mu), \qquad (5.119)$$

has a root in $(0,1)$, denoted by $\dfrac{\delta}{2}$,

and

for

$$\delta_0 = \frac{L\,\eta + 2\,\mu}{1 - L_0\,\eta}, \qquad\qquad (5.120)$$

$$\alpha = \frac{2\,(L - 2\,M)}{L + \sqrt{L^2 + 8\,L_0\,(L - 2\,M)}}, \qquad\qquad (5.121)$$

the following holds

$$\delta_0 \leq \delta \leq 2\,\alpha. \qquad\qquad (5.122)$$

Chapter 5. Convergence Using Recurrent Functions   91

Then, scalar sequence $\{t_n\}$ $(n \geq 0)$ given by

$$t_0 = 0, \quad t_1 = \eta,$$
$$t_{n+2} = t_{n+1} + \frac{L\,(t_{n+1} - t_n) + 2\,(M\,t_n + \mu)}{2\,(1 - L_0\,t_{n+1})}\,(t_{n+1} - t_n) \tag{5.123}$$

is increasing, bounded above by

$$t^{\star\star} = \frac{2\,\eta}{2 - \delta}, \tag{5.124}$$

and converges to its unique least upper bound $t^{\star} \in [0, t^{\star\star}]$.
Moreover the following estimates hold for all $n \geq 1$:

$$t_{n+1} - t_n \leq \frac{\delta}{2}\,(t_n - t_{n-1}) \leq \left(\frac{\delta}{2}\right)^n \eta, \tag{5.125}$$

and

$$t^{\star} - t_n \leq \frac{2\,\eta}{2 - \delta}\,\left(\frac{\delta}{2}\right)^n. \tag{5.126}$$

Proof. We shall show using induction on the integer m:

$$\begin{aligned}
0 \;<\; t_{m+2} - t_{m+1} &= \frac{L\,(t_{m+1} - t_m) + 2\,(M\,t_m + \mu)}{2\,(1 - L_0\,t_{m+1})}\,(t_{m+1} - t_m) \\
&\leq \frac{\delta}{2}\,(t_{m+1} - t_m),
\end{aligned} \tag{5.127}$$

and

$$L_0\,t_{m+1} < 1. \tag{5.128}$$

If (5.127), and (5.128) hold, we have (5.125) holds, and

$$\begin{aligned}
t_{m+2} &\leq t_{m+1} + \frac{\delta}{2}\,(t_{m+1} - t_m) \\
&\leq t_m + \frac{\delta}{2}\,(t_m - t_{m-1}) + \frac{\delta}{2}\,(t_{m+1} - t_m) \\
&\leq \eta + \left(\frac{\delta}{2}\right)\eta + \cdots + \left(\frac{\delta}{2}\right)^{m+1}\eta \\
&= \frac{1 - \left(\frac{\delta}{2}\right)^{m+2}}{1 - \frac{\delta}{2}}\,\eta < \frac{2\,\eta}{2 - \delta} = t^{\star\star} \qquad \text{by (5.124)}.
\end{aligned}$$

Estimates (5.127) and (5.128) hold by the initial conditions for $m = 0$. Indeed (5.127), and (5.128) become:

$$\begin{aligned}
0 < t_2 - t_1 &= \frac{L\,(t_1 - t_0) + 2\,(M\,t_0 + \mu)}{2\,(1 - L_0\,t_1)}\,(t_1 - t_0) \\
&= \frac{L\,\eta + 2\,\mu}{2\,(1 - L_0\,\eta)}\,(t_1 - t_0) = \frac{\delta_0}{2}\,(t_1 - t_0) \leq \frac{\delta}{2}\,(t_1 - t_0),
\end{aligned}$$

$$L_0\,\eta < 1,$$

which are true by the choices of δ_0, and δ, (5.122), (5.123), and the initial conditions. Let us assume (5.125), (5.127), and (5.128) hold for all $m \leq n+1$.

Estimate (5.127) can be re–written as

$$L\left(t_{m+1} - t_m\right) + 2\left(M\, t_m + \mu\right) \leq \left(1 - L_0\, t_{m+1}\right)\delta$$

or

$$L\left(t_{m+1} - t_m\right) + 2\left(M\, t_m + \mu\right) + \delta\, L_0\, t_{m+1} \leq 0, \qquad (5.129)$$

or

$$L\left(\frac{\delta}{2}\right)^m \eta + 2\left(M\, \frac{1 - \left(\frac{\delta}{2}\right)^m}{1 - \frac{\delta}{2}}\, \eta + \mu\right) + \delta\, L_0\, \frac{1 - \left(\frac{\delta}{2}\right)^{m+1}}{1 - \frac{\delta}{2}}\, \eta - \delta \leq 0. \qquad (5.130)$$

Replace $\frac{\delta}{2}$ by s, and define functions f_m on $[0, +\infty)$ $(m \geq 1)$:

$$\begin{aligned}
f_m(s) &= L\, s^m\, \eta + 2\left(M\left(1 + s + s^2 + \cdots + s^{m-1}\right)\eta + \mu\right) + \\
&\quad 2\, s\, L_0\left(1 + s + \cdots + s^m\right)\eta - 2\, s.
\end{aligned} \qquad (5.131)$$

We need to find a relationship between two consecutive f_m:

$$\begin{aligned}
&f_{m+1}(s) \\
&= L\, s^{m+1}\, \eta + 2\left(M\left(1 + s + s^2 + \cdots + s^{m-1} + s^m\right)\eta + \mu\right) + \\
&\quad 2\, s\, L_0\left(1 + s + \cdots + s^m + s^{m+1}\right)\eta - 2\, s \\
&= L\, s^{m+1}\, \eta - L\, s^m\, \eta + L\, s^m\, \eta + \\
&\quad 2\left(M\left(1 + s + s^2 + \cdots + s^{m-1}\right)\eta + \mu\right) + \\
&\quad 2\, M\, s^m\, \eta + 2\, s\, L_0\left(1 + s + \cdots + s^m\right)\eta + 2\, s\, L_0\, s^{m+1}\, \eta - 2\, s \\
&= f_m(s) + L\, s^{m+1}\, \eta - L\, s^m\, \eta + 2\, M\, s^m\, \eta + 2\, s\, L_0\, s^{m+1}\, \eta \\
&= f_m(s) + g(s)\, s^m\, \eta,
\end{aligned} \qquad (5.132)$$

where,

$$g(s) = 2\, L_0\, s^2 + L\, s + 2\, M - L. \qquad (5.133)$$

Note that in view of (5.118), function g has a unique positive zero α given by (5.121), and

$$g(s) < 0 \qquad s \in (0, \alpha). \qquad (5.134)$$

Estimate (5.130) certainly holds, if

$$f_m\left(\frac{\delta}{2}\right) \leq 0 \qquad (m \geq 1). \qquad (5.135)$$

Clearly, (5.135) holds for $m = 1$ as equality. We then get by (5.131), (5.132)–(5.134):

$$f_2\left(\frac{\delta}{2}\right) = f_1\left(\frac{\delta}{2}\right) + g\left(\frac{\delta}{2}\right)\frac{\delta}{2}\eta = g\left(\frac{\delta}{2}\right)\frac{\delta}{2}\eta \leq 0,$$

Chapter 5. Convergence Using Recurrent Functions 93

Assume (5.135) holds for all $k \leq m$. We shall show (5.135) for m replaced by $m+1$. Indeed, we have

$$f_{m+1}\left(\frac{\delta}{2}\right) = f_m\left(\frac{\delta}{2}\right) + g\left(\frac{\delta}{2}\right)\left(\frac{\delta}{2}\right)^m \eta \leq 0,$$

which shows (5.135) for all m.

Moreover, we obtain

$$f_\infty\left(\frac{\delta}{2}\right) = \lim_{m\to\infty} f_m\left(\frac{\delta}{2}\right) \leq 0.$$

That completes the induction.

Moreover, estimate (5.126) follows from (5.125) by using standard majorization techniques [155], [477].

Finally, sequence $\{t_n\}$ is increasing, bounded above by $t^{\star\star}$, and as such it converges to its unique least upper bound $t^\star$.

That completes the proof of Lemma 5.7. $\diamondsuit$

Remark 5.6. The hypotheses of Lemma 5.7 have been left as uncluttered as possible. Note that these hypotheses involve only computations only at the initial point x_0. Below, we shall provide some simpler but stronger hypotheses under which the hypotheses of Lemma 5.7 hold.

We can also show the following alternative to Lemma 5.7.

Lemma 5.8. *Let $L > 0$, $M > 0$, $\mu > 0$, with $0 < L_0 \leq L$, and $\eta > 0$, be such that:*

$$\mu < \alpha, \quad 2M < L,$$

and

$$0 < h_A = a\eta \leq \frac{1}{2}, \tag{5.136}$$

where,

$$a = \frac{1}{4(\alpha - \mu)} \max\{2L_0\alpha^2 + 2L_0\alpha + L\alpha + 2M, L + 2\alpha L_0\}. \tag{5.137}$$

Then, the following hold:

$$f_1 \text{ has a positive root } \frac{\delta}{2},$$

$$\max\{\delta_0, \delta\} \leq 2\alpha,$$

and

the conclusions of Lemma 5.7 hold, with α replacing $\frac{\delta}{2}$.

Proof. It follows from (5.132), (5.136) that

$$f_m(\alpha) = f_1(\alpha) \leq 0, \quad (m \geq 1), \tag{5.138}$$

which together with $f_1(0) = 2(M\eta + \mu) > 0$, imply that there exists a positive root $\dfrac{\delta}{2}$ of polynomial f_1, satisfying

$$\delta \leq 2\alpha. \tag{5.139}$$

It also follows from (5.120), and (5.136) that

$$\delta_0 \leq 2\alpha, \tag{5.140}$$

and (5.135) holds, with α replacing $\dfrac{\delta}{2}$ (by (5.138)). Note also $\alpha \in (0,1)$ by (5.118).

That completes the proof of Lemma 5.8 $\qquad\qquad\qquad \diamondsuit$

We shall show the following semilocal convergence theorem for (5.2).

Theorem 5.3. *Let* $F : \mathcal{D} \subseteq X \longrightarrow \mathcal{Y}$ *be a Fréchet–differentiable operator,* $G : \mathcal{D} \longrightarrow X$ *be a continuous operator, and let* $A(x) \in L(X,\mathcal{Y})$ *given in (5.2) be an approximation of* $F'(x)$. *Assume that there exist an open convex subset* $\mathcal{D}$ *of* X, $x_0 \in \mathcal{D}$, *a bounded inverse* A_0^{-1} *of* A_0, *and constants* $L > 0$, $M > 0$, $L_0 \geq 0$, $\mu_0 \geq 0$, $\mu_1 \geq 0$, $\eta > 0$, *such that for all* $x,y \in \mathcal{D}$:

$$\| A_0^{-1}(F(x_0) + G(x_0)) \| \leq \eta, \tag{5.141}$$

$$\| A_0^{-1}(F'(x) - F'(y)) \| \leq L \, \| x - y \|, \tag{5.142}$$

$$\| A_0^{-1}(F'(x) - A(x)) \| \leq M \, \| x - x_0 \| + \mu_0, \tag{5.143}$$

$$\| A_0^{-1}(A(x) - A_0) \| \leq L_0 \, \| x - x_0 \|, \tag{5.144}$$

$$\| A_0^{-1}(G(x) - G(y)) \| \leq \mu_1 \, \| x - y \|, \tag{5.145}$$

$$\overline{U}(x_0, t^\star) = \{ x \in X, \, \| x - x_0 \| \leq t^\star \} \subseteq \mathcal{D}, \tag{5.146}$$

and

the hypotheses of Lemmas 5.7, or 5.8 hold with

$$\mu = \mu_0 + \mu_1.$$

Then, sequence $\{x_n\}$ $(n \geq 0)$ *generated by (5.2) is well defined, remains in* $\overline{U}(x_0, t^\star)$ *for all* $n \geq 0$, *and converges to a solution* $x^\star$ *of equation* $F(x) + G(x) = 0$ *in* $\overline{U}(x_0, t^\star)$.

Moreover, the following estimates hold for all $n \geq 0$:

$$\| x_{n+1} - x_n \| \leq t_{n+1} - t_n, \tag{5.147}$$

and

$$\| x_n - x^\star \| \leq t^\star - t_n, \tag{5.148}$$

where, sequence $\{t_n\}$ $(n \geq 0)$, *and* $t^\star$ *are given in Lemma 5.7.*

Furthermore, the solution $x^\star$ *of equation (5.1) is unique in* $\overline{U}(x_0, t^\star)$ *provided that:*

$$\left(\frac{L}{2} + M + L_0 \right) t^\star + \mu < 1.$$

Chapter 5. Convergence Using Recurrent Functions   95

Proof. We shall show using induction on $n \geq 0$:

$$\| x_{n+1} - x_n \| \leq t_{n+1} - t_n, \tag{5.149}$$

and

$$\overline{U}(x_{n+1}, t^\star - t_{n+1}) \subseteq \overline{U}(x_n, t^\star - t_n). \tag{5.150}$$

For every $z \in \overline{U}(x_1, t^\star - t_1)$,

$$\begin{aligned} \| z - x_0 \| &\leq \| z - x_1 \| + \| x_1 - x_0 \| \\ &\leq t^\star - t_1 + t_1 = t^\star - t_0, \end{aligned}$$

implies $z \in \overline{U}(x_0, t^\star - t_0)$. We also have

$$\| x_1 - x_0 \| = \| A_0^{-1}(F(x_0) + G(x_0)) \| \leq \eta = t_1 - t_0.$$

That is (5.149), and (5.150) hold for $n = 0$. Given they hold for $m \leq n$, then:

$$\begin{aligned} \| x_{n+1} - x_0 \| &\leq \sum_{i=1}^{n+1} \| x_i - x_{i-1} \| \\ &\leq \sum_{i=1}^{n+1} (t_i - t_{i-1}) = t_{n+1} - t_0 = t_{n+1}, \end{aligned}$$

and

$$\| x_n + \theta(x_{n+1} - x_n) - x_0 \| \leq t_n + \theta(t_{n+1} - t_n) \leq t^\star,$$

for all $\theta \in (0,1)$.

Using (5.144), and the induction hypotheses, we get:

$$\begin{aligned} \| A_0^{-1}(A_{n+1} - A_0) \| &\leq L_0 \| x_{n+1} - x_0 \| \\ &\leq L_0(t_{n+1} - t_0) = L_0 t_{n+1} < 1. \end{aligned} \tag{5.151}$$

It follows from (5.151), and the Banach lemma on invertible operators that A_{n+1}^{-1} exists, and

$$\| A_{n+1}^{-1} A_0 \| \leq (1 - L t_{n+1})^{-1}. \tag{5.152}$$

We have for $x \in \overline{U}(x_0, t^\star)$:

$$\| A_0^{-1}(F'(x) - A(x)) \| \leq M \| x - x_0 \| + \mu_0. \tag{5.153}$$

We obtain also the approximation:

$$\begin{aligned} &x_{n+2} - x_{n+1} \\ =\ & -A_{n+1}^{-1}(F(x_{n+1} + G(x_{n+1}))) \\ =\ & -A_{n+1}^{-1} A_0 A_0^{-1} \\ &\left(\int_0^1 (F'(x_{n+1} + \theta(x_n - x_{n+1})) - F'(x_n))(x_{n+1} - x_n)\, d\theta + \right. \\ &\left. (F'(x_n) - A_n)(x_{n+1} - x_n) + G(x_{n+1}) - G(x_n) \right) \end{aligned} \tag{5.154}$$

Using (5.142), (5.143), (5.145), (5.152)–(5.154), and the induction hypotheses, we obtain in turn:

$$\| x_{n+2} - x_{n+1} \|$$

$$\leq (1 - L_0 t_{n+1})^{-1} \left(\frac{L}{2} \| x_{n+1} - x_n \|^2 + \right.$$

$$\left. (M \| x_n - x_0 \| + \mu_0) \| x_{n+1} - x_n \| + \mu_1 \| x_{n+1} - x_n \| \right) \quad (5.155)$$

$$\leq (1 - L_0 t_{n+1})^{-1} \left(\frac{L}{2} (t_{n+1} - t_n) + M t_n + \mu \right) (t_{n+1} - t_n)$$

$$= t_{n+2} - t_{n+1},$$

which shows (5.149) for all $n \geq 0$.

Thus, for every $z \in \overline{U}(x_{n+2}, t^\star - t_{n+2})$, we have:

$$\begin{aligned} \| z - x_{n+1} \| &\leq \| z - x_{n+2} \| + \| x_{n+2} - x_{n+1} \| \\ &\leq t^\star - t_{n+2} + t_{n+2} - t_{n+1} = t^\star - t_{n+1}, \end{aligned}$$

which shows (5.150) for all $n \geq 0$.

Lemmas 5.7 or 5.8 imply that sequence $\{t_n\}$ is Cauchy. Moreover, it follows from (5.149) and (5.150), that $\{x_n\}$ ($n \geq 0$) is also a Cauchy sequence in a Banach space X, and as such it converges to some $x^\star \in \overline{U}(x_0, t^\star)$ (since $\overline{U}(x_0, t^\star)$ is a closed set).

By letting $n \longrightarrow \infty$ in (5.155), we obtain $F(x^\star) + G(x^\star) = 0$. Furthermore estimate (5.148) is obtained from (5.147) by using standard majorization techniques. Finally to show that $x^\star$ is the unique solution of equation (5.1) in $\overline{U}(x_0, t^\star)$, as in (5.154) and (5.155), we get in turn for $y^\star \in \overline{U}(x_0, t^\star)$, with $F(y^\star) + G(y^\star) = 0$, the estimation:

$$\| y^\star - x_{n+1} \|$$

$$\leq \| A_n^{-1} A_0 \|$$

$$\left\{ \left(\left(\int_0^1 \| A_0^{-1} (F'(x_n + \theta (y^\star - x_n)) - F'(x_n)) \| \, d\theta \right. \right. \right.$$

$$\left. + \| A_0^{-1} (F'(x_n) - A_n) \| \right) \| y^\star - x_n \| +$$

$$\left. \| A_0^{-1} (G(x_n) - G(y^\star)) \| \right\}$$

$$\leq (1 - L_0 t_{n+1})^{-1} \left(\frac{L}{2} \| y^\star - x_n \|^2 + \right. \quad (5.156)$$

$$\left. (M \| x_n - x_0 \| + \mu) \| y^\star - x_n \| \right)$$

$$\leq (1 - L_0 t_{n+1})^{-1} \left(\frac{L}{2} (t^\star - t_n) + M t_n + \mu \right) \| y^\star - x_n \|$$

$$\leq (1 - L_0 t^\star)^{-1} \left(\frac{L}{2} (t^\star - t_0) + M t^\star + \mu \right) \| x^\star - x_n \|$$

$$< \| y^\star - x_n \|,$$

by the uniqueness hypothesis.

It follows by (5.156) that $\lim_{n \longrightarrow \infty} x_n = y^\star$. But we showed $\lim_{n \longrightarrow \infty} x_n = x^\star$. Hence, we deduce $x^\star = y^\star$.

Chapter 5. Convergence Using Recurrent Functions

That completes the proof of Theorem 5.3.

Note that t^* can be replaced by t^{**} given in closed form by (5.124) in the uniqueness hypothesis provided that $\overline{U}(x_0, t^{**}) \subseteq \mathcal{D}$, or in all hypotheses of the theorem.

Chapter 6

ω–Convergence

In this chapter, we provide a semilocal convergence analysis for Newton–type methods using our idea of recurrent functions in a Banach space setting. As in [303], [764], [770], we use Zabrejko–Zincenko conditions. In particular, we show that the convergence domains given in [155] can be extended under the same computational cost. Numerical examples are also provided to show that we can solve equations in cases not covered before [303], [770].

We are concerned with the problem of approximating a locally unique solution $x^\star$ of equation

$$F(x) + G(x) = 0, \tag{6.1}$$

where F is a Fréchet–differentiable operator defined on a open convex subset $\mathcal{D}$ of a Banach space X with values in a Banach space $\mathcal{Y}$, and $G : \mathcal{D} \longrightarrow \mathcal{Y}$ is a continuous operator.

We shall use the Newton–type method (NTM):

$$y_{n+1} = y_n - A(y_n)^{-1} P(y_n) \quad (n \geq 0), \quad (y_0 \in \mathcal{D}),$$
$$P(x) = F(x) + G(x), \ (x \in \mathcal{D}), \tag{6.2}$$

where, $A(x) \in L(X, \mathcal{Y})$ to generate a sequence approximating $x^\star$.

(a) If

$$A(x) = F'(x) \qquad (x \in \mathcal{D}), \tag{6.3}$$

we obtain the Zabrejko–Nguen iteration [764]:

$$y_{n+1} = y_n - F'(y_n)^{-1} P(y_n) \quad (n \geq 0), \quad (y_0 \in \mathcal{D}). \tag{6.4}$$

(b) If

$$A(x) = F'(x) + [x, y; G] \qquad (x \in \mathcal{D}), \tag{6.5}$$

where $[x, y; F]$ is a divided difference of order one for operator G, then we obtain an iteration faster than (6.4), first considered by Cătinaș [290]:

$$y_{n+1} = y_n - (F'(y_n) + [y_n, y_{n-1}; G])^{-1} P(y_n) \quad (n \geq 0), \quad (y_{-1}, y_0 \in \mathcal{D}). \tag{6.6}$$

(c) If

$$A(x) = F'(x), \quad G(x) = 0 \qquad (x \in \mathcal{D}), \qquad (6.7)$$

(NTM) reduces to Newton's method:

$$x_{n+1} = x_n - F'(x_n)^{-1} F(x_n) \quad (n \geq 0), \quad (x_0 \in \mathcal{D}). \qquad (6.8)$$

Several other choices are possible [155], [190]. A local as well as a semilocal convergence analysis for all these methods has been provided by many authors under Lipschitz–type conditions [89], [137], [303], [770]. A survey of such results can be found in [155], and the references there. We also refer the reader to the elegant related works by Proinov [601], [602], and Ezquerro/Hernández [375] whose works are also improved here in at least the Newton's method case.

Let $x^0 \in \mathcal{D}$, and $R > 0$ be such that:

$$U(x^0, R) = \subseteq \mathcal{D}. \qquad (6.9)$$

Chen and Yamamoto [303] provided a semilocal convergence for (NTM) for $y^0 \in U(x^0, R)$ under the conditions:

$(\mathcal{C})$:

$A(x^0)$ exists, and for any $x, y \in \overline{U}(x^0, R)$:

$$\| A(x^0)^{-1} (A(x) - A(x^0)) \| \leq \overline{v_0}(\| x - x^0 \|) + a,$$

$$\| A(x^0)^{-1} (F'(x + t (x - y)) - A(x)) \|$$
$$\leq \overline{v}(\| x - x^0 \| + t \| y - x \|) - \overline{v_0}(\| x - x^0 \|) + b, \quad t \in [0, 1],$$

and

$$\| A(x^0)^{-1} (G(x) - G(y)) \| \leq \omega(r) \| x - y \|,$$

where $\overline{v}(r + t) - \overline{v_0}(r), t \geq 0$, and $\omega(r)$ are non–decreasing, non–negative functions with

$$\omega(0) = \overline{v_0}(0) = \overline{v}(0) = 0,$$

$\overline{v_0}(r)$ is differentiable, $\overline{v_0}'(r) > 0$ at every point of $[0, R]$, and the constants a, b satisfy $a \geq 0$, $b \geq 0$, and $a + b < 1$.

Set

$$\| A(x^0)^{-1} P(x^0) \| \leq \eta,$$

$$\phi(r) = \eta - r + \int_0^1 \overline{v}(t) \, dt,$$

$$\psi(r) = \int_0^r \omega(t) \, dt,$$

$$\chi(r) = \phi(r) + \psi(r) + (a + b) \, r.$$

Chapter 6. ω–Convergence

Further, assume
$$\chi(R) \leq 0,$$

and define scalar sequence $\{s_n\}$ by

$$s_0 \in [0, R], \qquad s_{n+1} = s_n + \frac{u(s_n)}{p(s_n)}, \qquad (n \geq 0)$$

where,
$$u(r) = \chi(r) - \chi^\star, \qquad p(r) = 1 - \overline{v_0}(r) - a,$$

$\chi^\star$ is the minimal value of $\chi(r)$ in $[0, R]$, and $s^\star$ denotes the minimal point. Moreover, $t^\star$ denotes the unique zero of χ in $(0, s^\star]$.

Under these assumptions, there exists a unique solution $x^\star$ in $\in \overline{U}(x^0, t^\star)$, such that:

$$\| y_{n+1} - y_n \| \leq s_{n+1} - s_n,$$

and
$$\| x^\star - y_n \| \leq s^\star - s_n.$$

We shall use the more general set of conditions:

$(\mathcal{H})$:

$(\mathcal{H}_1)$ $A(x^0)$ exists, and for any $x, y \in \overline{U}(x^0, R)$ $(0 < r \leq R)$:

$$\| A(x^0)^{-1} (A(x) - A(x^0)) \| \leq v_0(\| x - x^0 \|) + a,$$

$(\mathcal{H}_2)$

$$\| A(x^0)^{-1} (F'(x + t (x - y)) - A(x)) \|$$
$$\leq v(t \| y - x \|) + \omega_0(\| x - x^0 \|) + b, \; t \in [0, 1],$$

$(\mathcal{H}_3)$

$$\| A(x^0)^{-1} (G(x) - G(y)) \| \leq \omega(r) \; \| x - y \|,$$

and

$(\mathcal{H}_4)$

$$\overline{U}(x^0, R) \subseteq \mathcal{D},$$

where, v_0, v, and ω_0 are non–decreasing, non–negative functions on $[0, R]$ with

$$v_0(0) = v(0) = \omega_0(0) = \omega(0) = 0,$$

and constants a and b are non–negative. Some more hypotheses are given in Lemma 6.1.

A semilocal convergence analysis is provided in this chapter under the $(\mathcal{H})$ conditions. We also compare the two sets of hypotheses. Numerical examples are also provided in this chapter to show how we can solve equations in cases not covered before [89], [770].

We need to define some parameters, functions and sequences.

Definition 6.1. Let $y_0 \in U(x^0, r)$. Define parameters r_0, r_1, iteration $\{r_n\}$, functions f_n, h_n, p_n on $[0, 1)$, and $q \in I_q = [0, 1)^2 \times [0, r_1 - r_0] \times [r_1 - r_0, \frac{r_1 - r_0}{1 - s}]^3$ ($s \in [0, 1)$) by

$$r_0 \geq \| y_0 - x^0 \|, \quad r_1 > r_0 + \| A(y_0)^{-1} (F(y_0) + G(y_0)) \|,$$

$$r_{n+1} = r_n + \frac{\int_0^1 v(t(r_n - r_{n-1})) \, dt + \omega_0(r_n) + \omega(r_n) + b}{1 - a - v_0(r_n)} (r_n - r_{n-1}), \tag{6.10}$$

$$f_n(s) = \int_0^1 v(t \, s^{n-1} (r_1 - r_0)) \, dt + \omega_1((1 + s + \cdots + s^{n-1})(r_1 - r_0)) + c, \tag{6.11}$$

$$\begin{aligned} h_n(s) = \ & \int_0^1 (v(t \, s^n (r_1 - r_0)) - v(t \, s^{n-1}(r_1 - r_0)) \, dt + \\ & \omega_1((1 + s + \cdots + s^n)(r_1 - r_0)) - \\ & \omega_1((1 + s + \cdots + s^{n-1})(r_1 - r_0)), \end{aligned} \tag{6.12}$$

where,

$$c = b - \alpha(1 - a) \qquad \text{for some } \alpha \in (0, 1), \tag{6.13}$$

$$\omega_1(s) = \omega_0(s) + \omega(s) + \alpha \, v_0(s), \tag{6.14}$$

$$\begin{aligned} p_n(s) & \\ = \ & \int_0^1 (v(t \, s^{n+1}(r_1 - r_0)) + v(t \, s^{n-1}(r_1 - r_0)) - 2v(t \, s^n (r_1 - r_0)) \, dt + \\ & \omega_1((1 + s + \cdots + s^{n+1})(r_1 - r_0)) + \\ & \omega_1((1 + s + \cdots + s^{n-1})(r_1 - r_0)) - \\ & 2\omega_1((1 + s + \cdots + s^n)(r_1 - r_0)), \end{aligned} \tag{6.15}$$

$$\begin{aligned} q(t, s, \lambda, \beta_0, \beta, \gamma_0) = \ & \int_0^1 (v(t \lambda s^2) + v(t \lambda) - 2v(t \lambda s)) \, dt + \\ & \omega_1(\beta_0 + \beta + \gamma_0) + \omega_1(\beta) - 2\omega_1(\beta_0 + \beta). \end{aligned} \tag{6.16}$$

Define function f_∞ on $[0, 1)$ by

$$f_\infty(s) = \lim_{n \to \infty} f_n(s). \tag{6.17}$$

It then follows from (6.11), and (6.17) that

$$f_\infty(s) = \omega_1\left(\frac{r_1 - r_0}{1 - s}\right) + c. \tag{6.18}$$

It can also easily be seen from (6.11), (6.12), (6.15), and (6.16) that the following hold:

$$f_{n+1}(s) = f_n(s) + h_n(s), \tag{6.19}$$

$$h_{n+1}(s) = h_n(s) + p_n(s), \tag{6.20}$$

and

$$\begin{aligned} & q(t, s, s^{n-1}(r_1 - r_0), s^n(r_1 - r_0), (1 + s + \cdots + \\ & s^{n-1})(r_1 - r_0), s^{n+1}(r_1 - r_0)) = p_n(s). \end{aligned} \tag{6.21}$$

Chapter 6. ω–Convergence   103

We need the following result on majorizing sequences for (NTM).

Lemma 6.1. *Let constants a, b, parameters r_0, r_1, functions v_0, v, ω_0, ω be as in the Introduction, and parameters α, c, functions ω_1, f_n, h_n, p_n, q be as in Definition 6.1.*
Assume there exists $\alpha \in (0,1)$ such that:

$$v_0(r_1) + a < 1, \tag{6.22}$$

$$\frac{\int_0^1 v(t(r_1 - r_0))\, dt + \omega_0(r_1) + \omega(r_1) + b}{1 - a - v_0(r_1)} \le \alpha, \tag{6.23}$$

$$c < 0, \tag{6.24}$$

$$q(t, s, \lambda, \beta_0, \beta, \gamma_0) \ge 0 \quad \text{on } I_q, \tag{6.25}$$

$$h_1(\alpha) \ge 0, \tag{6.26}$$

and

$$f_\infty(\alpha) \le 0. \tag{6.27}$$

Then, scalar sequence $\{r_n\}$ $(n \ge 0)$ given by (6.10) is non–decreasing, bounded from above by

$$r^{\star\star} = \frac{r_1 - r_0}{1 - \alpha}, \tag{6.28}$$

and converges to its unique least upper bound $r^\star$ satisfying $r^\star \in [0, r^{\star\star}]$.
Moreover the following estimates hold for all $n \ge 0$:

$$0 \le r_{n+1} - r_n \le \alpha\,(r_n - r_{n-1}) \le \alpha^n\,(r_1 - r_0), \tag{6.29}$$

and

$$r^\star - r_n \le \frac{r_1 - r_0}{1 - \alpha}\,\alpha^n. \tag{6.30}$$

Proof. Estimate (6.29) is true, if

$$0 \le \frac{\int_0^1 v(t(r_n - r_{n-1}))\, dt + \omega_0(r_n) + \omega(r_n)}{1 - a - v_0(r_n)} \le \alpha \tag{6.31}$$

holds for all $n \ge 1$.

In view of (6.10), (6.22), (6.23), estimate (6.31) holds true for $n = 1$. We also have by (6.31) that $0 \le r_2 - r_1 \le \alpha(r_1 - r_0)$.

Let us assume that (6.29) and (6.31) hold for all $k \le n$. Then, we have:

$$r_n \le \frac{1 - \alpha^n}{1 - \alpha}\,(r_1 - r_0). \tag{6.32}$$

Using the induction hypotheses, and (6.32), estimate (6.31) is true, if

$$\int_0^1 v(t\,\alpha^{n-1}\,(r_1 - r_0))\, dt + \omega_1\left(\frac{1 - \alpha^n}{1 - \alpha}\,(r_1 - r_0)\right) + c \le 0, \tag{6.33}$$

where, c, and ω_1 are given by (6.13), and (6.14), respectively. Estimate (6.33) (for $s = \alpha$) motivates us to introduce function f_n given by (6.11), and show instead of (6.33):

$$f_n(\alpha) \leq 0 \qquad (n \geq 1). \tag{6.34}$$

We have by (6.19)–(6.21) (for $s = \alpha$), and (6.26) that

$$f_{n+1}(\alpha) \geq f_n(\alpha) \qquad (n \geq 1). \tag{6.35}$$

In view of (6.17), (6.18), and (6.35), estimate (6.34) shall holds, if (6.27) is true, since

$$f_n(\alpha) \leq f_\infty(\alpha) \qquad (n \geq 1). \tag{6.36}$$

The induction is completed. It follows that iteration $\{r_n\}$ is non–decreasing, bounded above by r^{**} (given by (6.28)), and as such it converges to r^*. Finally, estimate (6.21) follows from (6.29) by using standard majorization techniques.

That completes the proof of Lemma 6.1. $\qquad\qquad\qquad\qquad\qquad\qquad\qquad\diamondsuit$

The hypotheses $(\mathcal{H})$, and those of Lemma 6.1 will be called $(\mathcal{A})$.

We can show the main semilocal convergence result for (NTM).

Theorem 6.1. *Assume hypotheses $(\mathcal{A})$ hold. Then, sequence $\{y_n\}$ $(n \geq 0)$ generated by (NTM) is well defined, remains in $\overline{U}(x^0, r^*)$ for all $n \geq 0$, and converges to a solution $x^* \in \overline{U}(x^0, r^*)$ of equation $F(x) + G(x) = 0$.*

Moreover, the following estimates hold for all $n \geq 0$:

$$\| y_{n+1} - y_n \| \leq r_{n+1} - r_n, \tag{6.37}$$

and

$$\| y_n - x^* \| \leq r^* - r_n, \tag{6.38}$$

where, sequence $\{r_n\}$ $(n \geq 0)$, and r^* are given in Lemma 6.1.

Furthermore, if there exists

$$R_0 \in [r^*, R] \tag{6.39}$$

such that

$$\int_0^1 v(t(R_0 + r_0)) \, dt + \omega_0(r^*) + \omega(R_0) + v_0(r^*) + a + b \leq 1, \tag{6.40}$$

then, the solution x^* of equation (6.1) is unique in $U(x^0, R_0)$.

Proof. We shall show using induction:

$$\| y_n - y_{n-1} \| \leq r_n - r_{n-1}, \tag{6.41}$$

and

$$\| y_n - x^0 \| \leq r_n. \tag{6.42}$$

Estimates (6.41), and (6.42) hold for $n = 1$, by (6.4) and (6.10).

Assume (6.41), and (6.42) hold for $n \leq k$.

Using $(\mathcal{H}_1)$, and (6.22), we get

$$
\begin{aligned}
\| A(x_0)^{-1} [A(y_1) - A(x^0)] \| &\leq v_0(\| y_1 - x^0 \|) + a \\
&\leq v_0(r_1) + a < 1.
\end{aligned}
\tag{6.43}
$$

It follows from (6.43), and the Banach lemma on invertible operators that $A(y_1)^{-1}$ exists, and

$$
\| A(y_1)^{-1} A(x^0) \| \leq (1 - a - v_0(r_1))^{-1}.
\tag{6.44}
$$

We also showed in Lemma 6.1 that

$$
v_0(r_k) + a < 1.
\tag{6.45}
$$

It then follows as in (6.43) with r_k, y_k replacing r_1, y_1, respectively, that $A(y_k)^{-1}$ exists, and

$$
\| A(y_k)^{-1} A(x^0) \| \leq (1 - a - v_0(r_k))^{-1}.
\tag{6.46}
$$

Using (6.2), $(\mathcal{H}_2)$, $(\mathcal{H}_3)$, (6.10), (6.41), (6.42), and (6.46), we obtain in turn:

$$
\begin{aligned}
&\| y_{k+1} - y_k \| \\
={}& \| A(y_k)^{-1} (F(y_k) + G(y_k)) \| \\
\leq{}& \| A(y_k)^{-1} A(x^0) \| \, \| A(x^0)^{-1} \Big(F(y_k) + \\
& G(y_k) - A(y_{k-1})(y_k - y_{k-1}) - F(y_{k-1}) - G(y_{k-1}) \Big) \| \\
\leq{}& (1 - a - v_0(r_k))^{-1} \Big(\int_0^1 \| A(x^0)^{-1} [F'(y_{k-1} + \theta(y_k - y_{k-1})) - \\
& A(y_{k-1})] \| \, \| y_k - y_{k-1} \| \, d\theta + \| A(x^0)^{-1} (G(y_k) - G(y_{k-1})) \| \Big) \\
\leq{}& (1 - a - v_0(r_k))^{-1} \Big(\int_0^1 v(t \, \| y_k - y_{k-1} \|) \, dt + \omega_0(\| y_k - x^0 \|) + \\
& \omega(\| y_k - x^0 \|) + b \Big) \| y_k - y_{k-1} \| \\
\leq{}& (1 - a - v_0(r_k))^{-1} \Big(\int_0^1 v(t(r_k - r_{k-1})) \, dt + \omega_0(r_k) + \\
& \omega(r_k) + b \Big) (r_k - r_{k-1}) = r_{k+1} - r_k.
\end{aligned}
\tag{6.47}
$$

Moreover, we have

$$
\begin{aligned}
\| y_{k+1} - x^0 \| &\leq \| y_{k+1} - y_k \| + \| y_k - x^0 \| \\
&\leq (r_{k+1} - r_k) + r_k = r_{k+1} \leq r^\star.
\end{aligned}
\tag{6.48}
$$

The induction for (6.41), and (6.42) is completed.

In view of Lemma 6.1, (6.41), and (6.42), sequence $\{y_n\}$ ($n \geq 0$) is Cauchy in a Banach space X, and as such it converges to some $x^\star \in \overline{U}(x^0, r^\star)$ (since $\overline{U}(x^0, r^\star)$ is a closed set). Estimate (6.38) follows from (6.37) by using standard majorization techniques.

Using (6.47), we obtain

$$\begin{aligned}
&\| A(x^0)^{-1}(F(x_k)+G(x_k)) \| \\
&\leq \left(\int_0^1 \mathsf{v}(t(r_k-r_{k-1}))\,dt + \omega_0(r_k) + \omega(r_k) + b \right)(r_k-r_{k-1}).
\end{aligned} \tag{6.49}$$

By letting $k \longrightarrow \infty$ in (6.49), we obtain $F(x^\star)+G(x^\star)=0$. Finally to show that $x^\star$ is the unique solution of equation (6.1) in $\overline{U}(x^0,R_0)$, let $y^\star \in \overline{U}(x^0,R_0)$, with $F(y^\star)+G(y^\star)=0$.
Using the approximation

$$\begin{aligned}
y^\star - y_{k+1} = \ & y^\star - y_k + A(y_k)^{-1}(F(y_k)+G(y_k)) - \\
& A(y_k)^{-1}(F(y^\star)+G(y^\star)),
\end{aligned} \tag{6.50}$$

as in (6.47), we obtain in turn:

$$\begin{aligned}
& \| y^\star - y_{k+1} \| \\
\leq \ & (1-a-\mathsf{v}_0(r_k))^{-1} \left(\int_0^1 \| A(x^0)^{-1}[F'(y_k+\theta(y^\star-y_k)) - \right. \\
& \left. A(y_k)] \| \, \| y^\star - y_k \| \, d\theta + \| A(x^0)^{-1}(G(y^\star)-G(y_k)) \| \right) \\
\leq \ & (1-a-\mathsf{v}_0(r_k))^{-1} \left(\int_0^1 \mathsf{v}(t\|y^\star-y_k\|)\,dt + \omega_0(\|y_k-x^0\|) + \right. \\
& \left. \omega(\|y_k-x^0\|) + b \right) \| y^\star - y_k \| \\
< \ & (1-a-\mathsf{v}_0(r^\star))^{-1} \left(\int_0^1 \mathsf{v}(t(R_0+r_0))\,dt + \omega_0(r^\star) + \right. \\
& \left. \omega(R_0) + b \right) \| y_k - y^\star \| \leq \| y_k - y^\star \| \quad \text{(by (6.40))}.
\end{aligned} \tag{6.51}$$

It follows by (6.51) that $\lim\limits_{k \longrightarrow \infty} y_k = y^\star$. But we showed $\lim\limits_{k \longrightarrow \infty} y_k = x^\star$. Hence, we deduce $x^\star = y^\star$.

That completes the proof of Theorem 6.1. $\qquad\qquad\qquad\qquad\qquad \diamond$

Application 6.1. (Newton's method). Let $A(x)=F'(x)$, and $G(x)=0$ $(x \in \mathcal{D})$.
Then, in the case of the (C) conditions, we have:

$$y_0 = x^0 = x_0, \quad s_0=0, \quad s_1=\eta, \quad a=b=0,$$

$$\overline{\mathsf{v}}_0(r)=L_0\,r, \quad \overline{\mathsf{v}}(r)=L\,r, \quad \omega(r)=0,$$

$$\chi(r)=\phi(r)=\frac{L}{2}r^2-r+\eta, \quad \chi^\star=\frac{2L\eta-1}{2L},$$

$$s^\star=\frac{1}{L}, \quad t^\star=\frac{1-\sqrt{1-2L\eta}}{L}, \quad R \in [t^\star,s^\star]$$

$$u(r)=\frac{1}{2L}(Lr-1)^2, \quad p(r)=1-L_0\,r,$$

$$h_{CY}=L\eta \leq \frac{1}{2}, \tag{6.52}$$

and

$$s_0 = 0, \quad s_1 = \eta,$$
$$s_{n+1} = s_n + \frac{(Ls_n - 1)^2}{2L(1 - L_0 s_n)} = s_n + \frac{L(s_n - s_{n-1})^2}{2(1 - L_0 s_n)}, \quad (n \geq 1), \tag{6.53}$$

Moreover, in the case of the $(\mathcal{A})$ conditions, we have:

$$y_0 = x^0 = x_0, \quad r_0 = 0, \quad r_1 = \eta, \quad a = b = 0,$$

$$v_0(r) = L_0 r, \quad v(r) = L r, \quad \omega(r) = \omega_0(r) = 0,$$

$$\alpha = \frac{2L}{L + \sqrt{L^2 + 8L_0 L}}$$

$$h_{AH} = \overline{L}\eta \leq \frac{1}{2}, \tag{6.54}$$

where,

$$\overline{L} = \frac{1}{8}\left(L + 4L_0 + \sqrt{L^2 + 8L_0 L}\right),$$

and

$$s_0 = 0, \quad s_1 = \eta,$$
$$s_{n+1} = s_n + \frac{L(s_n - s_{n-1})^2}{2(1 - L_0 s_n)}, \quad (n \geq 1). \tag{6.55}$$

Note that

$$L_0 \leq L \tag{6.56}$$

holds in general, and $\dfrac{L}{L_0}$ can be arbitrarily large [155], [190].

Let us now compare the results. It follows from (6.52), (6.54), and (6.56)

$$h_{CY} \leq \frac{1}{2} \implies h_{AH} \leq \frac{1}{2}, \tag{6.57}$$

but not necessarily vice versa unless if $L_0 = L$.

Hence the convergence domains approach in [303] do not necessarily produce the weakest sufficient convergence conditions not even in the simplest possible case of a Newton–like method which is Newton's method (6.8). The recurrent functions approach produces sufficient convergence (6.54) that can always replace (6.52). Hence, the applicability of Newton's method has been extended, under the same hypotheses and computational cost as in [303], [477]. Note that the results in [375], [601], [602] are also improved in at least the Newton's method case, since their conditions also lead to (6.52) instead of (6.54).

In the Newton's method case, although the majorizing sequences $\{s_n\}$, and $\{r_n\}$ coincide the convergence domains approach fails to take advantage of the relationship between L_0 and L, since L_0 does not appear in (6.52). The same is happening in the general case, since function χ does not depends on $\overline{v_0}$. However, our approach depends on v_0 for the derivation of the sufficient convergence conditions. Under our method the ratio "α" of convergence for $\{s_n\}$ is known, but this is not true for iteration $\{r_n\}$. We also refer the reader to Chapter 1 for some applications and examples.

Chapter 7

Gauss–Newton Method

The famous for its simplicity and clarity Newton–Kantorovich hypothesis of Newton's method has been used for a long time as the sufficient convergence condition for solving nonlinear equations. Recently, in the elegant study by N. Hu, W. Shen and C. Li [455], a Kantorovich–type convergence analysis for the Newton–Gauss method (NGM) was given improving earlier results by Häußler [423], and extending some results by Argyros [145], [151] to hold for systems of equations with constant rank derivatives. In this chapter, we use our new idea of recurrent functions to extend the applicability of (GNM) by replacing existing conditions by weaker ones. Finally, we provide numerical example to solve equations in cases not covered before [423], [455], [477].

We are concerned with the problem of approximating a locally unique solution $x^\star$ of equation

$$F(x) = 0, \tag{7.1}$$

where, F is a Fréchet–differentiable operator defined on a open, nonempty, convex subset $\mathcal{D}$ of $\mathbb{R}^i$ in $\mathbb{R}^j$, where $i, j \in \mathbb{N}^\star$.

We shall use the Gauss–Newton method (GNM)

$$x_{n+1} = x_n - F'(x_n)^+ F(x_n) \quad (n \geq 0), \quad (x_0 \in \mathcal{D}), \tag{7.2}$$

to generate a sequence $\{x_n\}$ approximating a solution $x^\star$ of equation

$$F'(x)^+ F(x) = 0, \tag{7.3}$$

where, $F'(x)^+$ is Moore–Penrose–pseudoinverse of matrix $F'(x)$, $(x \in \mathcal{D})$ (see Definition 7.1).

There is an extensive literature on local convergence results for (GNM) depending on the information around the least square solution of F (see, e.g. Shub and Smale in [656], Dedieu and Shub in [330], Dedieu and Kim in [329], Li et al. in [508]). A survey on local as well as semilocal convergence results for Newton–type methods can be found in [151], [192] (see also [136], [740]).

Häußler considered in [423] a special class of singular nonlinear systems using the hypothesis

$$\| F'(y)^+ (I - F'(x) F'(x)^+ F(x)) \| \leq \kappa \| x - y \|, \quad \text{for all } x, y \in \mathcal{D}$$

where, $\kappa \in [0,1)$, and I is the identity matrix. He provided a Kantorovich–type semilocal convergence analysis using Lipschitz continuity of F'. Hu, Shen and Li in the elegant study [455] used an idea by Argyros [136], [157], where the center Lipschitz condition is used to provide upper bounds on the inverses of the linear operators involved. This way they provided on the one hand a finer convergence analysis than in [423], and on the other hand they extended the result of Argyros [145].

In this chapter, we use our new idea of recurrent functions to provide a converegence analysis further extending the applicability of (GNM) by providing weaker sufficient convergence conditions than before [455], [423], [329] under the same hypotheses, and computational cost. Numerical example is also provided to show that our results apply to solve equation (7.3) but not earlier ones.

We need the following results on majorizing sequences for (GNM).

Lemma 7.1. *Assume there exist constants $L > 0$, $L_0 > 0$, $\eta > 0$, and $\kappa \in [0,1)$ such that:*

(i) *Quadratic polynomial f_1 given by*

$$f_1(s) = 2\,L_0\,\eta\,s^2 - \left(2\,(1 + L_0\,\eta\,\kappa) - (L + 2\,L_0)\,\eta\right)s + 2\,\kappa\,(1 - L_0\,\eta), \qquad (7.4)$$

has a root in $(0,1)$, denoted by $\dfrac{\delta}{2}$;

(ii) *For*

$$\delta_0 = \frac{L\,\eta + 2\,\kappa\,(1 - L_0\,\eta)}{1 - L_0\,\eta}, \qquad (7.5)$$

$$\alpha = \frac{2\,L}{L - 2\,L_0\,\kappa + \sqrt{(L - 2\,L_0\,\kappa)^2 + 8\,L_0\,L}}, \qquad (7.6)$$

the following holds

$$0 < \max\{2\,\kappa,\,\delta_0\} \le \delta \le 2\,\alpha. \qquad (7.7)$$

Then, scalar sequence $\{t_n\}$ $(n \ge 0)$ given by

$$t_0 = 0, \quad t_1 = \eta,$$
$$t_{n+2} = t_{n+1} + \frac{L\,(t_{n+1} - t_n) + 2\,\kappa\,(1 - L_0\,t_{n+1})}{2\,(1 - L_0\,t_{n+1})}\,(t_{n+1} - t_n) \qquad (7.8)$$

is increasing, bounded from above by

$$t^{\star\star} = \frac{2\,\eta}{2 - \delta}, \qquad (7.9)$$

and converges to its unique least upper bound $t^\star \in [0, t^{\star\star}]$.

Moreover the following estimates hold for all $n \ge 1$:

$$t_{n+1} - t_n \le \frac{\delta}{2}\,(t_n - t_{n-1}) \le \left(\frac{\delta}{2}\right)^n \eta, \qquad (7.10)$$

and

$$t^\star - t_n \le \frac{2\,\eta}{2 - \delta}\left(\frac{\delta}{2}\right)^n. \qquad (7.11)$$

Chapter 7. Gauss–Newton Method

Proof. We shall show using induction on the integer m:

$$0 \;<\; t_{m+2} - t_{m+1} = \frac{L\,(t_{m+1} - t_m) + 2\,\kappa\,(1 - L_0\,t_{m+1})}{2\,(1 - L_0\,t_{m+1})}\,(t_{m+1} - t_m)$$
$$\leq \;\frac{\delta}{2}\,(t_{m+1} - t_m), \tag{7.12}$$

and

$$L_0\,t_{m+1} < 1. \tag{7.13}$$

If (7.12), and (7.13) hold, we have (7.10) holds, and

$$
\begin{aligned}
t_{m+2} \;&\leq\; t_{m+1} + \frac{\delta}{2}\,(t_{m+1} - t_m)\\
&\leq\; t_m + \frac{\delta}{2}\,(t_m - t_{m-1}) + \frac{\delta}{2}\,(t_{m+1} - t_m)\\
&\leq\; \eta + \left(\frac{\delta}{2}\right)\eta + \cdots + \left(\frac{\delta}{2}\right)^{m+1}\eta\\
&=\; \frac{1 - \left(\dfrac{\delta}{2}\right)^{m+2}}{1 - \dfrac{\delta}{2}}\,\eta \;<\; \frac{2\,\eta}{2 - \delta} = t^{\star\star} \qquad \text{by (7.9)}.
\end{aligned}
$$

Estimates (7.12) and (7.13) hold by the initial conditions for $m = 0$. Indeed (7.12), and (7.13) become:

$$
\begin{aligned}
0 < t_2 - t_1 \;&=\; \frac{L\,(t_1 - t_0) + 2\,\kappa\,(1 - L_0\,\eta)}{2\,(1 - L_0\,t_1)}\,(t_1 - t_0)\\
&=\; \frac{\delta_0}{2}\,(t_1 - t_0) \leq \frac{\delta}{2}\,(t_1 - t_0),
\end{aligned}
$$

$$L_0\,\eta < 1,$$

which are true by the choices of δ_0, and δ, (7.7), (7.8), and the initial conditions. Let us assume (7.10), (7.12), and (7.13) hold for all $m \leq n + 1$.

Estimate (7.12) can be re–written as

$$L\,(t_{m+1} - t_m) + 2\,\kappa\,(1 - L_0\,t_{m+1}) \leq (1 - L_0\,t_{m+1})\,\delta \tag{7.14}$$

or

$$L\left(\frac{\delta}{2}\right)^{m}\eta + (\delta - 2\,\kappa)\,L_0\,\frac{1 - \left(\dfrac{\delta}{2}\right)^{m+1}}{1 - \dfrac{\delta}{2}}\,\eta + 2\,\kappa - \delta \leq 0. \tag{7.15}$$

Replace $\dfrac{\delta}{2}$ by s, and define functions f_m on $[0, +\infty)$ $(m \geq 1)$:

$$f_m(s) \;=\; L\,s^m\,\eta + 2\,(s - \kappa)\,(1 + s + s^2 + \cdots + s^m)\,L_0\,\eta + 2\,(\kappa - s) \tag{7.16}$$

We need to find a relationship between two consecutive f_m:

$$
\begin{aligned}
f_{m+1}(s) &= L\,s^{m+1}\,\eta + 2\,(s-\kappa)\,(1+s+s^2+\cdots+s^{m+1})\,L_0\,\eta + \\
&\quad 2\,(\kappa-s) + 2\,(\kappa-s) - L\,s^m\,\eta - 2\,(s-\kappa)\,L_0\,\eta\,(1+s+ \\
&\quad s^2+\cdots+s^m) - 2\,(\kappa-s) \\
&= f_m(s) + g(s)\,s^m\,\eta,
\end{aligned}
\tag{7.17}
$$

where,

$$
g(s) = 2\,L_0\,s^2 + (L - 2\,L_0\,\kappa)\,s - L.
\tag{7.18}
$$

Note that function g has a unique positive zero α given by (7.6), and

$$
g(s) < 0 \qquad s \in (0,\alpha).
\tag{7.19}
$$

Estimate (7.15) certainly holds, if

$$
f_m\left(\frac{\delta}{2}\right) \le 0 \qquad (m \ge 1).
\tag{7.20}
$$

Clearly, (7.20) holds for $m = 1$ as equality. We then get by (7.16), (7.17)–(7.19):

$$
f_2\left(\frac{\delta}{2}\right) = f_1\left(\frac{\delta}{2}\right) + g\left(\frac{\delta}{2}\right)\frac{\delta}{2}\eta = g\left(\frac{\delta}{2}\right)\frac{\delta}{2}\eta \le 0,
$$

Assume (7.20) holds for all $k \le m$. We shall show (7.20) for m replaced by $m+1$. Indeed, we have

$$
f_{m+1}\left(\frac{\delta}{2}\right) = f_m\left(\frac{\delta}{2}\right) + g\left(\frac{\delta}{2}\right)\left(\frac{\delta}{2}\right)^m \eta \le 0,
$$

which shows (7.20) for all m.

Moreover, we obtain

$$
f_\infty\left(\frac{\delta}{2}\right) = \lim_{m\to\infty} f_m\left(\frac{\delta}{2}\right) \le 0.
$$

That completes the induction.

Moreover, estimate (7.11) follows from (7.10) by using standard majorization techniques.

Finally, sequence $\{t_n\}$ is increasing, bounded above by $t^{\star\star}$, and as such it converges to its unique least upper bound $t^\star$.

That completes the proof of Lemma 7.1. $\diamond$

Remark 7.1. The hypotheses of Lemma 7.1 have been left as uncluttered as possible. Note that these hypotheses involve only computations only at the initial point x_0. Below, we shall provide some simpler but stronger hypotheses under which the hypotheses of Lemma 7.1 hold.

We can also show the following alternative to Lemma 7.1.

Chapter 7. Gauss–Newton Method

Lemma 7.2. *Let $L > 0$, $L_0 > 0$, $\kappa \in [0, 1)$, with $0 < L_0 \leq L$, and $\eta > 0$, be such that:*

$$\kappa < \alpha,$$

and

$$0 < h_A = a\,\eta \leq \frac{1}{2}, \tag{7.21}$$

where,

$$a = \frac{1}{4\,(\alpha - \kappa)}\,\max\{2\,L_0\,\alpha\,(\alpha - \kappa) + 2\,L_0\,\alpha\,(1 - \kappa) + L\,\alpha, L + 2\,L_0\,(\alpha - \kappa)\}. \tag{7.22}$$

Then, the following hold:

$$f_1 \text{ has a positive root } \frac{\delta}{2},$$

$$\max\{\delta_0, \delta\} \leq 2\,\alpha,$$

and

the conclusions of Lemma 7.1 hold, with α replacing $\frac{\delta}{2}$.

Proof. It follows from (7.17), (7.21) that

$$f_m(\alpha) = f_1(\alpha) \leq 0, \quad (m \geq 1), \tag{7.23}$$

which together with $f_1(0) = 2\,(1 - L_0\,\eta)\,\kappa > 0$, imply that there exists a positive root $\frac{\delta}{2}$ of polynomial f_1, satisfying

$$\delta \leq 2\,\alpha. \tag{7.24}$$

It also follows from (7.5), and (7.21) that

$$\delta_0 \leq 2\,\alpha, \tag{7.25}$$

and (7.20) holds, with α replacing $\frac{\delta}{2}$ (by (7.23)).

That completes the proof of Lemma 7.2 $\qquad\qquad\qquad\qquad\qquad\qquad \diamond$

Remark 7.2. The conclusions of Lemma 7.1 also hold under the following set of hypotheses:

(a) (with δ_0 replacing δ) There exist constants $\kappa \in [0, 1)$, $L_0 > 0$, $L > 0$, with $L_0 \leq L$, and $\eta > 0$, such that

$$\kappa < \alpha, \tag{7.26}$$

$$h_{AH} = a_0\,\eta \leq \frac{1}{2} \tag{7.27}$$

and

$$f_1(\frac{\delta_0}{2}) \leq 0, \tag{7.28}$$

where polynomial f_1 is given by (7.4), and

$$a_0 = \frac{L + 2 L_0 (\alpha - \kappa)}{4 (\alpha - \kappa)}.$$

Indeed, estimate (7.20) holds for $m = 1$ by (7.28). Using (7.16)–(7.19), we get

$$f_2 \left(\frac{\delta_0}{2} \right) = f_1 \left(\frac{\delta_0}{2} \right) + g \left(\frac{\delta_0}{2} \right) \eta \frac{\delta_0}{2} \leq 0.$$

Assume (7.20) holds for all $k \leq m$. We shall show (7.20) for m replaced by $m + 1$. We can have:

$$f_{m+1} \left(\frac{\delta_0}{2} \right) = f_m \left(\frac{\delta_0}{2} \right) + g \left(\frac{\delta_0}{2} \right) \eta \left(\frac{\delta_0}{2} \right)^m \leq 0,$$

which show (7.20) for all m.

Finally, we obtain:

$$f_\infty \left(\frac{\delta_0}{2} \right) = \lim_{m \to \infty} f_m \left(\frac{\delta_0}{2} \right) \leq 0.$$

(b) (with α replacing $\frac{\delta}{2}$) There exist constants $\kappa \in [0, 1)$, $L_0 > 0$, $L > 0$, and $\eta > 0$, such that

$$\kappa < \alpha,$$

$$h_{AH} = \overline{a_0} \, \eta \leq \frac{1}{2}$$

where,

$$\overline{a_0} = \max \{ a_0, a_1 \}, \qquad a_1 = \frac{L_0}{2 (1 - \alpha)}.$$

Note that we have by (7.16) and (7.17)

$$f_1(\alpha) = f_m(\alpha) = f_\infty(\alpha)$$

and

$$f_\infty(\alpha) = 2 (\alpha - \kappa) \left(\frac{L_0 \, \eta}{1 - \alpha} - 1 \right).$$

Hence, estimate (7.20) holds if

$$f_\infty(\alpha) \leq 0,$$

which is true, since $\eta \leq \frac{1 - \alpha}{L_0}$. Moreover, we have $\delta_0 \leq 2 \, \alpha$ by the choice of a_0.

In practice, we shall test these sets of hypotheses to see if any of them is satisfied.

Chapter 7. Gauss–Newton Method 115

Lemma 7.3 ([189], [192]). *Assume there exist constants $L_0 \geq 0$, $L \geq 0$, and $\eta \geq 0$, with $L_0 \leq L$, such that:*

$$h_{AH} = \overline{L}\,\eta \begin{cases} \leq \dfrac{1}{2} & \text{if } L_0 \neq 0 \\[2mm] < \dfrac{1}{2} & \text{if } L_0 = 0, \end{cases} \tag{7.29}$$

where,

$$\overline{L} = \frac{1}{8}\left(L + 4\,L_0 + \sqrt{L^2 + 8\,L_0\,L}\right). \tag{7.30}$$

Then, sequence $\{t_n\}$ $(n \geq 0)$ given by

$$t_0 = 0, \quad t_1 = \eta, \quad t_{n+1} = t_n + \frac{L\,(t_n - t_{n-1})^2}{2\,(1 - L_0\,t_n)} \quad (n \geq 1), \tag{7.31}$$

is nondecreasing, bounded from above by $t^{\star\star}$, and converges to its unique least upper bound $t^\star \in [0, t^{\star\star}]$, where

$$t^{\star\star} = \frac{2\,\eta}{2 - \delta}, \tag{7.32}$$

$$\delta = \frac{4\,L}{L + \sqrt{L^2 + 8\,L_0\,L}} < 1 \quad \text{for } L_0 \neq 0. \tag{7.33}$$

Moreover the following estimates hold:

$$L_0\,t^\star \leq 1, \tag{7.34}$$

$$0 \leq t_{n+1} - t_n \leq \frac{\delta}{2}\,(t_k - t_{k-1}) \leq \cdots \leq \left(\frac{\delta}{2}\right)^n \eta, \quad (n \geq 1), \tag{7.35}$$

$$t_{n+1} - t_n \leq \left(\frac{\delta}{2}\right)^n (2\,h_{AH})^{2^n - 1}\,\eta, \quad (n \geq 0), \tag{7.36}$$

$$0 \leq t^\star - t_n \leq \left(\frac{\delta}{2}\right)^n \frac{(2\,h_{AH})^{2^n - 1}\,\eta}{1 - (2\,h_{AH})^{2^n}}, \quad (2\,h_{AH} < 1), \quad (n \geq 0). \tag{7.37}$$

Definition 7.1. Let $\mathcal{M}$ be a matrix $i \times j$. The $j \times i$ matrix $\mathcal{M}^+$ is called the Moore–Penrose–pseudoinverse of $\mathcal{M}$ if the following four axioms hold:

$$(\mathcal{M}^+\,\mathcal{M})^T = \mathcal{M}^+\,\mathcal{M},$$

$$(\mathcal{M}\,\mathcal{M}^+)^T = \mathcal{M}\,\mathcal{M}^+,$$

$$\mathcal{M}^+\,\mathcal{M}\,\mathcal{M}^+ = \mathcal{M}^+,$$

and

$$\mathcal{M}\,\mathcal{M}^+\,\mathcal{M} = \mathcal{M}$$

where, $\mathcal{M}^T$ is the adjoint of $\mathcal{M}$.

In the case of a full rank (i, j) matrix $\mathcal{M}$, with rank $\mathcal{M} = j$, the pseudo–inverse is given by:

$$\mathcal{M}^+ = (\mathcal{M}^T\,\mathcal{M})^{-1}\,\mathcal{M}^T. \tag{7.38}$$

Note that if we denote by *Ker M* and *Im M* the kernel and image of M, respectively, and Π_E the projection onto a subspace E of $\mathbb{R}^i$, then

$$M^+ \, M = \Pi_{Ker \, M^\perp}$$

and

$$M \, M^+ = \Pi_{Im \, M}.$$

If M is full row rank, then $M \, M^+ = I$, where I is the identity matrix.

We need the following result on the pertubation bound of Moore–Penrose inverses [151], [728]:

Proposition 7.1. *Let A and N two $i \times j$ matrices. Assume that rank $A \leq$ rank N (rank $N \geq 1$), and $\| A - N \| \, \| N^+ \| < 1$. Then*

$$rank \, A = rank \, N$$

and

$$\| A^+ \| \leq \frac{N^+}{1 - \| A - N \| \, \| N^+ \|}.$$

Let $R > 0$, and $x \in \mathcal{D}$. We denote by $U(x,R)$ and $\overline{U}(x,R)$ the open and closed balls in $\mathbb{R}^i$ with center x and radius R.

We shall provide a semilocal convergence analysis for (GNM).

Theorem 7.1. *Let $F : \mathcal{D} \subseteq \mathbb{R}^i \longrightarrow \mathbb{R}^i$ be a Fréchet–differentiable operator.*

Assume:

there exist $x_0 \in \mathcal{D}$, constants $\kappa \in [0,1)$, $K > 0$, $K_0 > 0$, with $K_0 \leq K$, $\beta > 0$ and $\eta > 0$, such that for all $x, y \in \mathcal{D}$:

$$F'(x_0) \neq 0, \tag{7.39}$$

$$\| F'(x_0)^+ \, F(x_0) \| \leq \eta, \tag{7.40}$$

$$\| F'(x_0)^+ \| \leq \beta, \tag{7.41}$$

$$rank \, F'(x) \leq rank \, F'(x_0), \tag{7.42}$$

$$\| F'(x) - F'(x_0) \| \leq K_0 \, \| x - x_0 \|, \tag{7.43}$$

$$\| F'(x) - F'(y) \| \leq K \, \| x - y \|, \tag{7.44}$$

$$\| F'(y)^+ \, (I - F'(x) \, F'(x)^+) \, F(x) \| \leq \kappa \, \| x - y \|, \tag{7.45}$$

$$\overline{U}(x_0, t^\star) \subseteq \mathcal{D}, \tag{7.46}$$

where, $t^\star$ is given in Lemma 7.1,

and

hypotheses of Lemmas 7.1 or 7.2 or 7.3 hold, for

$$L_0 = \beta \, K_0, \quad \text{and} \quad L = \beta \, K. \tag{7.47}$$

Then, sequence $\{x_n\}$ ($n \geq 0$) generated by (GNM) is well defined, remains in $\overline{U}(x_0, t^\star)$ for all $n \geq 0$, and converges to a solution $x^\star$ of $F'(x)^+ \, F(x) = 0$ in $\overline{U}(x_0, t^\star)$.

Moreover, the following estimates hold for all $n \geq 0$:

$$\| x_{n+1} - x_n \| \leq t_{n+1} - t_n, \tag{7.48}$$

and

$$\| x_n - x^\star \| \leq t^\star - t_n, \tag{7.49}$$

where, sequence $\{t_n\}$ is given in Lemma 7.1.

Proof. We shall show using induction on m:

$$\| x_{m+1} - x_m \| \leq t_{m+1} - t_m, \tag{7.50}$$

and

$$\overline{U}(x_{m+1}, t^\star - t_{m+1}) \subseteq \overline{U}(x_m, t^\star - t_m). \tag{7.51}$$

For every $z \in \overline{U}(x_1, t^\star - t_1)$,

$$\begin{aligned} \| z - x_0 \| &\leq \| z - x_1 \| + \| x_1 - x_0 \| \\ &\leq t^\star - t_1 + t_1 = t^\star - t_0, \end{aligned}$$

implies $z \in \overline{U}(x_0, t^\star - t_0)$. We also have

$$\| x_1 - x_0 \| = \| F'(x_0)^+ F(x_0) \| = \eta = t_1 - t_0.$$

That is (7.50), and (7.51) hold for $m = 0$. Let us assume (7.50), and (7.51) hold for all $k \leq m$, then we have:

$$\| x_{m+1} - x_0 \| \leq \sum_{i=1}^{m+1} \| x_i - x_{i-1} \| \leq \sum_{i=1}^{m+1} (t_i - t_{i-1}) = t_{m+1} - t_0 = t_{m+1},$$

and

$$\| x_m + \theta (x_{m+1} - x_m) - x_0 \| \leq t_m + \theta (t_{m+1} - t_m) \leq t^\star,$$

for all $\theta \in (0, 1)$.

Let $x \in U(x_0, t^\star)$. It follows from (7.41), (7.43), (7.46), (7.47), and the proofs of the Lemmas 7.1, 7.2, 7.3 that:

$$\| F'(x_0)^+ \| \| F'(x) - F'(x_0) \| \leq \beta K_0 \| x - x_0 \| < \beta K_0 t^\star \leq 1. \tag{7.52}$$

Using (7.44), (7.45), (7.53), and the definition of iteration $\{t_n\}$, we get in turn:

$$\begin{aligned} & \| x_{m+2} - x_{m+1} \| \\ \leq\ & \| F'(x_{m+1})^+ \| \int_0^1 \| F'(x_m + \theta(x_{m+1} - x_m)) - F'(x_m) \| \| x_{m+1} - x_m \| \, d\theta + \\ & \| F'(x_{m+1})^+ (I - F'(x_m)) F'(x_m)^+ F(x_m) \| \\ \leq\ & \frac{\beta K}{2 (1 - \beta K_0 \| x_{m+1} - x_0 \|)} \| x_{m+1} - x_m \|^2 + \kappa \| x_{m+1} - x_m \|. \end{aligned}$$

It follows from (7.52), and the Banach lemma on invertible operators that for each $x \in U(x_0, t^\star)$:

$$rank\ F'(x) = rank\ F'(x_0),$$

118   Ioannis K. Argyros, Saïd Hilout and Mohammad A. Tabatabai

and

$$\| F'(x)^+ \| \le \frac{\beta}{1 - \beta\, K_0 \, \| x - x_0 \|}. \tag{7.53}$$

We have $x_0, x_1 \in U(x_0, t^\star)$. Let us assume $x_m, x_{m+1} \in U(x_0, t^\star)$. Using (7.2), we obtain the identity

$$
\begin{aligned}
x_{m+2} \;=\; & x_{m+1} + F'(x_{m+1})^+ \times \\
& \int_0^1 \left(F'(x_m + \theta\,(x_{m+1} - x_m)) - F'(x_{m+1}) \right)(x_{m+1} - x_m)\, d\theta + \\
& F'(x_{m+1})^+ \left(I - F'(x_m) \right) F'(x_m)^+ F(x_m).
\end{aligned} \tag{7.54}
$$

We also have:

$$\frac{\beta\, K}{2\,(1 - \beta\, K_0\, t_{m+1})}\,(t_{m+1} - t_m)^2 + \kappa\,(t_{m+1} - t_m) = t_{m+2} - t_{m+1}, \tag{7.55}$$

which shows (7.50) for all m. Moreover, for every $w \in \overline{U}(x_{m+2}, t^\star - t_{m+2})$, we have:

$$
\begin{aligned}
\| w - x_{m+1} \| \;\le\;& \| w - x_{m+2} \| + \| x_{m+2} - x_{m+1} \| \\
\le\;& t^\star - t_{m+2} + t_{m+2} - t_{m+1} = t^\star - t_{m+1},
\end{aligned} \tag{7.56}
$$

which completes the induction.

Lemmas 7.1, 7.2, 7.3 implies that sequence $\{t_n\}$ is a complete sequence. Moreover, in view of (7.50) and (7.51), $\{x_n\}$ ($n \ge 0$) is a complete sequence too, and as such it converges to some $x^\star \in \overline{U}(x_0, t^\star)$ (since $\overline{U}(x_0, t^\star)$ is a closed set).

By letting $m \longrightarrow \infty$ in the estimate

$$
\begin{aligned}
& \| F'(x^\star)^+ F(x_m) \| \\
\le\;& \| F'(x^\star)^+ \left(I - F'(x_m)\, F'(x_m)^+ \right) F(x_m) \| + \\
& \| F'(x^\star)^+ \| \, \| \, \| F'(x_m)\, F'(x_m)^+ F(x_m) \| \\
\le\;& \kappa\, \| x_m - x^\star \| + \| F'(x^\star)^+ \| \, \| F'(x_m) \| \, \| x_{m+1} - x_m \|,
\end{aligned} \tag{7.57}
$$

we obtain $x^\star$ is a solution of $F'(x)^+ F(x) = 0$. Finally, estimate (7.49) follows from (7.50) by using standard majorization techniques.

That completes the proof of Theorem 7.1. $\diamond$

Note that $t^{\star\star}$ given in closed form by (7.9) can replace $t^\star$ in Theorem 7.1.

Remark 7.3.   (a) Note that

$$K_0 \le K \tag{7.58}$$

holds in general, and $\dfrac{K}{K_0}$ can be arbitrarily large.

(b) In view of (7.54), and (7.55) for $m = 0$, we see that $\{r_n\}$ given by:

$$
\begin{aligned}
& r_0 = 0, \quad r_1 = \eta, \\
& r_{n+2} = r_{n+1} + \frac{\beta\, K_1\,(r_{n+1} - r_n)^2 + 2\,\kappa\,(r_{n+1} - r_n)}{2\,(1 - \beta\, K_0\, r_{n+1})} \quad (n \ge 0)
\end{aligned} \tag{7.59}
$$

is also majorizing for $\{x_k\}$, where,

$$K_1 = \begin{cases} K_0, & if \quad n = 0 \\ K, & if \quad n > 0 \end{cases}. \tag{7.60}$$

If strict inequality holds in (7.58), then, we have:

$$r_n < t_n, \qquad (n \geq 2), \tag{7.61}$$

$$r_{n+1} - r_n < t_{n+1} - t_n, \qquad (n \geq 1), \tag{7.62}$$

$$r^\star - r_n \leq t^\star - t_n, \qquad (n \geq 2), \tag{7.63}$$

and

$$r^\star \leq t^\star, \tag{7.64}$$

where,

$$r^\star = \lim_{n \to \infty} r_n.$$

That is $\{r_n\}$ is a tighter majorizing sequence for $\{x_n\}$ than $\{t_n\}$ (under the same hypotheses).

Let us compare our results with the corresponding ones in [455], which in turn have improved earlier results by Häußler [423].

We need to define scalar sequence $\{s_n\}$ by

$$s_0 = 0, \qquad s_{n+1} = s_n - \frac{g_q(t_n)}{g_1'(t_n)} \qquad (n \geq 0), \tag{7.65}$$

where,

$$g_q(t) = \frac{p}{2} t^2 - q t + \eta, \tag{7.66}$$

$$p = \frac{\beta K}{1 + (K - K_0) \beta \eta}, \tag{7.67}$$

and

$$q = 1 - \frac{(1 - \beta \eta K_0) \kappa}{1 + (K - K_0) \beta \eta}. \tag{7.68}$$

If

$$h_{HSL} = \frac{p \eta}{q^2} \leq \frac{1}{2}$$

then, sequence $\{s_n\}$ is nondecreasing and converges to

$$s^\star = \frac{q - \sqrt{q^2 - 2 p \eta}}{p}. \tag{7.69}$$

Moreover, under the additional hypotheses (7.39)–(7.46), the conclusions of Theorem 7.1 hold with $\{s_n\}$, $s^\star$ replacing $\{t_n\}$, $t^\star$, respectively [455].

120   Ioannis K. Argyros, Saïd Hilout and Mohammad A. Tabatabai

In particular it was also shown (see Lemma 7.1, and (3.18) in [455, Theorem 3.1]) that:

$$
\begin{aligned}
\| x_{n+2} - x_{n+1} \| &\leq \frac{\beta K (s_{n+1} - s_n)^2}{2 (1 - \beta K_0 s_{n+1})} + \kappa (s_{n+1} - s_n) \\
&\leq -\frac{g_q(s_{n+1})}{g_1'(s_{n+1})} = s_{n+2} - s_{n+1}.
\end{aligned}
\tag{7.70}
$$

It then follows from the definition of sequences $\{t_n\}$, $\{s_n\}$, (7.70), and a simple inductive argument that under the hypotheses of our Theorem 7.1 and Theorem 3.1 in [455], we have in turn:

$$
t_{n+1} - t_n < s_{n+1} - s_n, \qquad (n \geq 1),
\tag{7.71}
$$

and consequently

$$
t^\star - t_n \leq s^\star - s_n, \qquad (n \geq 2),
\tag{7.72}
$$

$$
t^\star \leq s^\star.
\tag{7.73}
$$

That is $\{t_n\}$ is an at least as tight majorizing sequence as $\{s_n\}$.

We provide a numerical example, where the hypotheses of Theorem 7.1 and Lemma 7.2 are satisfied, but not earlier ones [423], [455].

Example 7.1 ([455]). Let $i = j = 2$, and $\mathbb{R}^2$ equipped with the ℓ_1-norm. Choose:

$$
x_0 = (.2505, 1)^T, \quad \mathcal{D} = \{x = (v, w)^T : -1 < v < 1 \text{ and } -1 < w < 1\}.
$$

Define function F on $U(x_0, \sigma) \subseteq \mathcal{D}$ ($\sigma = .72$) by

$$
F(x) = (v - w, .5 (v - w)^2), \qquad x = (v, w)^T \in \mathcal{D}.
\tag{7.74}
$$

Then, for each $x = (v, w)^T \in \mathcal{D}$, the Fréchet–derivative of F at x, and the Moore–Penrose–pseudoinverse of $F'(x)$ are given by

$$
F'(x) = \begin{bmatrix} 1 & -1 \\ v - w & w - v \end{bmatrix}
\tag{7.75}
$$

and

$$
F'(x)^+ = \frac{1}{2 (1 + (v - w)^2)} \begin{bmatrix} 1 & v - w \\ -1 & w - v \end{bmatrix}
\tag{7.76}
$$

respectively.

Let $x = (v_1, w_1)^T \in \mathcal{D}$ and $y = (v_2, w_2)^T \in \mathcal{D}$. By (7.75), we have

$$
\| F'(x) - F'(y) \| = |(v_1 - v_2) - (w_1 - w_2)| \leq \| x - y \|.
$$

That is $K = K_0 = 1$.

Using (7.75), (7.76) and (7.45), we obtain:

$$
\begin{aligned}
& \| F'(y)^+ (I - F'(x) F'(x)^+) F(x) \| \\
&= \frac{(v_1 - w_1)^2}{4 (1 + (v_2 - w_2)^2) (1 + (v_1 - w_1)^2)} |(v_1 - v_2) - (w_1 - w_2)| \\
&\leq \frac{(v_1 - w_1)^2}{2(1 + (v_1 - w_1)^2)} \| x - y \| \leq \frac{2}{5} \| x - y \|.
\end{aligned}
$$

Chapter 7. Gauss–Newton Method

Hence the constant κ in hypothesis (7.45) is given by:

$$\kappa = .4.$$

Using hypotheses of Theorem 7.1, (7.74)–(7.76), we obtain:

$$\eta = .243104595, \qquad \beta = 1.176664037.$$

Then, we have:

$$\beta \, \eta \, K = .286052435 > \frac{9}{50} = \frac{(1-\kappa)^2}{2}.$$

Theorem 2.4 in [423] is not applicable.
Moreover, we get:

$$\beta \, \eta \, K = .286052435 > .243060906 = \frac{9}{19+5\sqrt{13}} = \frac{(1-\kappa)^2}{\kappa^2 - \kappa + 1 + \sqrt{2\kappa^2 - 2\kappa + 1}}.$$

That is Theorem 3.1 in [455] cannot apply to solve the equation.
However, the hypotheses of Lemma 7.2 are satisfied, since

$$a = 2.034776751, \quad \alpha = .658872344,$$

$$h_{AH} = .494663578 < .5, \quad \text{and} \quad t^{**} = .712649915 < \sigma.$$

That is our Theorem 7.1 can apply to solve the equation.

Let us now consider the case $\kappa = 0$ in condition (7.45). Theorem 3.1 in [455] gives

$$h_{SHL} = \frac{K_0 + K}{2} \, \beta \, \eta \leq \frac{1}{2}, \tag{7.77}$$

whereas our Lemma 7.3 provides:

$$h_{AH} = \frac{1}{8} \, (K + 4\,K_0 + \sqrt{K^2 + 8\,K_0\,K}) \, \beta \, \eta \leq \frac{1}{2}. \tag{7.78}$$

Condition (7.77) was derived by us in [136] when $F'(x_0)$ is invertible.
It follows from (7.77) and (7.78) that:

$$h_{SHL} \leq \frac{1}{2} \implies h_{AH} \leq \frac{1}{2} \tag{7.79}$$

but not vice versa unless if $K = K_0$. If $K_0 = K$, (7.77), and (7.78) reduce to the famous Kantorovich hypothesis:

$$h_K = K \, \beta \, \eta \leq \frac{1}{2}. \tag{7.80}$$

Hence, we have extended the applicability of (GNM), since (7.78) can always replace (7.77).

Other applications, where $K_0 < K$ can be found in [136], [192] (see, also chapter 1).

Chapter 8

Convergence on $\mathcal{K}$–Normed Spaces

In this chapter, we use our new idea of recurrent functions to provide a new semilocal convergence result for a Newton–type method (NTM) for solving a nonlinear operator equation in a $\mathcal{K}$–normed space setting. Using more precise majorizing sequences than before [289], [135], we show how to expand the convergence domain of (NTM) under the same computational cost as before [289], [135].

We are concerned with the problem of approximating a locally unique solution $x^\star$ of equation

$$F(x) + G(x) = 0, \tag{8.1}$$

where, F, G are defined on a closed ball $\overline{U}(x_0, R)$ centered at some point x_0 of a Banach space X with $R > 0$, and with values in X. Operator F is differentiable, whereas the differentiability of G is not assumed.

We propose the Newton–type method (NTM)

$$x_{n+1} = x_n - F'(x_n)^{-1} (F(x_n) + G(x_n)) \quad (n \geq 0) \tag{8.2}$$

to generate a sequence approximating $x^\star$. A survey of local as well as semilocal convergence results for (NTM), under Lipschitz or Hölder type continuity conditions can be found in [135], [155], [289] (see also [60], [764]).

This chapter is motivated by the elegant works in [289], [762], and optimization considerations, where X is a real Banach space ordered by a closed convex cone $\mathcal{K}$. Note that passing from scalar majorants to vector majorants enlarges the range of applications, since the latter uses the spectral radius which is usually smaller than its norm used by the former.

In [135], Argyros used tighter vector majorants than before [289] and provided under the same hypotheses:

(a) Sufficient convergence conditions which are always weaker than before.

(b) Tighter error bounds on the distances involved, and an at least as precise information on the location of the solution $x^\star$ are provided.

Some applications are also provided in [135]. In particular Argyros showed as a special case that the famous Newton–Kantorovich hypothesis is weakened. In this chapter, we extend the convergence domain of (NTM) even further than [289], [135] using our new idea of recurrent functions.

124   Ioannis K. Argyros, Saïd Hilout and Mohammad A. Tabatabai

In order to make the chapter as self–contained as possible we need to reintroduce some concepts involving $\mathcal{K}$–normed spaces [135], [289], [762].

Let X be a real Banach space ordered by a closed convex cone $\mathcal{K}$. We say that cone $\mathcal{K}$ is regular if every increasing sequence

$$\lambda_1 \leq \lambda_2 \leq \lambda_n \leq \cdots$$

which is bounded above, converges in norm. Moreover, If

$$\lambda_n^0 \leq \lambda_n \leq \lambda_n^1 \quad \text{and} \quad \lim_{n \to \infty} \lambda_n^0 = \lim_{n \to \infty} \lambda_n^1 = \lambda^\star$$

then the regularity of $\mathcal{K}$ implies $\lim_{n \to \infty} \lambda_n = \lambda^\star$.

Let $\alpha, \beta \in X$, then we define the conic segment $\langle \alpha, \beta \rangle = \{\lambda : \alpha \leq \lambda \leq \beta\}$. An operator Q in X is called positive if $Q(\lambda) \in \mathcal{K}$ for all $\lambda \in \mathcal{K}$. Denote by $L(X, X)$ the space of all bounded linear operators in X, and $\mathcal{L}_{\text{sym}}(X^2, X)$ the space of bilinear, symmetric, bounded operators from X^2 to X. Using the standard linear isometry between $L(X^2, X)$, and $L(X, L(X, X))$, we consider the former embedded into the latter.

Let $\mathcal{D}$ be a linearly connected subset of $\mathcal{K}$, and φ be a continuous operator from $\mathcal{D}$ into $L(X, X)$ or $L(X, L(X, X))$. We say that the line integral of φ is independent of the path if for every polygonal line L in $\mathcal{D}$, the line integral depends only on the initial and final point of L. We define

$$\int_{r_0}^{r} \varphi(t)\, dt = \int_0^1 \varphi((1-s)\, r_0 + s\, r)\, (r - r_0)\, ds. \tag{8.3}$$

We need the definition of $\mathcal{K}$–normed space:

Definition 8.1. Let X be a real linear space. Then X is said to be $\mathcal{K}$–normed if operator $]\cdot[: X \to X$ satisfies:

$$]x[\, \geq 0 \quad (x \in X);$$

$$]x[\, = 0 \Leftrightarrow x = 0;$$

$$]\mu\, x[\, = |\mu|\,]x[\quad (x \in X, \mu \in \mathbb{R}); \tag{8.4}$$

$$]x + y[\, \leq\,]x[\, +\,]y[\quad (x, y \in X).$$

Definition 8.2. Let $x_0 \in X$ and $r \in \mathcal{K}$. Then we denote

$$\overline{U}(x_0, r) = \{x \in X :\,]x - x_0[\, \leq r\}. \tag{8.5}$$

Using $\mathcal{K}$–norm we can define convergence on X. A sequence $\{y_n\}$ $(n \geq 0)$ in X is said to be

(a) convergent to a limit $y \in X$ if

$$\lim_{n \to \infty}]y_n - y[\, = 0 \quad \text{in } X \tag{8.6}$$

and we write

$$(X) - \lim_{n \to \infty} y_n = y;$$

Chapter 8. Convergence on $\mathcal{K}$–Normed Spaces

(b) a Cauchy sequence if

$$\lim_{m,n\to\infty}]y_m - y_n[= 0.$$

The space X is complete if every Cauchy sequence is convergent.

We use the following conditions:

F is differentiable on the $\mathcal{K}$–ball $U(x_0,R)$, and for every $r \in S = \langle 0,R \rangle$, there exist positive operators $\omega_0(r), \overline{\omega}(r) \in \mathcal{L}_{\mathrm{sym}}(X^2, X)$ such that for all $z \in X$ and for all $x, y \in \overline{U}(x_0, r)$:

$$](F'(x) - F'(x_0))(z)[\leq \omega_0(r)(]x - x_0[,]z[) \tag{8.7}$$

and

$$](F'(x) - F'(y))(z)[\leq \overline{\omega}(r)(]x - y[,]z[) \tag{8.8}$$

where operators $\omega_0, \overline{\omega} : S \to \mathcal{L}_{\mathrm{sym}}(X^2, X)$ are increasing, with $\omega_0(0) = \overline{\omega}(0) = 0$. Moreover, the line integral of $\overline{\omega}$ (similarly for ω_0) is independent of the path, and the same is true for the operator $\omega : S \to L(X,X)$ given by

$$\omega(r) = \int_0^r \overline{\omega}(t)\, dt. \tag{8.9}$$

Note that in general

$$\omega_0(r) \leq \overline{\omega}(r) \quad \text{for all } r \in S, \tag{8.10}$$

and $\dfrac{\overline{\omega}}{\omega_0}$ can be arbitrarily large [155].

The Newton–Leibniz formula holds for F on $\overline{U}(x_0, R)$:

$$F(x) - F(y) = \int_x^y F'(z)\, dz, \tag{8.11}$$

for all segments $[x, y] \in \overline{U}(x_0, R)$; for every $r \in S$ there exists a positive operator $\omega_1(r) \in L(X,X)$ such that:

$$]G(x) - G(y)[\leq \omega_1(r)(]x - y[) \quad \text{for all } x, y \in \overline{U}(x_0, r), \tag{8.12}$$

where, $\omega_1 : S \to L(X,X)$, is increasing, $\omega_1(0) = 0$, and the line integral of ω_1 is independent of the path;

Operator $F'(x_0)$ is invertible and satisfies:

$$]F'(x_0)(y)[\leq b\,]y[\quad \text{for all} \quad y \in X \tag{8.13}$$

for some positive operator $b \in L(X,X)$.

Let

$$\eta =]F'(x_0)^{-1}(F(x_0) + G(x_0))[. \tag{8.14}$$

Define operator $f : S \to X$ by:

$$f(r) = \eta + b \int_0^r \omega(t)\, dt + b \int_0^r \omega_1(t)\, dt. \tag{8.15}$$

Using the monotonicity of operators ω, ω_1, we see that f is order convex, i.e., for all r, $\overline{r} \in S$, with $r \leq \overline{r}$,

$$f((1-s)r + s\overline{r}) \leq (1-s) f(r) + s\, f(\overline{r}) \quad \text{for all} \quad s \in [0,1]. \tag{8.16}$$

We will use the following results whose proofs can be found in [289]:

Lemma 8.1. (a) *If Lipschitz condition* (8.8) *holds then*

$$](F'(x+y) - F'(x))(z)[\le (\omega(r+]y[) - \omega(r))(]z[) \tag{8.17}$$

for all r, $r+]y[\in S$, $x \in \overline{U}(x_0, r)$, $z \in X$.

(b) *If Lipschitz condition* (8.12) *holds then*

$$]G(x+y) - G(x)[\le \int_r^{r+]y[} \omega_1(t)\, dt \tag{8.18}$$

for all r, $r+]y[\in S$, $x \in \overline{U}(x_0, r)$.

Lemma 8.2. *Denote by* $\mathrm{Fix}(f)$ *the set of all fixed points of the operator* f, *and assume:*

$$\mathrm{Fix}(f) \ne \emptyset. \tag{8.19}$$

Then there is a minimal element $r^\star$ *in* $\mathrm{Fix}(f)$, *which can be found by applying the method of successive approximations*

$$r = f(r) \tag{8.20}$$

with 0 *as the starting point.*
The set

$$B(f, r^\star) = \{r \in S : \lim_{n \to \infty} f^n(r) = r^\star\} \tag{8.21}$$

is the attracting zone of $r^\star$.

Remark 8.1 ([289]). Let $r \in S$. If

$$f(r) \le r \tag{8.22}$$

and

$$\langle 0, r \rangle \cap \mathrm{Fix}(f) = \{r^\star\} \tag{8.23}$$

then

$$\langle 0, r \rangle \subseteq B(f, r^\star). \tag{8.24}$$

Note that the successive approximations

$$\varepsilon_{n+r} = \delta(\varepsilon_n) \quad (\varepsilon_0 = r) \quad (n \in \mathcal{N}) \tag{8.25}$$

converges to a fixed point $\varepsilon^\star$ of f, satisfying $0 \le s^\star \le r$. Hence, we conclude $s^\star = r^\star$, which implies $r \in B(f, r^\star)$.

In particular, we have:

$$\langle 0, (1-s)\, r^\star + s\, r \rangle \subseteq B(f, r^\star) \tag{8.26}$$

for every $r \in \mathrm{Fix}(f)$, with $\langle 0, r \rangle \cap \mathrm{Fix}(f) = \{r^\star, r\}$, and for all $\lambda \in [0, 1)$.
In the scalar case $X = \mathbb{R}$, we have

$$B(f, r^\star) = [0, r^\star] \cup \{r \in S : r^\star < r, f(q) < q, (r^\star < q \le r)\}. \tag{8.27}$$

Chapter 8. Convergence on $\mathcal{K}$–Normed Spaces

We will also use the notation

$$E(r^\star) = \bigcup_{r \in B(f, r^\star)} \overline{U}(x_0, r). \tag{8.28}$$

Returning back to method (8.2), we consider the sequences of approximations

$$r_{n+1} = r_n - (b\,\omega_0(r_n) - I)^{-1}\,(f(r_n) - r_n) \quad (r_0 = 0,\ n \geq 0) \tag{8.29}$$

and

$$\overline{r}_{n+1} = \overline{r}_n - (b\,\omega(\overline{r}_n) - I)^{-1}\,(f(\overline{r}_n) - \overline{r}_n) \quad (\overline{r}_0 = 0,\ n \geq 0) \tag{8.30}$$

for the majorant equation (8.20).

Lemma 8.3 ([135]). *If operators*

$$I - b\,\omega_0(r), \quad r \in [0, r^\star) \tag{8.31}$$

are invertible with positive inverses, then sequence $\{r_n\}$ ($n \geq 0$) given by (8.29) is well defined for all $n \geq 0$, monotonically increasing and convergent to $r^\star$.

Remark 8.2. If equality holds in (8.10), then sequence $\{\overline{r}_n\}$ becomes $\{r_n\}$ ($n \geq 0$) and Lemma 8.3 reduces to [289, Lemma 3, p. 555].

Moreover as it can easily be seen using induction on n

$$r_{n+1} - r_n \leq \overline{r}_{n+1} - \overline{r}_n \tag{8.32}$$

and

$$r_n \leq \overline{r}_n \tag{8.33}$$

for all $n \geq 0$. Furthermore if strict inequality holds in (8.10) so does in (8.32) and (8.33). If $\{r_n\}$ ($n \geq 0$) is a majorizing sequence for method (8.2), then (8.32) shows that the error bounds on the distances $\| x_{n+1} - x_n \|$ are tighter. It turns out that this is indeed the case.

We can show the semilocal convergence theorem for method (8.2).

Theorem 8.1. *Assume hypotheses (8.8), (8.9), (8.11)–(8.13), (8.19) hold, and operators (8.31) are invertible with positive inverses.*

Then sequence $\{x_n\}$ ($n \geq 0$) generated by (NTM) is well defined, remains in the $\mathcal{K}$–ball $U(x_0, r^\star)$ for all $n \geq 0$, and converges to a solution $x^\star$ of equation (8.1) in $E(r^\star)$, where $E(r^\star)$ is given by (8.28).

Moreover the following error bounds hold for all $n \geq 0$:

$$]x_{n+1} - x_n[\leq r_{n+1} - r_n, \tag{8.34}$$

and

$$]x^\star - x_n[\leq r^\star - r_n, \tag{8.35}$$

where sequence $\{r_n\}$ is given by (8.29).

Proof. We first show (8.34) using induction on $n \geq 0$ (by (8.14)). For $n = 0$;

$$]x_1 - x_0[\, = \,]F'(x_0)^{-1}(F(x_0) + G(x_0))[\, = \eta = r_1 - r_0. \qquad (8.36)$$

Assume:

$$]x_k - x_{k-1}[\, \leq r_k - r_{k-1}, \qquad k = 1, 2, \ldots, n. \qquad (8.37)$$

Using (8.37) we get

$$]x_n - x_0[\, \leq \sum_{k=1}^{n}]x_k - x_{k-1}[\, \leq \sum_{k=1}^{n} (r_k - r_{k-1}) = r_n. \qquad (8.38)$$

Define operators $Q_n : X \to X$ by

$$Q_n = -F'(x_0)^{-1}(F'(x_n) - F'(x_0)). \qquad (8.39)$$

By (8.7) and (8.13) we get

$$
\begin{aligned}
]Q_n(z)[\quad &= \quad]F'(x_0)^{-1}(F'(x_n) - F'(x_0))(z)[\\
&\leq \quad b](F'(x_n) - F'(x_0))(z)[\, \leq b \, \omega_0(r_n)(]z[),
\end{aligned} \qquad (8.40)
$$

and

$$]Q_n^i(z)[\, \leq (b \, \omega_0(r_n))^i(]z[) \quad (i \geq 1). \qquad (8.41)$$

Hence

$$\sum_{i=0}^{\infty}]Q_n^i(z)[\, \leq \sum_{j=0}^{\infty} (b \, \omega_0(r_n))^i \, (]z[). \qquad (8.42)$$

That is series $\displaystyle\sum_{i=0}^{\infty} Q_n^i(z)$ is convergent in X. Hence operator $I - Q_n$ is invertible, and

$$](I - Q_n)^{-1}(z)[\, \leq (I - b \, \omega_0(r_n))^{-1} (]z[). \qquad (8.43)$$

Operator $F'(x_n)$ is invertible for all $n \geq 0$, since $F'(x_n) = F'(x_0) \, (I - Q_n)$, and for all $x \in X$ we have:

$$
\begin{aligned}
]F'(x_n)^{-1}(x)[\quad &= \quad](I - Q_n)^{-1} \, F'(x_0)^{-1}(x)[\\
&\leq \quad (I - b \, \omega_0(r_n))^{-1}(]F'(x_0)^{-1}(x)[) \\
&\leq \quad (I - b \, \omega_0(r_n))^{-1}(b \,]x[).
\end{aligned} \qquad (8.44)
$$

Using (8.5) we obtain the approximation

$$
\begin{aligned}
]x_{n+1} - x_n[\quad &= \quad]F'(x_n)^{-1} \, (F(x_n) + G(x_n)) - \\
&\quad F'(x_n)^{-1} \, (F'(x_{n-1}) \, (x_n - x_{n-1}) + \\
&\quad F(x_{n-1}) + G(x_{n-1}))[.
\end{aligned} \qquad (8.45)
$$

Chapter 8. Convergence on $\mathcal{K}$–Normed Spaces 129

It now follows from (8.7)–(8.13), (8.15), (8.29) and (8.45)

$$
\begin{aligned}
&]x_{n+1} - x_n[\\
\leq\ &]F'(x_n)^{-1}(F(x_n) - F(x_{n-1}) - F'(x_{n-1})(x_n - x_{n-1})[+ \\
&]F'(x_n)^{-1}(G(x_n) - G(x_{n-1})) \leq (I - b\omega_0(r_n))^{-1} \Big\{ \\
& b\Big]\int_0^1 (F'((1-\lambda)x_{n-1} + \lambda x_n) - F'(x_{n-1}))\,(x_n - x_{n-1})d\lambda\Big[\Big\}+ \\
& (I - b\,\omega_0(r_n))^{-1}\,(b\,]G(x_n) - G(x_{n-1})[) \\
\leq\ & (I - b\,\omega_0(r_n))^{-1}\Big\{ b\int_0^1 (\omega((1-\lambda)r_{n-1} + \lambda r_n) - \\
& \omega(r_{n-1}))(r_n - r_{n-1})d\lambda \Big\}+ \\
& (I - b\,\omega_0(r_n))^{-1}\left(b\int_{r_{n-1}}^{r_n} \omega_1(t)\,dt \right) \\
=\ & (I - b\,\omega_0(r_n))^{-1}\Big\{ b\int_{r_{n-1}}^{r_n} \omega(t)\,dt - b\,\omega(r_{n-1})(r_n - r_{n-1}) + \\
& b\int_{r_{n-1}}^{r_n} \omega_1(t)\,dt \Big\} \\
=\ & (I - b\,\omega_0(r_n))^{-1}(f(r_n) - f(r_{n-1}) - b\,\omega(r_{n-1})(r_n - r_{n-1})) \\
=\ & (I - b\,\omega_0(r_n))^{-1}((f(r_n) - r_n) - (f(r_{n-1}) - r_{n-1}) - \\
& (b\,\omega(r_{n-1}) - I)\,(r_n - r_{n-1})) \\
=\ & (I - b\,\omega_0(r_n))^{-1}((f(r_n) - r_n) - (f(r_{n-1}) - r_{n-1}) - \\
& (b\,\omega(r_{n-1}) - I)(r_n - r_{n-1})) \\
\leq\ & (I - b\,\omega_0(r_n))^{-1}((f(r_n) - r_n) - (f(r_{n-1}) - r_{n-1}) - \\
& (b\,\omega_0(r_{n-1}) - I)(r_n - r_{n-1})) \\
=\ & (I - b\,\omega_0(r_n))^{-1}(f(r_n) - r_n) = r_{n+1} - r_n.
\end{aligned}
\tag{8.46}
$$

By Lemma 8.7, sequence $\{r_n\}$ $(n \geq 0)$ converges to $r^\star$. Hence $\{x_n\}$ is a convergent sequence, and its limit is a solution of equation (8.1). Therefore x_n converges to $x^\star$.

Finally (8.35) follows from (8.34) by using standard majorization techniques. The uniqueness part is omitted since it follows exactly as in [289, Theorem 2].

That completes the proof of Theorem 8.1. $\diamond$

Remark 8.3. It follows immediately from (8.46) that sequence

$$
\begin{aligned}
& t_0 = t_0, \quad t_1 = \eta, \\
& t_{n+1} - t_n = (I - b\,\omega_0(t_n))^{-1}\Big\{ b\int_{t_{n-1}}^{t_n} \omega(t)\,dt - b\,\omega(t_{n-1})(t_n - t_{n-1}) + \\
& \qquad\qquad b\int_{t_{n-1}}^{t_n} \omega_1(t)dt \Big\} \quad (n \geq 1)
\end{aligned}
\tag{8.47}
$$

is also a tighter majorizing sequence of $\{x_n\}$ $(n \geq 0)$ and converges to some $t^\star$ in $\langle 0, r^\star \rangle$.

The proof of Theorem 8.1 was also essentially given in [135], but the iteration (8.47) uses $\omega_1(t_n)\,(t_n - t_{n-1})$ instead of $\int_{t_{n-1}}^{t_n} \omega_1(t)\,dt$.

Moreover the following hold for all $n \geq 0$

$$
]x_1 - x_0[\leq t_1 - t_0 = r_1 - r_0,
\tag{8.48}
$$

$$]x_{n+1} - x_n [\leq t_{n+1} - t_n \leq r_{n+1} - r_n, \tag{8.49}$$

$$]x^\star - x_n [\leq t^\star - t_n \leq r^\star - r_n, \tag{8.50}$$

$$t_n \leq r_n, \tag{8.51}$$

and

$$t^\star \leq r^\star. \tag{8.52}$$

That is $\{t_n\}$ is a tighter majorizing sequence than $\{r_n\}$ and the information on the location of the solution $x^\star$ is more precise. Therefore, Argyros [135] remarks that if studying the convergence of $\{t_n\}$ without assuming (8.19) can lead to weaker sufficient convergence conditions for (NTM). In Theorem 8.1, Argyros responds to this question.

We need the following definition of some operators.

Definition 8.3. Define operators:

$$f_n, h_n, p_n : [0, 1) \longrightarrow X$$

and

$$q : I_q = \left[1, \frac{1}{1-\gamma}\right] \times [0,1)^4 \longrightarrow X, \quad \gamma \in [0, 1)$$

by

$$
\begin{aligned}
f_n(\gamma) = {} & b \Bigg\{ \int_0^1 \left(\omega\left(\left(\frac{1-\gamma^{n-1}}{1-\gamma} + t\,\gamma^{n-1}\right) \eta \right) - \omega\left(\frac{1-\gamma^{n-1}}{1-\gamma}\,\eta \right) \right) dt + \\
& \omega_1\left(\frac{1-\gamma^{n-1}}{1-\gamma}\,\eta \right) + \gamma\,\omega_0\left(\frac{1-\gamma^n}{1-\gamma}\,\eta \right) \Bigg\} - \gamma,
\end{aligned}
\tag{8.53}
$$

$$
\begin{aligned}
h_n(\gamma) = {} & b \Bigg\{ \int_0^1 \left(\omega\left(\left(\frac{1-\gamma^n}{1-\gamma} + t\,\gamma^n\right) \eta \right) - \right. \\
& \left. \omega\left(\left(\frac{1-\gamma^{n-1}}{1-\gamma} + t\,\gamma^{n-1}\right) \eta \right) \right) dt + \\
& \left(\omega\left(\frac{1-\gamma^{n-1}}{1-\gamma}\,\eta \right) - \omega\left(\frac{1-\gamma^n}{1-\gamma}\,\eta \right) \right) + \\
& \left(\omega_1\left(\frac{1-\gamma^n}{1-\gamma}\,\eta \right) - \omega_1\left(\frac{1-\gamma^{n-1}}{1-\gamma}\,\eta \right) \right) + \\
& \gamma\left(\omega_0\left(\frac{1-\gamma^{n+1}}{1-\gamma}\,\eta \right) - \omega_0\left(\frac{1-\gamma^n}{1-\gamma}\,\eta \right) \right) \Bigg\},
\end{aligned}
\tag{8.54}
$$

$$
\begin{aligned}
\overline{p_n}(\gamma) = {} & \int_0^1 \left(\omega\left(\left(\frac{1-\gamma^{n+1}}{1-\gamma} + t\,\gamma^{n+1}\right) \eta \right) + \omega\left(\left(\frac{1-\gamma^{n-1}}{1-\gamma} + t\,\gamma^{n-1}\right) \eta \right) - \right. \\
& \left. 2\,\omega\left(\left(\frac{1-\gamma^n}{1-\gamma} + t\,\gamma^n\right) \eta \right) \right) dt + \\
& \left(2\,\omega\left(\frac{1-\gamma^n}{1-\gamma}\,\eta \right) - \omega\left(\frac{1-\gamma^{n-1}}{1-\gamma}\,\eta \right) - \omega\left(\frac{1-\gamma^{n+1}}{1-\gamma}\,\eta \right) \right) + \\
& \left(\omega_1\left(\frac{1-\gamma^{n+1}}{1-\gamma}\,\eta \right) + \omega_1\left(\frac{1-\gamma^{n-1}}{1-\gamma}\,\eta \right) - 2\,\omega_1\left(\frac{1-\gamma^n}{1-\gamma}\,\eta \right) \right) + \\
& \gamma\left(\omega_0\left(\frac{1-\gamma^{n+2}}{1-\gamma}\,\eta \right) + \omega_0\left(\frac{1-\gamma^n}{1-\gamma}\,\eta \right) - 2\,\omega_0\left(\frac{1-\gamma^{n+1}}{1-\gamma}\,\eta \right) \right),
\end{aligned}
$$

$$p_n(\gamma) = b\,\overline{p_n}(\gamma), \tag{8.55}$$

$$
\begin{aligned}
\overline{q}(v_1, v_2, v_3, v_4, \gamma) \;=\; & \int_0^1 \Big(\omega((v_1 + v_2 + v_3 + t\,v_4)\,\eta) + \omega((v_1 + t\,v_2)\,\eta) - \\
& 2\,\omega((v_1 + v_2 + t\,v_3)\,\eta)\Big)\,dt + \\
& \Big(2\,\omega((v_1 + v_2)\,\eta) - \omega(v_1\,\eta) - \omega((v_1 + v_2 + v_3)\,\eta)\Big) + \\
& \Big(\omega_1((v_1 + v_2 + v_3)\,\eta) + \omega_1(v_1\,\eta) - 2\,\omega_1((v_1 + v_2)\,\eta)\Big) + \\
& \gamma\Big(\omega_0((v_1 + v_2 + v_3 + v_4)\,\eta) + \omega_0((v_1 + v_2)\,\eta) - \\
& 2\,\omega_0((v_1 + v_2 + v_3)\,\eta)\Big),
\end{aligned}
$$

$$q(v_1, v_2, v_3, v_4, \gamma) = b\,\overline{q}(v_1, v_2, v_3, v_4, \gamma), \tag{8.56}$$

where, η is given by (8.14).

Moreover, define function $f_\infty : [0, 1) \longrightarrow X$ by

$$f_\infty(\gamma) = \lim_{n \longrightarrow \infty} f_n(\gamma). \tag{8.57}$$

It then follows from (8.53), and (8.57) that

$$f_\infty(\gamma) = b\left(\omega\left(\frac{\eta}{1-\gamma}\right) + \gamma\,\omega_0\left(\frac{\eta}{1-\gamma}\right)\right) - \gamma. \tag{8.58}$$

It can also easily be seen from (8.53)–(8.56) that the following identities hold:

$$f_{n+1}(\gamma) = f_n(\gamma) + h_n(\gamma), \tag{8.59}$$

$$h_{n+1}(\gamma) = h_n(\gamma) + p_n(\gamma), \tag{8.60}$$

and for

$$v_1 = \sum_{i=0}^{n-2} \gamma^i, \quad v_2 = \gamma^{n-1}, \quad v_3 = \gamma^n, \quad v_4 = \gamma^{n+1}, \tag{8.61}$$

we have

$$q(v_1, v_2, v_3, v_4, \gamma) = p_n(\gamma). \tag{8.62}$$

We need the following result on majorizing sequences for (NTM).

Lemma 8.4. *Assume:*

Operator $I - b\,\omega_0(\eta)$ is positive, invertible, and with a positive inverse; there exists $\alpha \in (0, 1)$, such that:

$$\frac{\eta}{1-\alpha} \le R; \tag{8.63}$$

$$0 \le \left(I - b\,\omega_0(\eta)\right)^{-1} b\left(\int_0^1 \omega(t\,\eta)\,dt - \omega(0) + \omega_1(0)\right) \le \alpha\,I; \tag{8.64}$$

$$q(v_1, v_2, v_3, v_4, \gamma) \geq 0 \quad \text{on} \quad I_q, \tag{8.65}$$

$$h_1(\alpha) \geq 0, \tag{8.66}$$

and

$$f_\infty(\alpha) \leq 0, \tag{8.67}$$

where, 0 and I is the zero endomorphism and the identity opertor on X, respectively.

Then iteration $\{t_n\}$ ($n \geq 0$) given by (8.47) is non–decreasing, bounded from above by

$$t^{\star\star} = \frac{\eta}{1-\alpha}, \tag{8.68}$$

and converges to its unique least upper bound $t^\star$ satisfying

$$t^\star \in \langle 0, t^{\star\star} \rangle. \tag{8.69}$$

Moreover the following error bounds hold for all $n \geq 0$:

$$0 \leq t_{n+1} - t_n \leq \alpha \, (t_n - t_{n-1}) \leq \alpha^n \, \eta, \tag{8.70}$$

and

$$t^\star - t_n \leq \frac{\eta}{1-\alpha} \, \alpha^n. \tag{8.71}$$

Proof. Estimate (8.70) is true, if

$$0 \leq \left(I - b \, \omega_0(\eta) \right)^{-1} b \left(\int_0^1 \omega(t_{n-1} + t \, (t_n - t_{n-1})) \, dt - \omega(t_{n-1}) + \omega_1 \, (t_{n-1}) \right) \leq \alpha \, I \tag{8.72}$$

hold for all $n \geq 1$.

In view of (8.64), and (8.68), estimate (8.72) holds for $n = 1$. We also have by (8.47), and (8.72) that

$$0 \leq t_2 - t_1 \leq \alpha \, (t_1 - t_0).$$

Let us assume that (8.70), and (8.72) hold for all $k \leq n$. Then, we have

$$t_n \leq \frac{1 - \alpha^n}{1 - \alpha} \, \eta. \tag{8.73}$$

Using the induction hypotheses, and (8.72), we have by Lemma 8.2 that $(I - b \, \omega_0(t_n))^{-1}$ exists, and is positive. Moreover, (8.70) and (8.72) shall hold if

$$b \left\{ \int_0^1 \left(\omega \left((\frac{1-\alpha^{n-1}}{1-\alpha} + t \, \alpha^{n-1}) \, \eta \right) - \omega \left(\frac{1-\alpha^{n-1}}{1-\alpha} \, \eta \right) \right) \, dt + \omega_1 \left(\frac{1-\alpha^{n-1}}{1-\alpha} \, \eta \right) + \alpha \, \omega_0 \left(\frac{1-\alpha^n}{1-\alpha} \, \eta \right) \right\} - \alpha \leq 0. \tag{8.74}$$

Estimate (8.74) motivates us to define functions f_n (for $\gamma = \alpha$), and show instead

$$f_n(\alpha) \leq 0. \tag{8.75}$$

Chapter 8. Convergence on $\mathcal{K}$–Normed Spaces

We have by (8.59)–(8.62), (8.65) and (8.66) that

$$f_{n+1}(\alpha) \geq f_n(\alpha). \tag{8.76}$$

In view of (8.57) and (8.76), estimate (8.75) shall hold, if (8.67) is true. The induction is completed.

It follows that iteration $\{t_n\}$ is non–decreasing, bounded from above by t^{**} (given by (8.68)), and as such it converges to t^* satisfying (8.69).

Finally, estimate (8.71) follows from (8.70) by using standard majorizing techniques. That completes the proof of Lemma 8.4. $\diamond$

We also state a result from [135], so we can compare with Lemma 8.4.

Lemma 8.5 ([135])**.** *Assume there exist parameters* $\eta \geq 0$, $\delta \in [0,2)$ *such that*

(I) *Operators*

$$I - b\,\omega_0\left(2\,(2\,I - \delta\,I)^{-1}\left(I - \left(\frac{\delta I}{2}\right)^{n+1}\right)\eta\right) \tag{8.77}$$

    be positive, invertible, and with positive inverses for all $n \geq 0$;

(II)

$$2(I - b\,\omega_0(\eta))^{-1}\left(b\,\omega_1(\eta) + b\int_0^1 \omega(s\eta)\,ds - b\,\omega(0)\right) \leq \delta\,I; \tag{8.78}$$

(III)

$$\begin{aligned}
2\,b\int_0^1 &\omega\left(2\,(2\,I - \delta\,I)^{-1}\left(I - \left(\frac{\delta\,I}{2}\right)^{n+1}\right)\eta + \right.\\
&\left. s\left(\frac{\delta\,I}{2}\right)^{n+1}\eta\right)ds - \\
2\,b\,\omega&\left(2\,(2\,I - \delta\,I)^{-1}\left(I - \left(\frac{\delta\,I}{2}\right)^{n+1}\right)\eta\right) + \\
2\,b\,\omega_1&\left(2\,(2\,I - \delta\,I)^{-1}\left(I - \left(\frac{\delta\,I}{2}\right)^{n+1}\eta\right) + \right.\\
\delta\,b\,\omega_0&\left(2\,(2\,I - \delta I)^{-1}\left(I - \left(\frac{\delta\,I}{2}\right)^{n+1}\right)\eta\right) \\
&\leq 2\,b\int_0^1 \omega(s\,\eta)\,ds - 2\,b\,\omega(0) + 2\,b\,\omega_1(\eta) + \delta\,b\,\omega_0(\eta),
\end{aligned} \tag{8.79}$$

    for all $n \geq 0$.

Then iteration $\{t_n\}$ *(*$n \geq 0$*) given by (8.47) is non–decreasing, bounded above by*

$$t^{**} = 2\,(2\,I - \delta\,I)^{-1}\,\eta, \tag{8.80}$$

converges to some t^*, *such that*

$$0 \leq t^* \leq t^{**}. \tag{8.81}$$

Moreover the following error bounds hold for all $n \geq 0$:

$$0 \leq t_{n+2} - t_{n+1} \leq \frac{\delta\,I}{2}\,(t_{n+1} - t_n) \leq \left(\frac{\delta\,I}{2}\right)^{n+1}\eta. \tag{8.82}$$

134   Ioannis K. Argyros, Saïd Hilout and Mohammad A. Tabatabai

We can show the main semilocal convergence theorem for (NTM).

Theorem 8.2. *Assume:*
hypotheses (8.7)–(8.9), (8.11)–(8.13), *hypotheses of Lemma* 8.4, (8.77)–(8.79) *hold, and*

$$t^{\star\star} \leq R, \tag{8.83}$$

where $t^{\star\star}$ is given by (8.68).

Then sequence $\{x_n\}$ $(n \geq)$ generated by (NTM) is well defined, remains in the $\mathcal{K}$–ball $U(x_0, t^\star)$ for all $n \geq 0$ and converges to a solution x^* of equation (8.1), *which is unique in $E(t^\star)$.*

Moreover the following error bounds hold for all $n \geq 0$:

$$]x_{n+1} - x_n[\leq t_{n+1} - t_n \tag{8.84}$$

and

$$]x^* - x_n[\leq t^* - t_n, \tag{8.85}$$

where sequence $\{t_n\}$ $(n \geq 0)$ and t^* are given by (8.47) *and* (8.81) *respectively.*

Proof. The proof is identical to Theorem 8.1 with sequence t_n replacing r_n until the derivation of (8.47). But then the right hand side of (54) with these changes becomes $t_{n+1} - t_n$. By Lemma 8.5: $\{t_n\}$ converges to $t^\star$. Hence $\{x_n\}$ is a convergent sequence, its limit converges to a solution of equation (8.1). Therefore, $\{x_n\}$ converges to $x^\star$. Estimate (8.85) follows from (8.84) by using standard majorization techniques. The uniqueness part is omitted since it follows exactly as in [289, Theorem 2].

That completes the proof of Theorem 8.2. $\diamond$

Remark 8.4. The hypotheses of Lemma 8.4 are easier to verify than Lemma 8.5.

Application 8.1. Assume operator $][$ is given by a norm $\| \cdot \|$, and set $G(x) = 0$ for all $x \in \overline{U}(x_0, R)$. Choose for all $r \in \mathcal{S}$:

$$\overline{\omega}(r) = \ell\, r, \tag{8.86}$$

$$\omega_0(r) = \ell_0\, r \tag{8.87}$$

and

$$\omega_1(r) = 0. \tag{8.88}$$

That is we are consider Lipschitz and center–Lipschitz conditions of the form:

$$\| F'(x) - F'(y) \| \leq \ell \, \| x - y \|, \tag{8.89}$$

and

$$\| F'(x) - F'(x_0) \| \leq \ell_0 \, \| x - x_0 \|, \tag{8.90}$$

for all $x, y \in \overline{U}(x_0, R)$.

Chapter 8. Convergence on $\mathcal{K}$–Normed Spaces 135

Remark 8.5. The sufficient convergence condition in [289] using (8.15) resuces to the famous for its simplicity and clarity Newton–Kantorovich hypotheses for solving nonlinear equation [477]:

$$h_K = \beta \, \ell \, \eta \le \frac{1}{2}. \tag{8.91}$$

Moreover, the conditions of Lemma 8.5 becomes for $\delta = 1$ [135]:

$$h_A = \beta \, \frac{\ell + \ell_0}{2} \, \eta \le \frac{1}{2}. \tag{8.92}$$

Furthemore, the conditions of Lemma 8.4 give for

$$\alpha = \frac{4 \, \ell}{\ell + \sqrt{\ell^2 + 8 \, \ell_0 \, \ell}}, \tag{8.93}$$

$$h_{AH} = \beta \, \bar{\ell} \, \eta \le \frac{1}{2}, \tag{8.94}$$

where,

$$\bar{\ell} = \frac{1}{8} \left(\ell + 4 \, \ell_0 + \sqrt{\ell^2 + 8 \, \ell_0 \, \ell} \right). \tag{8.95}$$

It follows that

$$h_K \le \frac{1}{2} \implies h_A \le \frac{1}{2} \implies h_{AH} \le \frac{1}{2} \tag{8.96}$$

but not vice versa unless if $\ell_0 = \ell$.

Hence, we have expanded the applicability of (NM) under the same computational cost as in [135], [289].

Note that in practice the computation of ℓ requires that of ℓ_0. Hence, (8.90) is not an additional hypothesis.

Remark 8.6. The results obtained here hold under even weaker conditions. Indeed, since (8.8) is not "directly" used in the proofs above, it can be replaced by the weaker condition (8.17) throughout this chapter. As we showed in Lemma 8.1

$$(8.8) \implies (8.17)$$

but not necessarily vice versa unless if operator ω is convex [764, p. 674].

Chapter 9

Newton's Method on Spaces with Convergence Structure

In this chapter, we present a semilocal convergence theorem for Newton's method (NM) on spaces with a convergence structure. Using our new idea of recurrent functions, we provide a tighter analysis, with weaker hypotheses than before, and under the same computational cost [73], [87], [151], [521], [523].

We are concerned with the problem of approximating a solution of a nonlinear equation

$$F(x) = 0. \tag{9.1}$$

using Newton's method (NM).

Our chapter is motivated by the elegant works by Meyer [523], Ezquerro–Hernández [375], Proinov [601], [602], and optimization considerations (see, also [73], [87], [151], [521], [522]). Meyer provided in [523] a general structure for the convergence analysis of (NM), which enables us to consider under the same framework, Kantorovich–type semilocal convergence results and theorems based on monotonicity considerations. Here, we use our new idea of recurrent functions to provide a tighter semilocal analysis of (NM) with weaker conditions, and under the same computational cost as before [73], [87], [151], [375], [521], [523], [601], [602]. Other approaches on the Kantorovich–type semilocal convergence of (NM) can also be found in [137], [216], [476], [601], [602].

The chapter is organized as follows: the first part contains the needed terms and concepts; the second part contains two different semilocal convergence approaches for (NM). Finally, we develop special case.

In order for us to make the chapter as self contained as possible, we re–introduce concepts that can be found in [73], [87], [151], [521], [523]:

Definition 9.1. A Banach space X with convergence structure is a triple (X, V, W) satisfying:

(1) $(X, \| . \|)$ is a real Banach space.

(2) $(V, \mathbb{C}, \| . \|_V)$ is a real Banach space which is partially ordered by the closed convex cone $\mathbb{C}$; the norm $\| . \|_V$ is assumed to be monotone on $\mathbb{C}$.

(3) $\mathcal{W}$ is a closed convex cone in $X \times \mathcal{V}$ satisfying $\{0\} \times \mathbb{C} \subset \mathcal{W} \subset X \times \mathbb{C}$.

(4) The following map $/./ : \mathcal{D} \longrightarrow \mathbb{C}$ is well defined:

$$/x/ := \inf\{p \in \mathbb{C} : (x,p) \in \mathcal{W}\} \tag{9.2}$$

where,

$$\mathcal{D} := \{x \in X : \exists\, p \in \mathbb{C},\, (x,p) \in \mathcal{W}\}.$$

(5) For every $x \in \mathcal{D}$, we have

$$\|x\| \leq \| /x/ \|_{\mathcal{V}}. \tag{9.3}$$

We use standard properties of partial orderings on $\mathcal{V}$ and $X \times \mathcal{V}$ [263], [568]. The set $\mathcal{D}$ satisfies $\mathcal{D} + \mathcal{D} \subset \mathcal{D}$ and for $s > 0$: $s\mathcal{D} \subset \mathcal{D}$, whereas:

$$U(a) := \{x \in X : (x,a) \in \mathcal{W}\}$$

defines a generalized neighborhood of zero.

Definition 9.1 is motivated by the following examples [73], [87], [151], [523]:

Example 9.1. Let $X = \mathbb{R}^n$ equipped with the maximum–norm:

(a) $\mathcal{V} = \mathbb{R}$, $\mathcal{W} = \{(x,p) \in \mathbb{R}^{n+1} : \|x\|_\infty \leq p\}$. This case concerns classical convergence analysis in a Banach space.

(b) $\mathcal{V} = \mathbb{R}^n$, $\mathcal{W} = \{(x,p) \in \mathbb{R}^{2n} : |x| \leq p\}$, i.e., if $x = (x_i)_{i=1}^n$ and $p = (p_i)_{i=1}^n$, then $|x| \leq p \iff x_i \leq p_i$, for all $1 \leq i \leq n$. This case concerns componentwise analysis and error estimates.

(c) $\mathcal{V} = \mathbb{R}^n$, $\mathcal{W} = \{(x,p) \in \mathbb{R}^{2n} : 0 \leq x \leq p\}$. This case is used in monotone convergence analysis.

Remark 9.1. The convergence analysis is based on monotonocity considerations in $X \times \mathcal{V}$. Let $(x_n, p_n) \in \mathcal{W}^N$ be an increasing sequence, then:

$$(x_n, p_n) \leq (x_{n+m}, p_{n+m}) \implies 0 \leq (x_{n+m} - x_m, p_{n+m} - p_m).$$

If $p_n \longrightarrow p$, we obtain $0 \leq (x_{m+n} - x_n, p - p_n)$. Using (9.3) of Definition 9.1, we have

$$\|x_{n+m} - x_n\| \leq \|p - p_n\|_{\mathcal{V}} \xrightarrow[n \to \infty]{} 0.$$

Then $\{x_n\}$ is a Cauchy sequence. When deriving error estimates we shall use sequence $p_n = a_0 - a_n$ with a decreasing sequence $\{x_n\} \in \mathbb{C}$ to obtain estimate:

$$0 \leq (x_{m+n} - x_n, a_n - a_{n+m}) \leq (x_{m+n} - x_n, a_n).$$

If $x_n \longrightarrow x$ this implies the estimate $/x - x_n/ \leq a_n$.

Definition 9.2. We denote the space of multilinear, symmetric, bounded operators $A : X^n \longrightarrow X$ on a Banach space X by $L(X^n)$, and for an orderer Banach space $\mathcal{V}$:

$$L_+(\mathcal{V}^n) = \{L \in L(\mathcal{V}^n) : 0 \leq x_i \, (1 \leq i \leq n) \Longrightarrow 0 \leq L(x_1, \cdots, x_n)\}.$$

A map $L \in \mathbb{C}^1(\mathcal{V}_L \longrightarrow V)$ on a open subset V_L of an ordered Banach space $\mathcal{V}$ is defined to be order convex on an intervall $[a, b] \subset V_L$, if

$$c, d \in [a, b], \, c \leq d \Longrightarrow L'(d) - L'(c) \in L_+(\mathcal{V}).$$

Definition 9.3. As the set of bounds for an operator $A \in L(X^n)$, we define:

$$B(A) = \{L \in L_+(\mathcal{V}^n) : (x_i, p_i) \in W \Longrightarrow (A(x_1, \cdots, x_n), L(p_1, \cdots, p_n)) \in W\}.$$

Lemma 9.1 ([521]). *Let $A : [0, 1] \longrightarrow L(X^n)$, and $L : [0, 1] \longrightarrow L_+(\mathcal{V}^n)$ be continuous maps, then:*

$$\forall t \in [0, 1] : L(t) \in B(A(t)) \Longrightarrow \int_0^1 L(t) \, dt \in B\left(\int_0^1 A(t) \, dt\right)$$

which will used for the remainder of Taylor's formula.

Finally the following conventions are needed: Let $T : \mathcal{Y} \longrightarrow \mathcal{Y}$ be a map on a subset $\mathcal{Y}$ of a normed space. Then $T^n(x)$ denotes the result of n–fold application of T and in case of convergence:

$$T^\infty(x) = \lim_{n \longrightarrow \infty} T^n(x).$$

In particular, we define a right inverse through:

Definition 9.4. Let $A \in L(X)$ and $y \in X$ be given. Then we can write

$$A^\star y := z \Longleftrightarrow z \in T^\infty(0), T(x) := (I - A) \, x + y \Longleftrightarrow z = \sum_{j=0}^\infty (I - A)^j \, y$$

provided this limit exists.

We provide sufficient semilocal convergence conditions for (NM) to determine a zero $x^\star$ of operator G on Banach space. Our results are stated for the operator

$$F(x) = A \, G(x_0 + x), \tag{9.4}$$

where x_0 is the initial guess for (NM), and A is an approximation of $G'(x_0)^{-1}$. That is the following result is affine invariant in the sense of [348].

Theorem 9.1. *Let X be a Banach space with a convergence structure $(X, \mathcal{V}, \mathcal{W})$ with $\mathcal{V} = (\mathcal{V}, \mathbb{C}, \| \cdot \|_\mathcal{V})$, an operator $F \in \mathbb{C}^1(X_F \longrightarrow X)$ with $X_F \subseteq X$, an operator $L \in \mathbb{C}^1(\mathcal{V}_L \longrightarrow \mathcal{V})$ with $\mathcal{V}_L \subseteq \mathcal{V}$, an operator $L_0 \in \mathbb{C}^1(\mathcal{V}_{L_0} \longrightarrow \mathcal{V})$ with $\mathcal{V}_L \subseteq \mathcal{V}_{L_0}$, and a point $a \in \mathbb{C}$ such that the following hypotheses are satisfied:*

$$U(a) \subseteq X_F \quad \text{and} \quad [0, a] \subseteq \mathcal{V}_L; \tag{9.5}$$

L is order–convex on $[0,a]$, satisfying

$$L'(|x|+|y|)-L'(|x|) \in B(F'(x)-F'(x+y)) \tag{9.6}$$

for all x, $y \in U(a)$ with $|x|+|y| \leq a$;
L_0 is order–convex on $[0,a]$, satisfying $L_0' \leq L'$ on $\mathcal{V}_L$, and

$$L_0'(|x|)-L_0'(0) \in B(F'(0)-F'(x)) \tag{9.7}$$

for all $x \in U(a)$ with $|x| \leq a$;

$$L_0'(0) \in B(I-F'(0)) \quad \text{and} \quad (-F(0),L(0)) \in \mathcal{W}; \tag{9.8}$$

$$L(a) \leq a; \tag{9.9}$$

and

$$L'(a)^n a \longrightarrow 0 \quad \text{as} \quad n \longrightarrow \infty. \tag{9.10}$$

Then, sequence $\{x_n\}$ generated by (NM):

$$x_0 = 0, \quad x_{n+1} = x_n + F'(x_n)^\star (-F(x_n)) \tag{9.11}$$

is well defined, remains in $U(a)$ for all $n \geq 0$, and converges to the unique zero $x^\star$ of operator F in $U(a)$.
Moreover, the following estimates hold true for all $n \geq 0$:

$$/x_{n+1}-x_n/ \leq d_{n+1}-d_n, \tag{9.12}$$

$$/x_{n+1}-x^\star/ \leq b-d_n, \tag{9.13}$$

where,

$$b = L^\infty(0) \tag{9.14}$$

is the minimal fixed point of operator L in $[0,a]$, and sequence $\{d_n\}$ is given by

$$d_0 = 0, \quad d_{n+1} = L(d_n)+L_0'(|x_n|)c_n, \quad c_n = /x_{n+1}-x_n/. \tag{9.15}$$

Furthemore, sequence $(x_n,d_n) \in (X \times V)^N$ is well defined, remains in $\mathcal{W}^N$, and is monotone.

Remark 9.2. (a) If $L_0' = L'$ on $\mathcal{V}_L$, then hypotheses of Theorem 9.1 reduce to the ones in [523, Theorem 5]. However, note that in general

$$L_0' \leq L' \tag{9.16}$$

holds, and $\dfrac{L'}{L_0'}$ can be arbitrarily large.

Condition (9.7) is not an additional hypothesis. In practice, computing operator L' also requires determining L_0'. Moreover, operator L_0' always exists (if L' exists). The benefits out of introducing condition (9.7) are given in Remark 9.3.

Chapter 9. Newton's Method on Spaces with Convergence Structure 141

(b) Assume conditions (9.5)–(9.7) of Theorem 9.1 hold, and

$$\exists t \in (0,1) : L(a) \leq t\,a.$$

Then, there exists $a' \in [0, t\,a]$ satisfying all conditions of Theorem 9.1. The zero $z \in U(a')$ is unique in $U(a)$ [523].

Let $L \in \mathbb{C}^1(\mathcal{V}_L \to \mathcal{V})$ be a map satisfying the conditions of Theorem 9.1. Then L is monotone. Therefore, sequences $b_n = L^n(0)$ and $c_n = L^n(a)$ are monotone with

$$0 \leq b_n \leq b_{n+1} \leq c_{n+1} \leq c_n \leq a.$$

In view of (9.10), we conclude that the sequence $\{c_n - b_n\}$ converges to zero [521]. That is, we obtain $b = L^\infty(0)$ is well defined, and is the smallest solution of $L(p) \leq p$ in $[0, a]$.

In case of Remark 9.2(b), we obtain

$$0 \leq c_n - b_n \leq t^n\,a.$$

That is $b = L^\infty(0)$ is well defined, and the following inequalities hold:

$$L'(b)\,(a-b) \leq L(a) - L(b) \leq t\,(a-b)$$

$$b = L(b) \leq L(a) \leq t\,a \implies b \leq \frac{t}{1-t}\,(a-b),$$

so,

$$L'(b)^n\,b \leq \frac{t}{1-t}\,L'(b)^n\,(a-b) \leq \frac{t^{n+1}}{1-t}\,(a-b) \longrightarrow 0.$$

Therefore, $a' = b$ satisfies the additional hypothesis of Remark 9.2(b).

As a special case, we obtain the following result for affine maps:

Corollary 9.1 ([523])**.** *Let $L \in L_+(\mathcal{V})$, and a, $p \in \mathbb{C}$ be given such that:*

$$L\,p + a \leq p \quad \text{and} \quad L^n\,p \longrightarrow 0.$$

Then, the map

$$(I - L)^\star : [0, a] \longrightarrow [0, a]$$

is well defined and continuous.

As substitute for the Banach Lemma we use:

Lemma 9.2 ([523])**.** *Let $A \in L(X)$, $L \in B(A)$, $y \in \mathcal{D}$ and $p \in \mathbb{C}$ be given as in Theorem 9.1 such that:*

$$L\,p + /y/ \leq p \quad \text{and} \quad L^n\,p \longrightarrow 0.$$

Then, $x = (I - A)^\star y$ is well defined, $x \in \mathcal{D}$ and

$$/x/ \leq (I - L)^\star\,/y/ \leq p$$

The proof of Lemma 9.2 is easy. Simply note that the sequence $\{b_n\}$ defined by

$$b_0 = 0, \quad b_{n+1} = L b_n + /y/ \leq p$$

is well defined and converges to

$$b = (I - L)^\star \, /y/ \leq p.$$

If we consider $x_{n+1} = A x_n + y$, $x_0 = 0$, then the sequence (x_n, b_n) is monotone in $X \times \mathcal{V}$. Hence, the statement follows from the general principles seen earlier in this chapter.

Proof of Theorem 9.1. Conditions of Theorem 9.1 are satisfied for b replacing a.
We shall show that for each n, there exists x_{n+1} solving

$$p = (I - F'(x_n)) \, p + (-F(x_n)). \tag{9.17}$$

Let $n = 1$, and $p = b$. Using (9.6)–(9.9) we get:

$$\begin{aligned} |I - F'(0)| \, b + | - F(0)| &\leq L_0'(0) \, b + | - F(0)| \\ &\leq L'(0) \, b + | - F(0)| \\ &\leq L(b) - L(0) + L(0) = L(b) = b. \end{aligned} \tag{9.18}$$

Hence, x_1 is well defined, and $(x_1, b) \in \mathcal{W}$.
We also have

$$x_1 = (I - F'(0)) \, x_1 + (-F(0)), \tag{9.19}$$

So,

$$|x_1| \leq L_0'(0) \, |x_1| + L(0) = d_1, \tag{9.20}$$

and consequently,

$$d_1 = L_0'(0) \, |x_1| + L(0) \leq L'(0) \, b + L(0) \leq L(b) - L(0) + L(0) = L(b) = b. \tag{9.21}$$

Let us assume (x_k, d_k) is well defined, and monotone for all $k \leq n$, with

$$0 \leq (x_{k-1}, d_{k-1}) \leq (x_k, d_k), \quad d_k \leq b. \tag{9.22}$$

Using (9.7), and (9.8), we have:

$$\begin{aligned} |I - F'(x_k)| &\leq |(I - F'(0)) + (F'(0) - F'(x_k))| \\ &\leq |I - F'(0)| + |F'(0) - F'(x_k)| \\ &\leq L_0'(0) + L_0'(|x_k|) - L_0'(0) = L_0'(|x_k|), \end{aligned} \tag{9.23}$$

so,

$$L_0'(|x_k|) \in B(I - F'(x_k)). \tag{9.24}$$

In view of Lemma 9.2, we must solve for p:

$$L_0'(|x_k|) \, p + | - F(x_k)| \leq p. \tag{9.25}$$

Chapter 9. Newton's Method on Spaces with Convergence Structure 143

We need an estimate on $|-F(x_k)|$. By Taylor's theorem, (9.6), and Lemma 9.2, we obtain in turn:

$$
\begin{aligned}
|-F(x_k)| &= |-F(x_k) + F(x_{k-1}) + F'(x_{k-1})(x_k - x_{k-1})| \\
&\leq \int_0^1 \left[L'(|x_{k-1}| + t\, c_{k-1}) - L'(|x_{k-1}|) \right] c_{k-1}\, dt \\
&= L(|x_{k-1}| + c_{k-1}) - L(|x_{k-1}|) - L'(|x_{k-1}|)\, c_{k-1} \qquad (9.26) \\
&\leq L(d_{k-1} + d_k - d_{k-1}) - L(d_{k-1}) - L'(|x_{k-1}|)\, c_{k-1} \\
&\leq L(d_k) - L(d_{k-1}) - L_0'(|x_{k-1}|)\, c_{k-1} \\
&= L(d_k) - d_k
\end{aligned}
$$

Let $p = b - d_k$. Then, we have by (9.23), and (9.26):

$$
\begin{aligned}
L_0'(|x_k|)\, p + |-F(x_k)| + d_k &\leq L'(d_k)(b - d_k) + L(d_k) \\
&\leq L(b) - L(d_k) + L(d_k) = L(b) = b. \qquad (9.27)
\end{aligned}
$$

Hence, x_{k+1} is well defined by Lemma 9.2, and $c_k \leq b - d_k$. Therefore, d_{k+1} is well defined too, and we obtain

$$
\begin{aligned}
d_{k+1} &\leq L(d_k) + L_0'(d_k)(b - d_k) \\
&\leq L(d_k) + L'(d_k)(b - d_k) \qquad (9.28) \\
&\leq L(d_k) + L(b) - L(d_k) = L(b) = b.
\end{aligned}
$$

In view of the estimate

$$
\begin{aligned}
c_k + d_k &\leq L_0'(|x_k|)\, c_k + |-F(x_k)| + d_k \\
&\leq L_0'(|x_k|)\, c_k + L(d_k) = d_{k+1}, \qquad (9.29)
\end{aligned}
$$

we deduce the monotonicity

$$
(x_k, d_k) \leq (x_{k+1}, d_{k+1}), \qquad (9.30)
$$

which also implies (9.12).

It follows inductively from (9.15) that:

$$
L^k(0) \leq d_k \leq b, \qquad (9.31)
$$

which together with $L^k(0) \longrightarrow b$, imply $d_k \longrightarrow b$ as $k \longrightarrow \infty$.

Sequence $\{x_n\}$ converges to some $x^\star \in U(b)$. By setting $k \longrightarrow \infty$ in (9.26), we deduce that $x^\star$ is a zero of operator F.

Estimate (9.13) now follows from (9.12) by using standard majorization techniques [151], [189], [375].

The uniqueness statement is given in [521], where the modified Newton method

$$
x_{n+1} = x_n - F(x_n) \qquad (9.32)
$$

is considered. Clearly, sequence $(x_n, L^n(0))$ is monotone in $X \times \mathcal{V}$.

Moreover, if there exists a zero $y^\star \in U(a)$ of F, then, it was shown in [521] that

$$
|y^\star - x_n| \leq L^n(a) - L^n(0) \longrightarrow 0 \quad \text{as} \quad n \longrightarrow \infty, \qquad (9.33)
$$

which implies $\lim_{n \to \infty} x_n = y^\star$. But, we showed $\lim_{n \to \infty} x_n = x^\star$. Hence, we deduce

$$
x^\star = y^\star. \qquad (9.34)
$$

That completes the proof of Theorem 9.1. $\diamond$

Remark 9.3. If equality holds in (9.16), then our Theorem 9.1 reduces to [523, Theorem 5]. Otherwise (i.e., if $L_0' < L'$), the former Theorem improves the latter. Indeed, the majorizing sequence $\{q_n\}$ used in [523] is given by

$$q_0 = 0, \quad q_{n+1} = L(q_n) + L'(|x_n|)\, c_n. \tag{9.35}$$

In view of (9.15), and (9.35), a simple inductive argument shows:

$$d_n < q_n, \quad (n \geq 1) \tag{9.36}$$

and

$$d_{n+1} - d_n < q_{n+1} - q_n, \quad (n \geq 1). \tag{9.37}$$

Hence, sequence $\{d_n\}$ is a tighter majorizing sequence for $\{x_n\}$ than $\{q_n\}$. As already noted in Remark 9.2, these advantages are obtained under the same hypotheses and computational cost as in [523].

By simply replacing L' by L_0' in the definitions of the operators involved, we can also improve the a posteriori estimates given in [523] under the hypotheses of Theorem 9.1. More, precisely in order for us to obtain a posteriori estimates, we define:

$$R_n(p) = \left(I - L_0'(|x_n|)\right)^\star S_n(p) + c_n \tag{9.38}$$

where,

$$S_n(p) = L(|x_n| + p) - L(|x_n|) - L_0'(|x_n|)\, p. \tag{9.39}$$

Operator S_n is monotone on the interval $I_n = [0, a - |x_n|]$; Moreover, if there exists $p_n \in \mathbb{C}$ such that $|x_n| + p_n \leq a$, and

$$S_n(p_n) + L_0'(|x_n|)(p_n - c_n) \leq p_n - c_n, \tag{9.40}$$

then operator $R_n : [0, p_n] \longrightarrow [0, p_n]$ is well defined by Corollary 9.1, and monotone.

We then have

$$\begin{aligned} d_n + c_n \leq d_{n+1} &\Rightarrow L(a) - L(d_n) - L_0'(|x_n|)\, c_n \leq a - d_n - c_n \\ &\Rightarrow S_n(a - d_n) + L_0'(|x_n|)(a - d_n - c_n) \leq a - d_n - c_n, \end{aligned} \tag{9.41}$$

which implies $a - d_n$ is a suitable choice for p_n.

Other ways for choosing suitable p_n are given by the following:

Proposition 9.1. *Assume:*

$$R_n(p) \leq p \quad \text{for some} \quad p \in I_n. \tag{9.42}$$

Then we have:

$$c_n \leq R_n(p) = \overline{p} \leq p \tag{9.43}$$

and

$$R_{n+1}(\overline{p} - c_n) \leq \overline{p} - c_n. \tag{9.44}$$

Chapter 9. Newton's Method on Spaces with Convergence Structure 145

Proof. Using (9.26), the order convexity of L_0, L, and the estimate

$$S_n(p) + L_0'(|x_n|)(\overline{p} - c_n) = \overline{p} - c_n, \tag{9.45}$$

we get

$$S_{n+1}(\overline{p} - c_n) + |-F(x_{n+1})| + L_0'(|x_{n+1}|)(\overline{p} - c_n) \leq \overline{p} - c_n. \tag{9.46}$$

That completes the proof of Proposition 9.1.

Proposition 9.2. *Assume that conditions of Theorem 9.1 hold, and there exists a solution $p_n \in I_n$ satisfying*

$$R_n(p) \leq p. \tag{9.47}$$

Define sequence

$$a_n = p_n, \quad a_{m+1} = R_m(a_m) - c_m \quad (m \geq n). \tag{9.48}$$

Then, the following a posteriori estimate holds

$$/x^\star - x_m/ \leq a_m. \tag{9.49}$$

Proof. An induction argument shows

$$R_m(a_m) \leq a_m, \tag{9.50}$$

so,

$$a_{m+1} + c_m \leq a_m, \tag{9.51}$$

and consequently, we deduce the monotonicity of $(x_m, a_n - a_m)$ in $X \times \mathcal{V}$.
That completes the proof of Proposition 9.2. $\diamondsuit$

The properties of R_n imply the existence of $R_n^\infty(0)$, which is a suitable choice for p_n in Proposition 9.2. Hence we arrive at:

Corollary 9.2. *Assume that conditions of Theorem 9.1 hold, and there exists $p \in I_n$ satisfying*

$$R_n(p) \leq p.$$

Then the following a posterioiri estimates holds:

$$|x^\star - x_n| \leq R_n^\infty(0) \leq p. \tag{9.52}$$

As already noted in [523], in view of the estimate

$$S_n(p) + /-F(x_n)/ + L_0'(|x_n|)p \leq p \Longrightarrow R_n(p) \leq p, \tag{9.53}$$

one may consider further majorization:

$$Q_n(p) = L(|x_{n-1}| + c_{n-1} + p) - L(|x_{n-1}|) - L_0'(|x_{n-1}|)c_{n-1}. \tag{9.54}$$

Note that an application to a two point boundary value problem was given in [523] in the case $L_0' = L_0$. Further discussion on applications, and practical aspects can be found in [87], [151], [521], [522], [706].

Remark 9.4. If $L_0' = L'$, then our a posteriori estimates reduce to the ones in [523]. However, if $L_0' < L'$, then our a posteriori estimates are tighter. So far, we showed how to improve on the error estimates given in [523]. We are now wondering if we can also weaken the sufficient convergence conditions (9.9) and (9.10).

It turns out that this can be done, using our new idea of recurrent functions [151], [192].

We need to define some operator sequences.

Definition 9.5. Let $\eta \in \mathbb{C}$. Define operators:

$$f_n, h_n, \beta_n : [0,1) \longrightarrow \mathcal{X}$$

and

$$\delta : I_\delta = \left[1, \frac{1}{1-\gamma}\right] \times [0,1)^4 \longrightarrow \mathcal{X}, \quad \gamma \in [0,1)$$

by

$$
\begin{aligned}
f_n(\gamma) &= \left\{ \int_0^1 \left(L'\left(\left(\frac{1-\gamma^{n-1}}{1-\gamma} + t\gamma^{n-1} \right) \eta \right) - L'\left(\frac{1-\gamma^{n-1}}{1-\gamma} \eta \right) \right) dt + \right. \\
&\quad \left. \gamma L_0'\left(\frac{1-\gamma^n}{1-\gamma} \eta \right) \right\} - \gamma,
\end{aligned}
\tag{9.55}
$$

$$
\begin{aligned}
h_n(\gamma) &= \left\{ \int_0^1 \left(L'\left(\left(\frac{1-\gamma^n}{1-\gamma} + t\,\gamma^n \right) \eta \right) - \right. \right. \\
&\quad \left. L'\left(\left(\frac{1-\gamma^{n-1}}{1-\gamma} + t\,\gamma^{n-1} \right) \eta \right) \right) dt + \\
&\quad \left(L'\left(\frac{1-\gamma^{n-1}}{1-\gamma} \eta \right) - L'\left(\frac{1-\gamma^n}{1-\gamma} \eta \right) \right) + \\
&\quad \left. \gamma \left(L_0'\left(\frac{1-\gamma^{n+1}}{1-\gamma} \eta \right) - L_0'\left(\frac{1-\gamma^n}{1-\gamma} \eta \right) \right) \right\},
\end{aligned}
\tag{9.56}
$$

$$
\begin{aligned}
\overline{\beta_n}(\gamma) &= \int_0^1 \left(L'\left(\left(\frac{1-\gamma^{n+1}}{1-\gamma} + t\,\gamma^{n+1} \right) \eta \right) + \right. \\
&\quad L'\left(\left(\frac{1-\gamma^{n-1}}{1-\gamma} + t\,\gamma^{n-1} \right) \eta \right) - \\
&\quad 2 L'\left(\left(\frac{1-\gamma^n}{1-\gamma} + t\,\gamma^n \right) \eta \right) \right) dt + \\
&\quad \left(2 L'\left(\frac{1-\gamma^n}{1-\gamma} \eta \right) - L'\left(\frac{1-\gamma^{n-1}}{1-\gamma} \eta \right) - L'\left(\frac{1-\gamma^{n+1}}{1-\gamma} \eta \right) \right) + \\
&\quad \left. \gamma \left(L_0'\left(\frac{1-\gamma^{n+2}}{1-\gamma} \eta \right) + L_0'\left(\frac{1-\gamma^n}{1-\gamma} \eta \right) - 2 L_0'\left(\frac{1-\gamma^{n+1}}{1-\gamma} \eta \right) \right) \right),
\end{aligned}
$$

$$\beta_n(\gamma) = \overline{\beta_n}(\gamma),
\tag{9.57}$$

Chapter 9. Newton's Method on Spaces with Convergence Structure 147

$$\bar{\delta}(v_1, v_2, v_3, v_4, \gamma) = \int_0^1 \Big(L'((v_1 + v_2 + v_3 + t\, v_4)\, \eta) + L'((v_1 + t\, v_2)\, \eta) - 2\, L'((v_1 + v_2 + t\, v_3)\, \eta) \Big)\, dt +$$

$$\Big(2\, L'((v_1 + v_2)\, \eta) - L'(v_1\, \eta) - L'((v_1 + v_2 + v_3)\, \eta) \Big) +$$

$$\gamma \Big(L_0'((v_1 + v_2 + v_3 + v_4)\, \eta) + L_0'((v_1 + v_2)\, \eta) - 2\, L_0'((v_1 + v_2 + v_3)\, \eta) \Big),$$

$$\delta(v_1, v_2, v_3, v_4, \gamma) = \bar{\delta}(v_1, v_2, v_3, v_4, \gamma). \tag{9.58}$$

Moreover, define function $f_\infty : [0, 1) \longrightarrow X$ by

$$f_\infty(\gamma) = \lim_{n \longrightarrow \infty} f_n(\gamma). \tag{9.59}$$

It then follows from (9.55), and (9.59) that

$$f_\infty(\gamma) = b \left(L'(\frac{\eta}{1 - \gamma}) + \gamma L_0'(\frac{\eta}{1 - \gamma}) \right) - \gamma. \tag{9.60}$$

It can also easily be seen from (9.55)–(9.58) that the following identities hold:

$$f_{n+1}(\gamma) = f_n(\gamma) + h_n(\gamma), \tag{9.61}$$

$$h_{n+1}(\gamma) = h_n(\gamma) + \beta_n(\gamma), \tag{9.62}$$

and for

$$v_1 = \sum_{i=0}^{n-2} \gamma^i, \quad v_2 = \gamma^{n-1}, \quad v_3 = \gamma^n, \quad v_4 = \gamma^{n+1}, \tag{9.63}$$

we have

$$\delta(v_1, v_2, v_3, v_4, \gamma) = \beta_n(\gamma). \tag{9.64}$$

Finally, let us define sequence $\{t_n\}$ by

$$t_0 = 0, \quad t_1 = L_0(0) + L_0'(0)\, \eta,$$
$$t_{n+1} = t_n + L_0'(t_n)\, (t_n - t_{n-1}) +$$
$$\int_0^1 (L'(t_{n-1} + t\, (t_n - t_{n-1})) - L'(t_{n-1}))\, dt\, (t_n - t_{n-1}). \tag{9.65}$$

We need the following result on majorizing sequences for (NM).

Lemma 9.3. *Assume that there exist* η, $a \in \mathbb{C}$, *and* $\alpha \in (0, 1)$ *such that*

$$\frac{\eta}{1 - \alpha} \in [0, a]; \tag{9.66}$$

$$0 \le L_0'(t_1) + \int_0^1 (L'(t\, t_1) - L'(0))\, dt \le \alpha\, I; \tag{9.67}$$

$$\delta(v_1, v_2, v_3, v_4, \gamma) \ge 0 \quad \text{on} \quad I_\delta, \tag{9.68}$$

$$h_1(\alpha) \geq 0, \tag{9.69}$$

and

$$f_\infty(\alpha) \leq 0, \tag{9.70}$$

where, 0 and I is the zero endomorphism and the identity opertor on X, respectively.
Then iteration $\{t_n\}$ $(n \geq 0)$ given by (9.65) is non–decreasing, bounded from above by

$$t^{\star\star} = \frac{\eta}{1 - \alpha}, \tag{9.71}$$

and converges to its unique least upper bound $t^\star$ satisfying

$$t^\star \in \langle 0, t^{\star\star} \rangle. \tag{9.72}$$

Moreover the following error bounds hold for all $n \geq 0$:

$$0 \leq t_{n+1} - t_n \leq \alpha \, (t_n - t_{n-1}) \leq \alpha^n \, \eta, \tag{9.73}$$

and

$$t^\star - t_n \leq \frac{\eta}{1 - \alpha} \, \alpha^n. \tag{9.74}$$

Proof. Estimate (9.73) is true, if

$$0 \leq L'_0(t_n) + \int_0^1 \left(L'(t_{n-1} + t \, (t_n - t_{n-1})) - L(t_{n-1}) \right) dt \leq \alpha \, I \tag{9.75}$$

holds for all $n \geq 1$.

In view of (9.67), and (9.71), estimate (9.75) holds for $n = 1$. We also have by (9.65), and (9.75) that

$$0 \leq t_2 - t_1 \leq \alpha \, (t_1 - t_0).$$

Let us assume that (9.73), and (9.75) hold for all $k \leq n$. Then, we have

$$t_n \leq \frac{1 - \alpha^n}{1 - \alpha} \, \eta. \tag{9.76}$$

Moreover, (9.73) and (9.75) shall hold if

$$\left\{ \int_0^1 \left(L' \left(\left(\frac{1 - \alpha^{n-1}}{1 - \alpha} + t \, \alpha^{n-1} \right) \eta \right) - L' \left(\frac{1 - \alpha^{n-1}}{1 - \alpha} \, \eta \right) \right) dt + \alpha \, L'_0 \left(\frac{1 - \alpha^n}{1 - \alpha} \, \eta \right) \right\} - \alpha \leq 0. \tag{9.77}$$

Estimate (9.77) motivates us to define functions f_n (for $\gamma = \alpha$), and show instead

$$f_n(\alpha) \leq 0. \tag{9.78}$$

We have by (9.61)–(9.64), (9.68) and (9.69) that

$$f_{n+1}(\alpha) \geq f_n(\alpha). \tag{9.79}$$

Chapter 9. Newton's Method on Spaces with Convergence Structure   149

In view of (9.59) and (9.79), estimate (9.78) shall hold, if (9.70) is true.
The induction is completed.

It follows that iteration $\{t_n\}$ is non–decreasing, bounded from above by t^{**} (given by (9.71)), and as such it converges to t^* satisfying (9.72).

Finally, estimate (9.74) follows from (9.73) by using standard majorizing techniques [151], [189].

That completes the proof of Lemma 9.3. $\diamondsuit$

We can show the following semilocal convergence result for (NM).

Theorem 9.2. *Let X be a Banach space with a convergence structure $(X, \mathcal{V}, \mathcal{W})$ with $\mathcal{V} = (\mathcal{V}, \mathbb{C}, \| . \|_{\mathcal{V}})$, an operator $F \in \mathbb{C}^1(X_F \longrightarrow X)$ with $X_F \subseteq X$, an operator $L \in \mathbb{C}^1(\mathcal{V}_L \longrightarrow \mathcal{V})$ with $\mathcal{V}_L \subseteq \mathcal{V}$, an operator $L_0 \in \mathbb{C}^1(\mathcal{V}_{L_0} \longrightarrow \mathcal{V})$ with $\mathcal{V}_L \subseteq \mathcal{V}_{L_0}$, and a point $a \in \mathbb{C}$ such that the following hypotheses are satisfied:*
(9.5)–(9.8);

$$L(\eta) \le \eta, \qquad \eta = L_0^\infty(0), \quad \eta \le a; \tag{9.80}$$

$$L'(\eta)^n \eta \longrightarrow 0 \quad as \quad n \longrightarrow \infty; \tag{9.81}$$

and hypotheses of Lemma 9.3 hold for $|x_1| \le \eta$, where, x_1 solves

$$p = (I - F'(0)) \, p + (-F(0)). \tag{9.82}$$

Then, sequence $\{x_n\}$ generated by (NM) is well defined, remains in $U(t^*)$ for all $n \ge 0$, and converges to a zero x^* of operator F in $U(t^*)$.

Moreover, the following estimates hold true for all $n \ge 0$:

$$/x_{n+1} - x_n/ \le t_{n+1} - t_n, \tag{9.83}$$

$$/x_{n+1} - x^*/ \le t^* - t_n. \tag{9.84}$$

Furthemore, sequence $(x_n, t_n) \in (X \times \mathcal{V})^N$ is well defined, remains in $\mathcal{W}^N$, and is monotone.

Proof. As in Theorem 9.1, we have b_0 is the smallest fixed point of operator L_0 in $[0, a]$ guaranteed to exist by (9.8), (9.80), (9.81), and Lemma 9.2 since

$$\begin{aligned} L_0'(0)\,\eta + |-F(0)| &\le L_0(\eta) - L_0(0) + L_0(0) \\ &= L_0(\eta) = \eta. \end{aligned} \tag{9.85}$$

Equation (9.82) is satisfied for $p = \eta$. Therfore, x_1 is well defined, and $(x_1, \eta) \in \mathcal{W}$. We also have

$$|x_1| \le L_0'(0)\,|x_1| + L(0) = L_0'(0)\,\eta + L_0(0) = t_1,$$

and

$$t_1 \le L_0(\eta) - L_0(0) + L_0(0) = L_0(\eta) \le \eta \le t^*.$$

We also have

$$L'_0(|x_1|)(t_1-t_0)+|-F(x_1)|$$

$$\leq\ L'_0(t_1)(t_1-t_0)+\int_0^1 (L'(|x_0|+t\,c_0)-L'(|x_0|))\,c_0\,dt$$

$$=\ L'_0(t_1)(t_1-t_0)+\int_0^1 (L'(t\,c_0)-L'(0))\,c_0\,dt$$

$$=\ t_2-t_1\leq\alpha\,(t_1-t_0),$$

which together with Lemma 9.2 implies x_2 is well defined, and (9.83) holds for $n=1$.

Let us assume (x_k,t_k) is well defined, and monotone for all $k\leq n$, i.e.,

$$0\leq(x_{k-1},t_{k-1})\leq(x_k,t_k),\quad\text{and}\quad t_k\leq t^\star,\quad k=1,\cdots,n.$$

In view of (9.7), (9.9), (9.23), (9.26), the definition of sequence $\{t_n\}$, we have in turn for $p=t_k-t_{k-1}$:

$$
\begin{aligned}
& L'_0(|x_k|)(t_k-t_{k-1})+|-F(x_k)| \\
\leq\ & L'_0(t_k)(t_k-t_{k-1})+ \\
& \int_0^1\left[L'(|x_{k-1}|+t\,c_{k-1})-L'(|x_{k-1}|)\right]c_{k-1}\,dt \\
\leq\ & L'_0(t_k)(t_k-t_{k-1})+ \\
& \int_0^1\left[L'(|t_{k-1}|+t\,(t_k-t_{k-1}))-L'(|t_{k-1}|)\right](t_k-t_{k-1})\,dt \\
=\ & t_{k+1}-t_k\leq\alpha\,(t_k-t_{k-1}).
\end{aligned}
\tag{9.86}
$$

It follows from (9.86) and Lemma 9.2 that x_{k+1} is well defined, and (9.83) holds for all n. Sequence t_{k+1} is well defined too, and bounded above by $t^\star$. According to what we saw earlier in this chapter, $\{x_n\}$ converges to some $x^\star\in U(t^\star)$. By setting $k\longrightarrow\infty$ in the upper bound of $|-F(x_k)|$ given in (9.86), we obtain that $x^\star$ is a zero of operator F.

Estimate (9.84) follows from (9.83) as in Theorem 9.1.

That completes the proof of Theorem 9.2.

Remark 9.5. (a) Note that $t^{\star\star}$, given in closed form by (9.71), can replace $t^\star$ in Theorem 9.1.

(b) In view of the proof of Theorem 9.2, it follows that sequence $\{s_n\}$ given by

$$s_0=0,\quad s_1=L_0(0)+L'_0(0)\,|x_1|,$$

$$s_{n+1}=s_n+L'_0(|x_n|)c_n+\int_0^1 (L'(|x_{n-1}|+t\,c_{n-1})-L'(|x_{n-1}|))\,c_{n-1}\,dt$$

is also a finer majorizing sequence for $\{x_n\}$ than t_n.

(c) **The monotone case.** This is a particular case of Theorem 9.1 (or Theorem 9.2) but is omitted here, since it follows along the lines of Theorem 13 in [523], where X is itself partially ordered, and satisfies the conditions for $\mathcal{V}$ in Definition 9.1. We set $X=\mathcal{V}$, $\mathcal{D}=\mathbb{C}^2$, and $/./=I$ (see Case 3 in Example 9.1). The (NM) is given by

$$u_0=u,\quad u_{n+1}=u_n+(A\,G'(u_n))^\star(-A\,G(u_n)),$$

Chapter 9. Newton's Method on Spaces with Convergence Structure 151

$$G \in \mathbb{C}^1(\mathcal{V}_G \longrightarrow \mathcal{Y}), \quad A \in L(\mathcal{Y} \longrightarrow X),$$

$$u, v \in \mathcal{V}, \quad L_0(p) = p - F(p), \quad [u, v] \subseteq \mathcal{V}_G, \quad \text{and} \quad a = v - u.$$

Application 9.1. Let X be a Banach space with real norm $\| \cdot \|$. We shall check the conditions of Theorem 9.1, and [523, Theorem 5]. Let us assume for simplicity that $F'(0) = I$, and that there exists a monotone operator $E : [0, a] \longrightarrow \mathbb{R}$, such that

$$\| F'(x_0)^{-1} (F'(x) - F'(y)) \| \leq E(\| x - y \|) \| x - y \| \tag{9.87}$$

for all $x, y \in U(a)$.

Define L by

$$L(p) = \eta + \int_0^p ds \int_0^s dt \, E(t), \quad \eta \geq \| F'(x_0)^{-1} F(x_0) \| . \tag{9.88}$$

We have to solve (9.9) for $E(t) \leq E(a) = \ell$, i.e.,

$$\eta + \frac{1}{2} \ell \, a^2 \leq a, \tag{9.89}$$

which is possible if

$$h_K = \ell \, \eta \leq \frac{1}{2}. \tag{9.90}$$

Condition (9.90) is the famous for its simplicity, and clarity Kantorovich sufficient convergence hypothesis for (NM) [476, Chapter 12], [87], [189], [568].

In view of (9.87), there exists a monotone operator $E_0 : [0, a] \longrightarrow \mathbb{R}$, such that

$$\| F'(x_0)^{-1} (F'(x) - F'(x_0)) \| \leq E_0(\| x - x_0 \|) \| x - x_0 \| \tag{9.91}$$

for all $x \in U(a)$.

Define operator L_0 by

$$L_0(p) = \eta + \int_0^p ds \int_0^s dt \, E_0(t). \tag{9.92}$$

Set

$$E_0(a) = \ell_0.$$

Then it can easily be seen that hypotheses of Lemma 9.3, and Theorem 9.2 hold, if

$$h_{AH} = \bar{\ell} \, \eta \leq \frac{1}{2}, \tag{9.93}$$

where

$$\bar{\ell} = \frac{1}{8} \left(\ell + 4 \, \ell_0 + \sqrt{\ell^2 + 8 \, \ell_0 \, \ell} \right), \tag{9.94}$$

and

$$\alpha = \frac{2 \, \ell}{\ell + \sqrt{\ell^2 + 8 \, \ell_0 \, \ell}}. \tag{9.95}$$

In view of (9.90), and (9.93), we have

$$h_K \leq \frac{1}{2} \implies h_{AH} \leq \frac{1}{2}, \tag{9.96}$$

but not necessarily vice versa, unless if $\ell_0 = \ell$.

Hence, in this special case, our Theorem 9.2 is weaker than [523, Theorem 5].

For examples where (9.90) is violated but (9.93) is satisfied, and $\ell_0 < \ell$, the reader can refer to chapter 1. More applications can be found in [73], [192], [521], [523].

Part II

MATHEMATICAL MODELLING

Chapter 10

Newton's Method and Interior Point Techniques

In this chapter, We use a weaker Newton–Kantorovich theorem for solving equations, introduced in [157] to analyze interior point methods. Our approach requires less number of steps than before [594] to achieve a certain error tolerance for both Newton's and Modified Newton's methods.

We are concerned with the problem of approximating a locally unique solution $x^\star$ of equation

$$F(x) = 0, \tag{10.1}$$

where F is a differentiable operator defined on a convex domain $\mathcal{D}$ of $\mathbb{R}^i$ (i an integer) with values in $\mathbb{R}^i$.

It has already been shown in [594] that the Newton–Kantorovich theorem can be used to construct and analyze optimal–complexity path following algorithms for linear complementary problems. Potra has chosen to apply this theorem to linear complementary problems because such problems provide a convenient framework for analyzing primal–dual interior point methods. Theoretical and experimental work conducted over the past decade has shown that primal–dual path following algorithms are among the best solution methods for linear programming (LP), quadratic programming (QP), and linear complementary problems (LCP) (see for example [600], [736]). Primal–dual path following algorithms are the basis of the best general–purpose practical methods, and they have important theoretical properties [640], [736], [752]. Potra, using the Newton–Kantorovich theorem, in particular showed how to construct path–following algorithms for LCP that have $O(\sqrt{n}L)$ iteration complexity [594].

Given a point x that approximates a point $x(\tau)$ on the central path of the LCP with complementarity gap τ, the algorithms compute a parameter $\theta \in (0,1)$ so that x satisfies the Newton–Kantorovich hypothesis (10.5) for the equation defining $x((1-\theta)\tau)$. It is proven that θ is bounded below by a multiple of $n^{-1/2}$. Since (10.5) is satisfied, the sequence generated by Newton's method or by the modified Newton method (take $F'(x_n) = F'(x_0), n \geq 0$) with starting x, will converge to $x((1-\theta)\tau)$. He showed that the number of steps required to obtain an acceptable approximation of $x((1-\theta)\tau)$ is bounded above by a number independent of n. Therefore, a point with complementarity less than ε can be obtained in at most $O\left(\sqrt{n}\log\left(\frac{\varepsilon}{\varepsilon_0}\right)\right)$ steps (for both methods), where ε_0 is the complementary gap of the starting point. For linear complementarity problems with rational input data of bit length L, this implies that an exact solution can be obtained in at most $O(\sqrt{n}L)$ iterations plus a rounding procedure including $O(n^3)$ arithmetic operations [736].

The differences between Potra's work and the earlier works by Renegar [613] and Renegar and Shub [614] have been stated in the introduction of this chapter and further analyzed in [594]. We also refer the reader to the excellent monograph of Nesterov and Nemirovskii [558] for an analysis of the construction of interior point methods for a larger class of problems than that considered in [594].

The famous Newton–Kantorovich theorem [477] has been used extensively to solve equation (10.1). A survey of such results can be found in [137] and the references there. Recently, in [137], [157], Argyros improved the Newton–Kantorovich theorem. Here we use this development to show that the results obtained in the elegant work in [594] in connection with interior point methods can be improved if our convergence conditions simply replace the stronger ones given there. Finally a numerical example is provided to show that fewer iterations than the ones suggested in [594] are needed to achieve the same error tolerance.

Let $\|\cdot\|$ be a given norm on $\mathbb{R}^i$, and x_0 be a point of $\mathcal{D}$ such that $\overline{U}(x_0, r)$ the closed ball of radius r centred at x_0, is included in $\mathcal{D}$, i.e.

$$\overline{U}(x_0, r) \subseteq \mathcal{D}. \tag{10.2}$$

We assume that the Jacobian $F'(x_0)$ is nonsingular and that the following affine–invariant Lipschitz condition is satisfied:

$$\|F'(x_0)^{-1}[F'(x) - F'(y)]\| \leq \omega\|x - y\| \tag{10.3}$$

for all $x, y \in \overline{U}(x_0, r)$.

The famous Newton–Kantorovich Theorem [477] states that if the quantity

$$\alpha := \|F'(x_0)^{-1}F(x_0)\| \tag{10.4}$$

together with ω satisfy

$$k = \alpha\,\omega \leq \frac{1}{2}, \tag{10.5}$$

then there exists $x^\star \in \overline{U}(x_0, r)$ with $F(x^\star) = 0$. Moreover the sequences produced by Newton's method

$$x_{n+1} = x_n - F'(x_n)^{-1}F(x_n) \quad (n \geq 0), \tag{10.6}$$

and by the modified Newton method

$$y_{n+1} = y_n - F'(y_0)^{-1}F(y_n), \quad y_0 = x_0 \quad (n \geq 0) \tag{10.7}$$

are well defined and converge to $x^\star$.

In [137], [157], Argyros introduced the center–Lipschitz condition

$$\|F'(x_0)^{-1}[F'(x) - F'(x_0)]\| \leq \omega_0\|x - x_0\| \tag{10.8}$$

Chapter 10. Newton's Method and Interior Point Techniques 157

for all $x \in \overline{U}(x_0, r)$ and provided a finer local and semilocal convergence analysis of method (10.6) by using the combination of conditions (10.3) and (10.8) given by

$$k^0 = \alpha \, \overline{\omega} \leq \frac{1}{2}, \tag{10.9}$$

where,

$$\overline{\omega} = \frac{1}{8} \left(\eta + 4 \, \eta_0 + \sqrt{\eta^2 + 8 \, \eta_0 \, \eta} \right) \quad [157]. \tag{10.10}$$

In general

$$\overline{\omega} \leq \omega \tag{10.11}$$

holds, and $\dfrac{\overline{\omega}}{\omega}$ can be arbitrarily large. Note also that

$$k \leq \frac{1}{2} \Rightarrow k^0 \leq \frac{1}{2} \tag{10.12}$$

but not vice versa unless if $\overline{\omega} = \omega$. Examples where weaker condition (10.9) holds but (10.5) fails have been also given in [137], [157].

Similarly by simply replacing ω with ω_0 (since (10.8) instead of (10.3) is actually needed in the proof) and condition (10.5) by the weaker

$$k^1 = \alpha \, \omega_0 \leq \frac{1}{2} \tag{10.13}$$

in the proof of [594, Theorem 1] we show that method (10.7) also converges to $x^\star$ and the improved bounds

$$\|y_n - x^\star\| \leq \frac{2 \, \beta_0 \, \lambda_0^2}{1 - \lambda_0^2} \, \xi_0^{n-1} \quad (n \geq 1) \tag{10.14}$$

where

$$\beta_0 = \frac{\sqrt{1 - 2k^1}}{}, \quad \lambda_0 = \frac{1 - \sqrt{1 - 2k^1} - h^1}{k^1} \quad \text{and} \quad \xi_0 = 1 - \sqrt{1 - 2k^1}, \tag{10.15}$$

hold. In case $\omega_0 = \omega$ (10.14) reduces to (10.11) in [594]. Otherwise our error bounds are finer. Note also that

$$k \leq \frac{1}{2} \Rightarrow k^1 \leq \frac{1}{2} \tag{10.16}$$

but not vice versa unless if $\omega_0 = \omega$.

The above suggest that all results on interior point methods obtained in [594] for Newton's method using (10.5) can now be rewritten using only (10.9). The same holds true for the modified Newton's Method where (10.13) also replaces (10.5).

For example k_1 and k_2 can be replaced by k_1^0, k_2^0 ($k_2^0 < .5$) in the case of Newton's method (10.6) and by k_1^1, k_2^1 ($k_2^1 < .5$) in the case of the modified Newton method (10.7) respectively in all the results in [594] where they appear.

Since $\dfrac{k_1}{k_1^0}, \dfrac{k_2}{k_2^0}, \dfrac{k_1}{k_1^1}, \dfrac{k_2}{k_2^1}$ can be arbitrarily large [137] for a given triplet α, ω and ω_0, the choices

$$k_1^0 = k_1^1 = .12, \quad k_2^0 = k_2^1 = .24 \quad \text{when } k_1 = .21 \text{ and } k_2 = .42$$

and

$$k_1^0 = k_1^1 = .24, \quad k_2^0 = k_2^1 = .48 \ \text{ when } \ k_1 = .245 \text{ and } k_2 = .49$$

are possible. As in [594] denote by N the number of Newton steps, by S the number of the modified Newton steps and by χ the parameter appearing in [594, Corollary 4] or [151, Chapter 10]. Then by using formulas (10.8), (10.9) and (10.10) in [594, Corollary 4 and Theorem 2] we obtain the following tables:

(a) If the HLCP is monotone and only Newton directions are performed, then:

Potra	Argyros
$\chi(.21, .42) > .17$	$\chi^0(.12, .24) > .1$
$\chi(.245, .49) > .199$	$\chi^0(.24, .48) > .196$

Potra	Argyros
$N(.21, .42) = 2$	$N^0(.12, .24) = 1$
$N(.245, .49) = 4$	$N^0(.24, .48) = 3$

(b) If the HLCP is monotone and Modified Newton directions are performed:

Potra	Argyros
$\chi(.21, .42) > .149$	$\chi^1(.12, .24) > .098$
$\chi(.245, .49) > .164$	$\chi^1(.24, .48) > .162$

Potra	Argyros
$S(.21, .42) = 5$	$S^1(.12, .24) = 1$
$S(.245, .49) = 18$	$S^1(.24, .48) = 12$

All the above improvements are obtained under weaker hypotheses and the same computational cost (in the case of Newton's method) or less computational cost (in the case of the modified Newton method) since in practice the computation of ω requires that of ω_0 and in general the computation of ω_0 is less expensive than that of ω.

Chapter 11

Finite Element Methods

In this chapter, we are concerned with the problem of approximating a locally unique solution of the nonlinear equation

$$F(x) = 0, \tag{11.1}$$

where F is a Fréchet–differentiable operator defined on an open convex subset D of a Banach space A with values in a Banach space B.

The famous Newton–Kantorovich theorem has been used to show existence and uniqueness of exact solutions of equation (11.1). Moreover, a priori and a posteriori estimates can be obtained as a direct consequence of the Newton–Kantorovich theorem. A survey of such recent results can be found in [157].

Tsuchiya in [693] used this theorem to show existence of finite element solutions of strongly nonlinear elliptic boundary value problems. However, it is possible that the basic condition in this theorem, the so–called the Newton–Kantorovich hypothesis [477] may be violated and still Newton's method converges [151], [157]. That is why we introduced a weaker hypothesis in [151] which can always replace the Newton–Kantorovich hypothesis and under the same computational cost. This way we can use Newton's method to solve a wider range of problem than before [693]. Moreover, finer estimates on the distances involved and a more precise information on the location of the solution are obtained in [151], [157]. Finally we provide examples of elliptic boundary value problems where our results apply.

First, we state the version of our main result in [157] needed in this chapter.

Theorem 11.1. *Let $F : D \subseteq A \to B$ be a nonlinear Fréchet differentiable operator. Assume:*
there exists a point $x_0 \in D$ such that the Fréchet–derivative $F'(x_0) \in L(A, B)$ is an isomorphism and $F(x_0) \neq 0$;
there exists positive constants ℓ_0 and ℓ such that the following center–Lipschitz and Lipschitz conditions are satisfied:

$$\left\| F'(x_0)^{-1} \left[F'(x) - F'(x_0) \right] \right\| \leq \ell_0 \, \|x - x_0\| \tag{11.2}$$

$$\left\| F'(x_0)^{-1} \left[F'(x) - F'(y) \right] \right\| \leq \ell \, \|x - y\| \tag{11.3}$$

for all $x, y \in D$;

Setting:

$$\eta = \left\| F'(x_0)^{-1} F(x_0) \right\|$$

and

$$h_1 = \frac{1}{4} \left(\ell + 4\,\ell_0 + \sqrt{\ell^2 + 8\,\ell_0\,\ell} \right) \eta,$$

we further assume

$$h_1 \leq 1; \tag{11.4}$$
$$\overline{U}(x_1, t^* - \eta) \subseteq D, \tag{11.5}$$

where, $x_1 = x_0 - F'(x_0)^{-1} F(x_0)$, and t^* a well defined point in $[\eta, 2\eta]$.

Then equation $F(x) = 0$ has a solution $x^* \in \overline{U}(x_1, t^* - \eta)$ and this solution is unique in $U(x_0, t^*) \cap D$, if $\ell_0 = \ell$ and $h_1 < 1$, and $\overline{U}(x_0, t^*) \cap D$, if $\ell_0 = \ell$ and $h_1 = 1$. If $\ell_0 \neq \ell$ the solution x^* is unique in $U(x_0, R)$ provided that $\frac{1}{2}(t^* + R)\ell_0 \leq 1$ and $U(x_0, R) \subseteq D$.

Moreover, we have the estimate

$$\|x^* - x_0\| \leq t^*. \tag{11.6}$$

We will simply use $\|\cdot\|$ if the norm of the element involved is well understood. Otherwise we will use $\|\cdot\|_X$ for the norm on a particular set X.

We assume the following:

(A_1) there exist Banach spaces $Z \subseteq X$ and $U \subseteq Y$ such that the inclusions are continuous, and the restriction of F to Z, denoted again by F, is a Fréchet differentiable operator from Z to U.

(A_2) For any $v \in Z$ the derivative $F'(v) \in L(Z.U)$ can be extended to $F'(v) \in L(X,Y)$ and it is:

* ★ Locally Lipschitz continuous on Z, i.e., for any bounded convex set $T \subseteq Z$ there exists a positive constant c_1 depending on T such that

$$\left\| F'(v) - F'(w) \right\| \leq c_1 \|v - w\|, \quad \text{for all } v.w \in T. \tag{11.7}$$

* ★ center locally Lipschitz continuous at a given $u_0 \in Z$, i.e., for any bounded convex set $T \subseteq Z$ with $u_0 \in T$ there exists a positive constant c_0 depending on u_0 and T such that

$$\left\| F'(v) - F'(u_0) \right\| \leq c_0 \|v - u_0\|, \quad \text{for all } v \in T. \tag{11.8}$$

(A_3) There are Banach spaces $V \subseteq Z$ and $W \subseteq U$ such that the inclusions are continuous. We suppose that there exists a subset $S \subseteq V$ for which the following holds: "if $F'(u) \in L(V,W)$ is an isomorphism between V and W at $u \in S$, then $F'(u) \in L(X,Y)$ is an isomporhism between X and Y as well".

To define discretized solutions of $F(u) = 0$, we introduce the finite dimensional subspaces $S_d \subseteq Z$ and $S_d \subseteq U$ parametrized by d, $0 < d < 1$ with the following properties:

(A_4) There exists $r \geq 0$ and a positive constant c_2 independent of d such that

$$\|v_d\|_Z \leq \frac{c_2}{d^r} \|v_d\|_X, \quad \text{for all } v_d \in S_d. \tag{11.9}$$

Chapter 11. Finite Element Methods   161

(A5) There exists projection $\Pi_d : X \to S_d$ for each S_d such that, if $u_0 \in S$ is a solution of $F(u) = 0$, then

$$\lim_{d \to 0} d^{-r} \|u_0 - \Pi_d u_0\|_X = 0 \tag{11.10}$$

and

$$\lim_{d \to 0} d^{-r} \|u_0 - \Pi_d u_0\|_Z = 0. \tag{11.11}$$

We can show the following result concerning the existence of locally unique solutions of discretized equations.

Theorem 11.2. *Assume that conditions* (A_1)–(A_5) *hold. Suppose* $F'(u_0) \in L(V,W)$ *is an isomorphism, and* $u_0 \in S$. *Moreover, assume* $F'(u_0)$ *can be decomposed into* $F'(u_0) = Q + R$, *where* $Q \in L(X,Y)$ *and* $R \in L(X,Y)$ *is compact. The discretized nonlinear operator* $F_d : Z \to U$ *is defined by*

$$F_d(u) = (I - P_d)Q(u) + P_d F(u) \tag{11.12}$$

where I is the identity of Y, and $P_d : Y \to S_d$ *is a projection such that*

$$\lim_{d \to 0} \|v - P_d v\|_Y = 0, \ \text{for all } v \in Y, \tag{11.13}$$

and

$$(I - P_d)Q(v_d) = 0, \ \text{for all } v_d \in S_d. \tag{11.14}$$

Then, for sufficiently small $d > 0$, *there exists* $u_d \in S_d$ *such that* $F_d(u_d) = 0$, *and* u_d *is locally unique.*

Moreover the following estimate holds

$$\|u_d - \Pi_d(u_0)\| \leq \ell_1 \|u_0 - \Pi_d(u_0)\| \tag{11.15}$$

where ℓ_1 *is a positive constant independent of d.*

Proof. The proof is similar to the corresponding one in [693, Th. 2.1, p. 126]. However, there are some crucial differences where weaker (11.8) is used (needed) instead of stronger condition (11.7).

Step 1. We claim that there exists a positive constant c_3, independent of d, such that, for sufficiently small $h > 0$,

$$\|F_d'(\Pi_d(u_0))v_d\|_Y \geq c_3 \|v_d\|_X, \ \text{for all } v_d \in S_d. \tag{11.16}$$

From (A_3) and $u_0 \in S$, $F'(u_0) \in L(X,Y)$ is an isomorphism. Set $B_0 = \left\|F'(u_0)^{-1}\right\|$.

We can have in turn

$$F_d'(\Pi_d(u_0))v_d = F'(u_0)v_d + P_d\left(F'(\Pi_d(u_0)) - F'(u_0)\right)v_d \tag{11.17}$$
$$- (I - P_d)\left(-Q + F'(u_0)\right)v_d.$$

Since $-Q + F'(u_0) \in L(X,Y)$ is compact we get by (11.13) that

$$\lim_{d \to 0} \left\|(I - P_d)\left(-Q + F'(u_0)\right)\right\| = 0. \tag{11.18}$$

By (11.13) there exists a positive constant c_4 such that

$$\sup_{d>0} \|P_d\| \leq c_4. \tag{11.19}$$

That is, using (11.8) we get

$$\left\|P_d \left(F'\left(\Pi_d\left(u_0\right)\right) - F'\left(u_0\right)\right)\right\| \leq c_0 c_4 \left\|\Pi_d\left(u_0\right) - u_0\right\|. \tag{11.20}$$

Hence, by (11.11) we can have

$$\left\|F_d'\left(\Pi_d\left(u_0\right)\right) v_d\right\| \geq \left(\tfrac{1}{B_0} - \delta\left(d\right)\right) \|v_d\|, \tag{11.21}$$

where $\lim_{d\to 0} \delta\left(d\right) = 0$, and (11.16) holds with $c_3 = \dfrac{B_0^{-1}}{2}$.

Step 2. We shall show:

$$\lim_{d\to 0} d^{-r} \left\|F_d'\left(\Pi_d\left(u_0\right)\right)^{-1} F_d\left(\Pi_d\left(u_0\right)\right)\right\| = 0. \tag{11.22}$$

Note that

$$\begin{aligned}
\left\|F_d\left(\Pi_d\left(u_0\right)\right)\right\| &\leq c_4 \left\|F_d\left(\Pi_d\left(u_0\right)\right) - F_d\left(u_0\right)\right\| \\
&\leq c_4 \int_0^1 \|G_t\| \, dt \left\|\Pi_d\left(u_0\right) - u_0\right\| \\
&\leq c_4 c_5 \left\|\Pi_d\left(u_0\right) - u_0\right\|,
\end{aligned} \tag{11.23}$$

where

$$G_t = F'\left((1-t)u_0 + t\Pi_d\left(u_0\right)\right) \tag{11.24}$$

and we used

$$\begin{aligned}
\|G_t\| &\leq \left\|G_t - F'\left(u_0\right)\right\| + \left\|F'\left(u_0\right)\right\| \\
&\leq c_0 t \left\|\Pi_d\left(u_0\right) - u_0\right\| + \left\|F'\left(u_0\right)\right\| \leq c_5
\end{aligned} \tag{11.25}$$

where c_5 is independent of d.

The claim is proved.

Step 3. We use our modification of the Newton–Kantorovich theorem with the following choices:

$A = S_d \subseteq Z$, with norm $d^{-r} \|w_d\|_X$,

$B = S_d \subseteq U$ with norm $d^{-r} \|w_d\|_Y$,

$x_0 = \Pi_d\left(u_0\right)$,

$F = F_d$.

Notice that $\|S\|_{L(A,B)} = \|S\|_{L(X,Y)}$ for any linear operator $S \in L\left(S_d, S_d\right)$.

By Step 1, we know $F_d'\left(\Pi_d\left(u_0\right)\right) \in L\left(S_d, S_d\right)$ is an isomorphism.

It follows from (11.7) and (A_4) that for any $w_d, v_d \in S_d$,

$$\begin{aligned}
\left\|F_d'\left(w_d\right) - F_d'\left(v_d\right)\right\| &\leq c_1 c_4 \left\|w_d - v_d\right\|_Z \\
&\leq c_1 c_2 c_4 d^{-r} \left\|w_d - v_d\right\|_X
\end{aligned} \tag{11.26}$$

Chapter 11. Finite Element Methods

Similarly, we get using (11.8) and (A$_4$) that

$$\left\| F'_d(w_d) - F'_d(\Pi_d(u_0)) \right\| \leq c_1 c_2 c_4 d^{-r} \| w_d - x_0 \|_X.$$

Hence assumptions are satisfied with

$$\ell = c_1 c_2 c_3^{-1} c_4 \quad \text{and} \quad \ell_0 = c_0 c_2 c_3^{-1} c_4. \tag{11.27}$$

From Step 2, we may take sufficiently small $d > 0$ such that $(\ell_0 + \ell)\eta \leq 1$, where

$$\eta = d^{-r} \left\| F'_d(\Pi_d(u_0))^{-1} F_d(\Pi_d(u_0)) \right\|_X.$$

That is, assumption $h_1 \leq 1$ is satisfied.

Hence for sufficiently small $d > 0$ there exists a locally unique $u_d \in S_d$ such that $F_d(u_d) = 0$ and

$$\| u_d - \Pi_d(u_0) \|_X \leq 2d^r \eta \leq 2c_3^{-1} \| F_d(\Pi_d(u_0)) \|_Y$$
$$\leq 2c_3^{-1} c_4 c_5 \| u_0 - \Pi_d(u_0) \|_X.$$

It follows (11.15) holds with $\ell_1 = 2c_3^{-1} c_4 c_5$.

That completes the proof of the Theorem. $\diamond$

Remark 11.1. In general

$$c_0 \leq c_1 \quad (\text{i.e., } \ell_0 \leq \ell) \tag{11.28}$$

holds and $\dfrac{\ell}{\ell_0}$ can be arbitrarily large, where ℓ and ℓ_0 are given by (11.27).

If $\ell = \ell_0$ our Theorem 11.2 reduces to the corresponding Theorem 2.1 in [693, p. 126].

Otherwise our condition $h_1 \leq 1$ is weaker than the corresponding one in [693] using the Newton–Kantorovich hypothesis $h = 2\,\ell\,\eta \leq 1$.

That is,

$$h \leq 1 \implies h_1 \leq 1$$

but not vice versa.

As already shown in [157], finer error estimates on the distances $\| u_d - \Pi_d(u_0) \|$ and a more precise information on the location of the solution are provided here and under the same computational cost since in practice the evaluation of c_1 requires that of c_0.

Note also that our parameter d will be smaller than the corresponding one in [693] which in turn implies fewer computations and smaller dimension subspaces S_d are used to approximate u_d. This observation is very important in computational mathematics.

The above observations suggest that all results obtained in [693] can be improved if rewritten with weaker $h_1 \leq 1$ instead of stronger $h \leq 1$.

However we do not attempt this here (leaving this task to the motivated reader). Instead we provide examples of nonlinear problems already reported in [693] where finite element methods apply along the lines of our theorem above.

164 — Ioannis K. Argyros, Saïd Hilout and Mohammad A. Tabatabai

Example 11.1. Find $u \in H_0^1(J)$, $J = (b,c) \subseteq \mathbb{R}$ such that

$$\langle F(u), v \rangle = \int_J \left[g_0\left(x, u, u'\right) v' + g\left(x, u, u'\right) v \right] dx = 0, \text{ for all } v \in H_0^1(J) \qquad (11.29)$$

where g_0 and g_1 are sufficiently smooth functions from $J \times \mathbb{R} \times \mathbb{R}$ to $\mathbb{R}$.

Example 11.2. For the N–dimensional case $(N = 2, 3)$ let $D \subseteq \mathbb{R}^N$ be a bounded domain with a Lipschitz boundary. Then consider the problem: find $u \in H_0^1(D)$ such that

$$\langle F(u), v \rangle = \int_D \left[q_0\left(x, u, \nabla u\right) \cdot \nabla v + q\left(x, u, \nabla u\right) \cdot v \right] dx = 0,$$
$$\text{for all } v \in H_0^1(D), \qquad (11.30)$$

where $q_0 \in D \times \mathbb{R} \times \mathbb{R}^N$ to $\mathbb{R}$ are sufficiently smooth functions.

Example 11.3. Since equations (11.29) and (11.30) are defined in divergence form, their finite element solutions are defined in a natural way. Finite element methods applied to nonlinear elliptic boundary value problems have also been considered by other authors [382], [604].

Chapter 12

Convergence in Riemannian Manifolds

In this chapter, we provide a finer semilocal convergence analysis of Newton's method in Riemannian manifolds Using more precise majorizing sequences than before [14], [383], and under the same computational cost. We obtain the following advantages: larger convergence domain, finer error bounds on the distances involved, and a more precise information on the location of the singularity of the vector field. We refer the reader to [14], [151], [352] for some of the concepts introduced but not detailed here.

Let X be a C^1 vector field defined on a connected, complete and finite–dimensional Riemannian manifold (M, g). In this chapter, we are concerned with the following problem:

$$\text{find } p^* \in M \text{ such that } X(p^*) = o \in T_{p^*}M. \tag{12.1}$$

A point p^* satisfying (12.1) is called a singularity of X.

The most popular method for generating a sequence $\{p_n\}$ $(n \geq 0)$ approximating p^* is undoubtedly Newton's method, described here as follows:

Assume there exists an initial guess $p_0 \in M$ such that the covariant $X'(p_0)$ of X at p_0 given by

$$X'(p)v := \nabla_v X(p) = (\nabla_y X)(p), \quad v \in T_p M, \tag{12.2}$$

is invertible, at $p = p_0$, for each pair of continuously differentiable vector fields X, Y where the vector field $\nabla_y X$ stands for the covariant derivative of X with respect to Y.

Define the Riemannian–Newton method by

$$p_{n+1} = \exp_{p_n}[-X'(p_n)^{-1}X(p_n)], \tag{12.3}$$

where $\exp_p : T_p M \to M$ is the exponential map at p. A survey of local as well as semilocal convergence results for method (12.3) can be found in [14] and the references there.

We are motivated in this chapter by optimization considerations and the recent excellent study [14] of the Riemannian analogue of the property used by Zabreijko and Nguen [764]. All the above advantages are achieved because we use more precise error estimates on the distances involved than in [14] along the lines of our relevant works for Newton's mehtod for solving nonlinear equations on Banach spaces [137], [157]. We cover the local and the semilocal convergence of method (12.3) respectively.

166   Ioannis K. Argyros, Saïd Hilout and Mohammad A. Tabatabai

We need the definition [14], [151]:

Definition 12.1. Let $G_2(p_0, r)$ denote the class of all the piecewise geodesic curves $c \colon [0, T] \to M$ which satisfy:

(a) $c(0) = p_0$ and the length of c is no greater than r;

(b) there exists $\tau \in (0, T)$ such that $c|_{[0,\tau]}$ is a minimizing geodesic and $c|_{[\tau,T]}$ is a geodesic.

We can now introduce a Lipschitz as well as a center–Lipschitz–type continuity of X':

Let $R > 0$. We suppose there exist continuous and nondecreasing functions $\ell_0, \ell \colon [0, R] \to [0, +\infty)$ such that: for every $r \in [0, R]$ and $c \in G_2(p_0, r)$,

$$\left\| X'(p_0)^{-1} \left[P_{c,b,0} X'(c(b)) - P_{c,0,0} X'(c(0)) \right] \right\|_{p_0} \leq \ell_0(r) \int_0^b |\dot{c}|, \ 0 \leq b, \quad (12.4)$$

$$\left\| X'(p_0)^{-1} \left[P_{c,b,0} X'(c(b)) - P_{c,a,0} X'(c(a)) \right] \right\|_{p_0} \leq \ell(r) \int_a^b |\dot{c}|,$$
$$0 \leq a \leq b, \qquad (12.5)$$

where

$$P_{c,t,0} T(c(t)) = T(c(0)) + \int_0^t \left[P_{c,s,0} T'(c(s)) \dot{c}(s) \right] ds. \qquad (12.6)$$

Without loss of generality we assume $\ell_0(r) > 0$, $\ell(r) > 0$ on $(0, R]$.

Remark 12.1. In general

$$\ell_0(r) \leq \ell(r) \quad r \in [0, R] \qquad (12.7)$$

holds and $\dfrac{\ell(r)}{\ell_0(r)}$ can be arbitrarily large.

It is convenient to define parameter

$$\eta = |X'(p_0)^{-1} X(p_0)|_{p_0}, \qquad (12.8)$$

functions $v, w \colon [0, R] \to [-\infty, +\infty)$ by

$$w(r) = \eta - r + \int_0^r (r - s) \ell(s) ds, \qquad (12.9)$$

$$v(r) = -1 + \ell_0(r) \qquad (12.10)$$

and iterations $\{t_n\}$, $\{r_n\}$ $(n \geq 0)$ by

$$t_0 = 0, \ t_1 = \eta, \ t_{n+1} = t_n - \frac{w(t_n)}{v(t_n)}, \qquad (12.11)$$

$$r_0 = 0, \ r_{n+1} = r_n - \frac{w(r_n)}{w'(r_n)} \qquad (12.12)$$

for all $n \geq 0$.

Let us consider the assumption:
the function w given by (12.9) has a unique zero r^* in $[0, R]$ with

$$w(R) \leq 0. \qquad (12.13)$$

We showed in [137] (see also [157], [352]):

Chapter 12. Convergence in Riemannian Manifolds 167

Proposition 12.1. *Under hypothesis* (12.13) *iterations* $\{t_n\}$, $\{s_n\}$ $(n \geq 0)$ *are well defined, monotonically increasing and convergent to* t^*, r^* *respectively with*

$$t^* \leq r^*. \tag{12.14}$$

Moreover the following estimates hold for all $n \geq 0$

$$t_n \;\leq\; r_n, \tag{12.15}$$
$$t_{n+1} - t_n \;\leq\; r_{n+1} - r_n, \tag{12.16}$$

and

$$t^* - t_n \leq r^* - r_n. \tag{12.17}$$

Furthermore if (12.7) *holds as a strict inequality so do* (12.15) *and* (12.16) *for* $n \geq 1$. *Since we shall show both* $\{t_n\}$, $\{r_n\}$ *are majorizing sequences for* $\{p_n\}$, *it follows by Proposition* 12.1 *that the claims made in the introduction for the local convergence of method* (12.3) *hold true.*

We can now state the main local convergence result for method (12.3) which improves the corresponding Theorem 3.1 in [14]:

Theorem 12.1. *Under hypotheses* (12.4), (12.5) *and* (12.13) *the following hold true:*

(a) *the vector field* X *admits a unique singularity* p^* *in* $\overline{U}(p_0, R) = \{p \in X \mid \|p - p_0\| \leq R\}$ *which belongs to* $\overline{U}(p_0, t^*)$. *If* $\ell_0(t^*) < 0$ *then* $X'(p^*) \in GL(T_{p^*}M)$.

(b) *Sequence* $\{p_n\}$ $(n \geq 0)$ *generated by method* (12.3) *is well defined,* $p_n \in \overline{U}(p_0, t_n)$ *for all* $n \geq 0$, *and* $\lim_{n \to \infty} p_n = p^*$.

(c) *The following estimates hold true for all* $n \geq 0$:

$$d(p_{n+1}, p_n) \leq |X'(p_n)^{-1} X(p_n)|_{p_n} \leq t_{n+1} - t_n \leq r_{n+1} - r_n, \tag{12.18}$$

and

$$d(p_n, p^*) \leq t^* - t_n \leq r^* - r_n. \tag{12.19}$$

Proof. Uses the more accurate (i.e. the one really needed) condition (12.4) instead of condition (12.5) used in [14] for the computation of the inverses $X'(p_n)^{-1}$ $(n \geq 0)$. The rest of the proof is identical to [14] and is omitted. $\diamondsuit$

Remark 12.2. (a) If $\ell_0(r) = \ell(r)$ our Theorem 12.1 reduces to Theorem 3.1 in [14]. Otherwise (see (12.7)) it is an improvement over it as already shown in Proposition 12.1. Note also that the rest of the bounds obtained in Theorem 3.1 in [14] (see parts (iv)-(v) of this Theorem hold true with $\{t_n\}$ replacing sequence $\{r_n\}$ there but we decided not to include those bounds here to avoid repetitions.

(b) Theorem 12.1 remains valid for a C^1 vector field $X : D \subset M \to TM$ which is defined only on an open subset D of M provided $\overline{U}(p_0, R) \subseteq D$ [14], [151], [352].

(c) It follows from the proof of the theorem that the sharper (than $\{t_n\}$) scalar $\{\bar{t}_n\}$ $(n \geq 0)$ given by

$$\bar{t}_0 = 0, \quad \bar{t}_1 = \eta,$$

$$\bar{t}_{n+2} = \bar{t}_{n+1} + \frac{\displaystyle\int_0^1 \ell(\bar{t}_n + t(\bar{t}_{n+1} - \bar{t}_n))(\bar{t}_{n+1} - \bar{t}_n)dt}{1 - \ell_0(t_{n+1})} \quad (n \geq 0), \tag{12.20}$$

is also a majorizing sequence for $\{p_n\}$ $(n \geq 0)$.

Sufficient convergence conditions for sequence (12.20) which are weaker than (12.13) have already been given in [137], [352]. Note that

$$\bar{t}_n \quad \leq \quad t_n, \tag{12.21}$$
$$\bar{t}_{n+1} - \bar{t}_n \quad \leq \quad t_{n+1} - t_n, \tag{12.22}$$
$$\bar{t}^* - \bar{t}_n \quad \leq \quad t^* - t_n, \tag{12.23}$$

and

$$\bar{t}^* \leq t^*. \tag{12.24}$$

See also Remark 12.3.

An extension of the Newton–Kantorovich theorem to finite–dimensional and complete Riemannian manifolds has been given by Ferreira and Svaiter in [352] and has been improved by us in [144]. Moreover the results in [352] were shown to be special cases of Theorem 3.1 in [14]. Here we weaken these results using L–Lipschitz and L_0–center Lipschitz conditions for tensors.

Definition 12.2. A $(1,k)$–tensor T on M is said to be: L_0–center Lipschitz continuous on a subset S of M, if for all geodesic curve $\gamma \colon [0,1] \to M$ with endpoints in S and $p_0 \in M$

$$\|P_{\gamma,1,0}T(\gamma(1)) - T(p_0)\|_{p_0} \leq L_0 \int_0^1 |\dot{\gamma}(t)|dt, \tag{12.25}$$

and L–Lipschitz continuous on S, if

$$\|P_{\gamma,1,0}T(\gamma(1)) - T(\gamma(0))\|_{\gamma(0)} \leq L \int_0^1 |\dot{\gamma}(t)|dt. \tag{12.26}$$

We can show the following improvement of Theorem 5.1 in [14] for the semilocal convergence of method (12.3) (see [137, page 387, Case 3, for $\delta = \delta_0$]):

Theorem 12.2. *Under hypotheses (12.25) and (12.26) on $S = \overline{U}(x_0, R)$ further suppose there exists $p_0 \in M$ such that $X'(p_0) \in GL(T_{p_0}M)$.*
Set:

$$a = \|X'(p_0)^{-1}\|_{p_0}$$

and

$$\bar{L} = \frac{1}{8}\left(L + 4L_0 + \sqrt{L^2 + 8L_0L}\right).$$

Assume:

$$h_0 = a\,\eta\,\overline{L} \le \frac{1}{2}, \tag{12.27}$$

$$\overline{U}(p_0, s^*) \subseteq \overline{U}(p_0, R), \tag{12.28}$$

where,

$$s^* = \lim_{n\to\infty} s_n \le b\,\eta, \quad b = \frac{2}{2-b_0}, \quad b_0 = \frac{1}{2}\left[-\frac{L}{L_0} + \sqrt{\left(\frac{L}{L_0}\right)^2 + 8\frac{L}{L_0}}\,\right], \tag{12.29}$$

$$s_0 = 0, \quad s_1 = \eta, \quad s_{n+2} = s_{n+1} + \frac{a\,L(s_{n+1}-s_n)^2}{2\,(1-L_0 s_{n+1})} \quad (n \ge 0). \tag{12.30}$$

Then

(a) *scalar sequence $\{s_n\}$ generated by (12.30) is monotonically increasing, bounded above by $b\,\eta$ and converges to $s^* \in [\eta, b\,\eta]$.*

(b) *Sequence $\{p_n\}$ $(n \ge 0)$ generated by method (12.3) is well defined, remains in $\overline{U}(p_0, s^*)$ for all $n \ge 0$ and converges to a unique singularity of X in $\overline{U}(p_0, s^*)$.*

(c) *The following error bounds hold true for all $n \ge 0$:*

$$d(p_{n+1}, p_n) \le s_{n+1} - s_n, \tag{12.31}$$

and

$$d(p_n, p^*) \le s^* - p_n. \tag{12.32}$$

(d) *If there exists $R_1 \in [s^*, R]$ such that*

$$aL_0(s^* + R_1) \le 2, \tag{12.33}$$

then p^* is unique in $U(p_0, R_1)$.

Proof. As the proof in Theorem 5.1 in [14, p. 18] but we use condition (12.25) for the computation of the upper bounds of $X'(p_n)^{-1}$ $(n \ge 0)$ which is really needed instead of (12.26) used in [14]. $\diamond$

Remark 12.3. As in Remark 12.1

$$L_0 \le L \tag{12.34}$$

holds in general and $\dfrac{L}{L_0}$ can be arbitrarily large.

If $L_0 = L$ holds then our theorem reduces to Theorem 5.1 in [14]. Otherwise it is an improvement. Indeed the condition corresponding to (12.27) in [14] is given by

$$h = a\,\eta\,L \le \frac{1}{2}. \tag{12.35}$$

Note that

$$h \le \frac{1}{2} \Rightarrow h_0 \le \frac{1}{2} \tag{12.36}$$

but not vice versa.

Moreover the corresponding majorizing sequence is given by

$$z_0 = 0, \ z_{n+1} = z_n - \frac{w_1(z_n)}{w_1'(z_n)},$$ (12.37)

where,

$$w_1(r) = \eta - r + \frac{aLr^2}{2},$$ (12.38)

and again

$$s_n \leq z_n,$$ (12.39)

$$s_{n+1} - s_n \leq z_{n+1} - z_n,$$ (12.40)

$$s^* - s_n \leq z^* - z_n,$$ (12.41)

and

$$s^* \leq z^* = \lim_{n \to \infty} z_n = \frac{1 - \sqrt{1 - 2h}}{aL}$$ (12.42)

with (12.39), (12.40) holding as strict inequalities for $n \geq 1$ if $L_0 < L$. Finally note that in [137], we provided sufficient convergence conditions for iteration (12.30) that are even weaker than (12.27).

All the above justify the advantages of our approach already stated in the introduction of this chapter. These ideas can be used to improve the rest of the results stated in [14]. However we leave the details for the motivated reader.

Chapter 13

Convergence on Lie Groups

In this chapter, we are concerned with the problem of approximating a locally unique zero x^* of a map f defined on a Lie group. Numerical algorithms on manifolds are very important in computational mathematics [140], [352], [383], [720], [570], because they appear in connection to eigenvalue problems, minimization problems, optimization problems. A convergence analysis of Newton's method on Riemannian manifolds under various condition similar to the corresponding ones on Banach spaces [137], [157], [411], [477] has been given in [720], [570] and the references there. Here we are motivated in particular by the elegant work in [720], and optimization considerations with advantages over earlier works [383], [720], [570].

A Lie group $(G, \cdot)$ is a Hausdorff topological group with countable bases which also has the structure of a smooth manifold such that the group product and the inversion are smooth operations in the differentiable structure given on the manifold. The dimension of a Lie group is that of the underlying manifold, and we shall always assume that it is finite. The symbol e designates the identity element of G. Let g be the Lie algebra of the Lie group G which is the tangent space T_eG of G at e, equipped with Lie bracket $[\cdot, \cdot] : g \times g \to g$. In the sequel we will make use of the left translation of the Lie group G. We define for each $y \in G$

$$L_y : G \to G$$
$$z \to y \cdot z,$$

the left multiplication in the group. The differential of L_y at e denoted by $(dL_y)_e$ determines an isomorphism of $g = T_eG$ with the tangent space T_yG via the relation

$$(dL_y)_e(g) = T_yG$$

or, equivalently,

$$g = (dL_y)_e^{-1}(T_yG) = (dL_{y-1})_y(T_yG).$$

The exponential map is a map

$$\exp : g \to G$$
$$u \to \exp(u),$$

which is certainly the most important construct associated to G and g. Given $u \in g$, the left invariant vector field $X_u : y \to (dL_y)_e(u)$ determines an one–parameter subgroup of G $\sigma_u : \mathbb{R} \to G$ such that $\sigma_u(0) = e$ and

$$\sigma'_u(t) = X_u(\sigma_u(t)) = (dL_{\sigma_u(t)})_e(u).$$

The exponential map is then defined by the relation

$$\exp(u) = \sigma_u(1).$$

Note that the exponential map is not surjective in general. However, the exponential map is a diffeomorphism on an open neighborhood $\mathcal{N}(l)$ of $0 \in g$. Let

$$N(e) = \exp(\mathcal{N}(0)).$$

Then for each $y \in N(e)$, there exists $\upsilon \in \mathcal{N}(0)$ such that $y = \exp(\upsilon)$. Furthermore, if

$$\exp(u) = \exp(\upsilon) \in N(e)$$

for some $u, \upsilon \in \mathcal{N}(0)$, then $u = \upsilon$. If G is Abelian, exp is also a homomorphism from g to G, i.e.,

$$\exp(u + \upsilon) = \exp(u) \cdot \exp(\upsilon) \tag{13.1}$$

for all $u, \upsilon \in g = T_e G$. In the non–abelian case, exp is not a homomorphism and (13.1) must be replaced by

$$\exp(\omega) = \exp(u) \cdot \exp(\upsilon),$$

where ω is given by the Baker–Campbell–Hausdorff (BCH) formula

$$\omega = u + \upsilon + \frac{1}{2}[u, \upsilon] + \frac{1}{12}([u[u, \upsilon]] + [\upsilon[\upsilon, u]]) + \cdots,$$

for all u, υ in an open neighborhood of $0 \in g$. To analyse convergence, we need a Riemannian metric on the Lie group G.

Following [352] take an inner product $\langle, \rangle_e$ on g and define

$$\langle u, \upsilon \rangle_x = \langle (dL_{x-1})_x(u), (dL_{x-1})_x(\upsilon) \rangle_e, \quad \text{for each } x \in G \quad \text{and} \quad u, \upsilon \in T_x G.$$

This construction actually produces a Riemannian metric on the Lie group G, see for example [352]. Let $\|\cdot\|_x$ be associated norm, where the subscript x is sometimes omitted if there is no confusion. For any two distinct elements $x, y \in G$, let $c : [0, 1] \to G$ be a piecewise smooth curve connecting X and y. Then the arc–length of c is defined by $l(c) := \int_0^1 \|c'(t)\| dt$, and the distance from x to y by $d(x, y) := \inf_c l(c)$, where the infimum is taken over all piecewise smooth curves $c : [0, 1 \to G$ connecting x and y. Thus, we assume throughout the whole chapter that G is connected and hence (G, d) is a complete metric space. Since we only deal with finite dimensional Lie algebras, every linear mapping $\varphi : g \to g$ is bounded and we define its norm by

$$\|\varphi\| = \sup_{u \neq 0} \frac{\|\varphi(u)\|}{\|u\|} = \sup_{\|u\|=1} \|\varphi(u)\| < \infty.$$

Chapter 13. Convergence on Lie Groups

173

For $r > 0$ we introduced the corresponding ball of radius r around $y \in G$ defined by one parameter subgroups of G as

$$C_r(y) = \{z \in G : z = y \cdot \exp(u), \|u\| \leq r\}.$$

We give the following definition on convergence:

Definition 13.1. Let $\{x_n\}_{n\geq 0}$ be a sequence of G and $x \in G$. Then $\{x_n\}_{n\geq 0}$ is said to be

(i) convergent to x if for any $\varepsilon > 0$ there exists a natural number K such that $x^{-1} \cdot x_n \in N(e)$ and $\|\exp^{-1}(x^{-1} \cdot x_n)\| \leq \varepsilon$ for all $n \geq K$;

(ii) quadratically convergent to x if $\{\|\exp^{-1}(x^{-1} \cdot x_n)\|\}$ is quadratically convergent to 0; that is, $\{x_n\}_{n\geq 0}$ is convergent to x and there exists a constant q and an natural number K such that

$$\left\|\exp^{-1}(x^{-1}) \cdot x_{n+1})\right\| \leq q \left\|\exp^{-1}(x^{-1} \cdot x_n)\right\|^2 \quad \text{for all} \quad n \geq K.$$

Note that convergence of a sequence $\{x_n\}_{n\geq 0}$ in G to x in the sense of Definition 13.1 above is equivalent to that $\lim_{n\to+\infty} d(x_n, x) = 0$.

In the remainder of this chapter, let $f : G \to g = T_e G$ be a C^1–mapping. The differential of f at a point $x \in G$ is a linear map $f'_x : T_x G \to g$ defined by

$$f'_x(\triangle_x) = \frac{d}{dt} f(x \cdot \exp(t((d_{x-1})_x)(\triangle_x)))|_{t=0} \text{ for any } \triangle_x \in T_x G. \tag{13.2}$$

The differential f'_x can be expressed via a function $df_x : g \to g$ given by

$$df_x = (f \circ L_x)'_e = f'_x \circ (dL_x)_e.$$

Thus, by (13.8), it follows that

$$df_x(u) = f'_x((dL_x)_e(u)) = \frac{d}{dt} f(x \cdot \exp(tu))|_{t=0} \text{ for any } u \in g.$$

Therefore the following lemma is clear.

Lemma 13.1. *Let $x \in G$, $u \in g$ and $t \in \mathbb{R}$. Then*

$$\frac{d}{dt} f(x \cdot \exp(-tu)) = -df_{x \cdot \exp(-tu)}(u) \tag{13.3}$$

and

$$f(x \cdot \exp(tu)) - f(x) = \int_0^t df_{x \cdot \exp(su)}(u) ds. \tag{13.4}$$

As in [570] Newton's method for f with initial point $x_0 \in G$ is defined as follows

$$x_{n+1} = x_n \cdot \exp(-df_{x_n}^{-1} \circ f(x_n)) \quad (n \geq 0). \tag{13.5}$$

We also define the modified Newton's method by

$$x_{n+1} = x_n \cdot \exp(-df_{x_0}^{-1} \circ f(x_n)) \quad (n \geq 0). \tag{13.6}$$

We will use the following definition involving Lipschitz conditions:

Definition 13.2. Let $r > 0$, and let $x_0 \in G$ be such that $df_{x_0}^{-1}$ exists. Then $df_{x_0}^{-1}df$ is said to satisfy: the center–Lipschitz condition with constant $\ell_0 > 0$ in $C(x_0, r)$ if

$$\left\| df_{x_0}^{-1}(df_{x_0 \exp(u)} - df_{x_0}) \right\| \leq \ell_0 \|u\|, \quad \text{for each } u \in g \text{ with } \|u\| \leq r; \qquad (13.7)$$

the center Lipschitz condition with constant ℓ in $C(x_0, r)$ if

$$\left\| df_{x_0}^{-1}(df_{x \cdot \exp(u)} - df_x) \right\| \leq \ell \|u\| \qquad (13.8)$$

holds for any $u, v \in g$ and $x = x_0 \exp(\tau)$ with $\|u\| + \|v\| \leq r$.

Remark 13.1. In general

$$\ell_0 \leq \ell, \qquad (13.9)$$

holds, and $\dfrac{\ell}{\ell_0}$ can be arbitrarily large.

We can show the following local convergence results for Newton's method (13.5).

Theorem 13.1. *Assume that G is an Abelian group. Choose $r \in \left(0, \dfrac{2}{2\ell_0 + \ell} \right)$, and let $x^* \in G$ such that $f(x^*) = 0$ and $df_{x^*}^{-1}$ exists. Moreover assume $df_{x^*}^{-1}df$ satisfies condition (13.8).*
Then sequence $\{x_n\}$ generated by Newton's method (13.5) is well defined, remains in $C(x^*, r)$ for all $n \geq 0$, and converges quadratically to x^* provided that $x_0 \in C(x^*, r)$, with ratio α given by

$$\alpha = \frac{\ell}{2(1 - \ell_0 \|u_0\|)}, \qquad (13.10)$$

where $u_0 \in g$ with $\|u_0\| \leq r$, and $x_0 = x^* \exp(u_0)$.

Proof. Set $\alpha_0 = \alpha \|u_0\|$. In view of (13.10) $\alpha_0 \in [0, 1)$. We shall show using induction that for each $n \geq 0$, x_n is well-defined, remains in $C(x^*, r)$, and there exists $u_n \in g$ with $\|u_n\| \leq r$ such that

$$x_n = x^* \exp(u_n), \quad \text{and} \quad \|u_{n+1}\| \leq \alpha \|u_n\|^2 \leq \alpha_0^{2^{n+1}-1} \|v_0\|. \qquad (13.11)$$

Estimates (13.11) hold true for $n = 0$ by the initial conditions. Assume estimates (13.11) hold true for $n \leq k$, x_n is well defined and there exist $U_n \in g$ with $\|u_n\| \leq r$ such that (13.11) hold. Using (13.7) we get

$$\left\| df_{x^*}^{-1}(df_{x^* \exp(u_k)} - df_{x^*}) \right\| \leq \ell_0 \|u_k\| < 1. \qquad (13.12)$$

It follows from the Banach Lemma that $df_{x_k}^{-1}$ exists and

$$\left\| df_{x_k}^{-1}df_{x^*} \right\| \leq \frac{1}{1 - \ell_0 \|u_k\|}. \qquad (13.13)$$

That is x_{k+1} is well defined. Set

$$u_{k+1} = u_k - df_{x_k}^{-1}(f(x_k)). \qquad (13.14)$$

In view of (13.13) and (13.14) we get in turn

$$\begin{aligned}\|u_{k+1}\| &= \left\|u_k - df_{x_k}^{-1}(f(x_k) - f(x^*))\right\| \\ &\leq \|df_{x_k}^{-1} df_{x^*}\| \int_0^1 \left\|df_{x^*}^{-1}(df_{x_k} - df_{x^* \exp(tu_k)})u_k\right\| dt \\ &\leq \frac{1}{1-\ell_0\|u_k\|} \int_0^1 (1-t)l\|u_k\|^2 dt \\ &= \frac{2}{2(1-\ell_0\|u_k\|)}\|u_k\|^2 \leq \alpha\|u_k\|^2 \leq \alpha_0^{2^{k+1}-1}\|u_0\| = r,\end{aligned}$$ (13.15)

which establishes the quadratic convergence and $\|u_{k+1}\| \leq \tau$. Moreover since G is an Abelian group

$$x_{k+1} = x^* \exp(u_k) \exp[-df_{x_k}^{-1} f(x_k)] = x^* \exp(u_{k+1}).$$ (13.16)

That completes the induction and the proof of the theorem.

Remark 13.2. If $\ell_0 = \ell$, Theorem 13.1 reduces to Theorem 3.1 in [720]. Otherwise it is an improvement. Indeed let τ_{WL} be the corresponding radius of convergence in [720] selected in $(0, \frac{2}{3l})$. Then since $\left(0, \frac{2}{3l}\right) \subseteq \left(0, \frac{2}{2\ell_0 + l}\right)$ it follows our radius r_A is such that

$$r_{WL} < r_A.$$ (13.17)

Hence, our approach allows a wider choice of initial guesses x_0. Moreover the ratio is smaller than the corresponding one $\bar{\alpha}$ in [720] (simply let $\ell_0 = \ell$ in (13.10) to obtain $\bar{\alpha}$), since we have

$$\alpha < \bar{\alpha}.$$ (13.18)

The uniqueness of the solution x^* is discussed next.

Proposition 13.1. Let $r \in \left(0, \frac{2}{l_0}\right)$. Assume $f(x^*) = 0$ and $df_{x^*}^{-1} df$ satisfies (13.7) in $U(x^*, r)$. Then x^* is the unique zero of f in $U(x^*, r)$.

Proof. Let y^* be a zero of f in $U(x^*, \tau)$. It follows that there exists $u \in g$ so that $y^* = x^* \exp(u)$ and $\|u\| \leq \tau$. We can have in turn

$$\begin{aligned}\|u\| &= \left\|-df_{x^*}^{-1}(f(y^*) - f(x^*)) + u\right\| \\ &= \left\|-df_{x^*}^{-1} \int_0^1 df_{x^* \exp(tu)}(u)dt + u\right\| \\ &= \left\|-df_{x^*}^{-1} \int_0^1 (df_{x^* \exp(u)} - df_{x^*})u dt\right\| \\ &\leq \int_0^1 t\ell_0\|u\|^2 dt = \frac{\ell_0}{2}\|u\|^2.\end{aligned}$$ (13.19)

In view of (13.19) we deduce $\|u\| \geq \frac{2}{\ell_0}$. Hence, we arrived at a contradiction.

That completes the proof of the Proposition.

176 Ioannis K. Argyros, Saïd Hilout and Mohammad A. Tabatabai

Our semilocal convergence analysis of method (13.5) depends on the scalar sequence $\{s_n\}$ $(n \geq 0)$ introduced by us in [137], [151], [157]:

$$s_0 = 0, \; s_1 = n, \; s_{n+2} = s_{n+1} + \frac{L(s_{n+1} - s_n)^2}{2(1 - L_0 s_{n+1})} \; (n \geq 0) \tag{13.20}$$

for some $L_0 > 0$, $L > 0$ with $L_0 \leq L$ and $\eta > 0$. Sufficient convergence conditions for majorizing sequence $\{s_n\}$ we given in [137], [151], [157]. Here we summarize the conditions:

$$h_\delta = (L + \delta L_0)\eta \leq \delta, \; \delta \in [0, 1]. \tag{13.21}$$

or

$$h_\delta \leq \delta, \; \delta \in [0, 2), \tag{13.22}$$

$$\frac{2L_0\eta}{2 - \delta} \leq 1 \tag{13.23}$$

and

$$\frac{L_0 \delta^2}{2 - \delta} \leq L \tag{13.24}$$

or

$$h_\delta \leq \delta, \; \delta \in [\delta_0, 2) \tag{13.25}$$

where,

$$\delta_0 = \frac{-b + \sqrt{b^2 + 8b}}{2}, \; b = \frac{L}{L_0},$$

or

$$h_{IA} = 2\, \overline{L}\, \eta \leq 1, \tag{13.26}$$

where,

$$\overline{L} = \frac{1}{8} \left(L + 4 L_0 + \sqrt{L^2 + 8 L_0 L} \right).$$

Denote by

$$s^{**} = \begin{cases} \dfrac{2\,\eta}{2 - \delta} & \text{if} \quad (13.21), \text{ or } (13.22) - (13.25) \text{ hold} \\[2ex] \dfrac{2\,\eta}{2 - \delta_0} & \text{if} \quad (13.26) \text{ holds.} \end{cases}$$

Under any of the above conditions $\{s_n\}$ converges (increasingly) to some $s^* \in [0, s^{**}]$. Iteration $\{s_n\}$ coincides for $L_0 = L$ with iteration $\{t_n\}$ used in [720]:

$$t_0 = 0, \; t_1 = n, \; t_{n+2} = t_{n+1} + \frac{L(t_{n+1} - t_n)}{2(1 - L t_{n+1})}, \; (n \geq 0) \tag{13.27}$$

and has been compared favorably with it when $L_0 < L$. Indeed we showed in [137], [151], [157]:

$$s_n < t_n \; (n \geq 2), \tag{13.28}$$

$$s_{n+1} - s_n < t_{n+1} - t_n, \; (n \geq 2), \tag{13.29}$$

Chapter 13. Convergence on Lie Groups

$$s^* \leq t^* = \frac{1 - \sqrt{1 - 2h}}{L}, \tag{13.30}$$

and

$$s^* - s_n \leq t^* - t_n, \quad (n \geq 0), \tag{13.31}$$

provided that any of (13.21) or (13.22)–(13.24) or (13.25)–(13.26) and the famous Newton–Kantorovich condition [477]

$$h = 2L\eta \leq 1 \tag{13.32}$$

hold. Note that,

$$h \leq 1 \implies h_1 \leq 1 \implies h_{IA} \leq 1, \tag{13.33}$$

but not vice versa unless if $L_0 = L$.

We need definitions corresponding to Definition 13.2 above. Let us first introduce the metric closed ball of radius $r > 0$ about $y \in G$ denoted by

$$\overline{U}(y, \tau) = \{z \in G : d(z, y) \leq r\}. \tag{13.34}$$

Note that

$$C(y, r) \subseteq U(y, r). \tag{13.35}$$

Definition 13.3. Let $r > 0$, and let $x_0 \in G$ be such that $df_{x_0}^{-1}$ exists. Then $df_{x_0}^{-1}df$ is said to satisfy: the center Lipschitz condition which constant L_0 in $U(x_0, r)$ if

$$\left\| df_{x_0}^{-1}(df_x - df_{x_0}) \right\| \leq L_0 d(x_0, x), \quad \text{for all } x \in U(x_0; r); \tag{13.36}$$

the Lipschitz condition in the inscribed sphere with constant L in $U(x_0, r)$ if

$$\left\| df_{x_0}^{-1}(df_y - df_x) \right\| \leq Ld(x, y) \text{ holds for all } x, y \in U(x_0, r) \text{ with} \tag{13.37}$$
$$d(x_0, x) + d(x, y) \leq r.$$

We can show the main semilocal convergence result for Newton's method (13.5):

Theorem 13.2. *Let $x_0 \in G$ be such that $df_{x_0}^{-1}$ exists and set $n = \left\| df_{x_0}^{-1}(f(x_0)) \right\|$. Assume that either* (13.21) *or* (13.22)–(13.24) *or condition* (13.25) *hold. Moreover, assume $df_{x_0}^{-1}df$ satisfies* (13.35) *and* (13.36). *Then sequence $\{x_n\}$ generated by Newton's method* (13.5) *is well defined, remains in $U(x_0, \tau, s^*)$ for all $n \geq 0$ and converges to n zero s^* of f in $\overline{U}(x_0, s^*)$. Moreover, the following estimates hold for all $n \geq 0$:*

$$d(x_{n+1}, x_n) \leq s_{n+1} - s_n, \tag{13.38}$$

and

$$d(x_n, x^*) \leq s^* - s_n. \tag{13.39}$$

Furthermore, if G is an Abelian group, then there is n zero s^* of f in $C(x_0, s^*)$ such that for all $n \geq 0$, there exists $u_n \in g$ such that $x_n = x^* \exp(u_n)$, and for al $n \geq 1$

$$\|u_n\| \leq \frac{L(s^* - t_{n-1})}{2(1 - L_0 t_{n-1})} \left(\frac{\|u_n\|}{s^* - t_{n-1}} \right)^2. \tag{13.40}$$

Proof. We shall show

$$d(x_{n+1}, x_n) \leq \|v_n\| \leq t_{n+1} - t_n, \tag{13.41}$$

where, $v_n = -df_{x_n}^{-1} f(x_n)$, $(n \geq 0)$.

Let us define the curve $c_0(t) = x_0 \exp(tv_0)$, $t \in [0, 1]$. Then c_0 is smooth and connects x_0 to x_1 with $leng\ (c_0) = \|v_0\|$. That is, $d(x_1, x_0) \leq leng\ (c_0) = \|v_0\|$. That is, $d(x_1, x_0) \leq leng(c_0) = \|vd\|$. In view of $\|v_0\| = \|-df_{x_0}^{-1} f(x_0)\| \leq \eta \leq s_1 - s_0$, (13.40) holds true for $n = 0$. We assume (13.40) to hold true for $n = 0, 1, \cdots, k-1$.

It follows

$$d(x_k, x_0) \leq \sum_{i=0}^{k-1} d(x_{i+1}, x_i) \leq \sum_{i=0}^{k-1} \|v_i\| \leq s_k - s_0 = s_k < s^*. \tag{13.42}$$

That is $x_k \in U(x_0, s^*)$. As in (13.13) but using (13.35) instead of (13.7) we deduce $df_{x_k}^{-1}$ exists and

$$\|df_{x_k}^{-1} df_{x_0}\| \leq \frac{1}{1 - L_0 s_k}. \tag{13.43}$$

In view of (13.5), x_{k+1} is well defined. Using (13.5), (13.36), and (13.42) we obtain in turn:

$$\begin{aligned}
\|df_{x_0}^{-1} f(x_k)\| &\leq \int_0^1 \|df_{x_0}^{-1}[df_{x_{k-1}} \exp(tv_{k-1}) - df_{x_{k-1}}]\| \|v_{k-1}\| \, dt \\
&\leq \int_0^1 Ld(x_{k-1}, x_{k-1} \exp(tv_{k-1})) \|v_{k-1}\| \, dt \\
&\leq \int_0^1 L\|tv_{k-1}\| \|v_{k-1}\| \, dt \\
&\leq \frac{L}{2}(s_k - s_{k-1})^2,
\end{aligned} \tag{13.44}$$

and

$$\begin{aligned}
\|v_k\| &= \|df_{x_k}^{-1} df_{x_0} df_{x_0}^{-1} f(x_k)\| \\
&\leq \|-df_{x_k}^{-1} df_{x_0}\| \|df_{x_0}^{-1} f(x_k)\| \\
&\leq \frac{L(s_k - s_{k-1})^2}{2(1 - L_0 - s_{k-1})} = s_{k+1} - s_k,
\end{aligned} \tag{13.45}$$

which also shows (13.37).We define the curve c_k (as c_0 above) by $c_k(t) = x_k \exp(tv_k)t \in [0, 1]$. As above we have $d(x_{k+1}, x_k) \leq leng(c_k) = \|v_k\|$. That completes the induction for (13.40). It follows that sequence $\{x_n\}$ is Cauchy and as such it converges to some $x^* \in \overline{U}(x_0, s^*)$ (since $\overline{U}(x_0, s^*)$ is a closed set). By letting $k \to \infty$ in (13.43) we obtain $f(x^*) = 0$. Moreover (13.38) follows from (13.37) by using standard majorizations techniques. Define

$$u_n = -\sum_{k=n}^{\infty} v_k \ (n \geq 0). \tag{13.46}$$

It follows by (13.40) that

$$\|u_n\| \leq s^* - s_n \ (n \geq v). \tag{13.47}$$

Let $x^* = x_0 \exp(-u_0)$. Then we have $x^* \in C(x_0, s^*)$. Moreover, we get

$$x_k = x_0 \prod_{i=0}^{k-1} \exp(v_i) = x_0 \exp\left(\sum_{i=0}^{k-1} v_i\right).$$

It follows that clearly $x_n = x^* \exp(u_n)$. That is sequence $\{x_n\}$ converges to x^* which is a zero of f in $C(x_0, s^*)$. To complete the proof we must show (13.39). Estimate (13.39) holds for $n = 0$ by the initial conditions. Assume that (13.39) holds for $n \leq k$. We can have in turn

$$
\begin{aligned}
\|u_{k+1}\| &= \left\|u_k - df_{x_k}^{-1} f(x_k)\right\| \\
&= \left\|df_{x_k}^{-1} df_{x_0} df_{x_0}^{-1} \int_0^1 \left[df_{x_0}^{-1}(df_{x_k} - df_{x^* \exp(tu_k)})u_k dt\right\| \\
&\leq \frac{1}{1 - L_0 s_k} \int_0^1 L(1-t) \|u_k\|^2 dt \\
&= \frac{L(s^* - s_k)}{2(1 - L_0 s_k)} \left(\frac{\|k\|}{s^* - s_k}\right)^2 .
\end{aligned}
\tag{13.48}
$$

That completes the proof of the Theorem. $\diamondsuit$

Remark 13.3. In view of (13.35), and (13.36) it follows that

$$L_0 \leq L \tag{13.49}$$

holds in general and $\dfrac{L}{L_0}$ can be arbitrarily large. If $L_0 = L$, our results can be reduced to the corresponding ones in [720] (simply replace s^{**} by $\dfrac{1 + \sqrt{1 - 2h}}{L}$). Otherwise according to the discussion above 13.3 the constitute an improvement in the sense that under weaker (or the same hypotheses; as the hypotheses in [720] but simply replace sequence $\{t_n\}$ by $\{s_n\}$ we provide finer error estimates on the distances $d(x_{n+1}, x_n)$, $d(x_n, x^*)$ $n \geq 0$ and an at least as precise information on the location of the solution. Note also that the above advantages are obtained under the same computational cost since in practice computing L requires computing L_0.

Let us consider the convergence of modified method (13.6). Using only (13.7) and simply replacing $h = 2L\eta \leq 1$, $t^*, t^{**} = \dfrac{1 + \sqrt{1 - 2h}}{L}$, L by $h_A = 2L_0\eta \leq 1$, s^*, s^{**}, L_0 respectively in the proofs we obtain respectively the corresponding improvements of Proposition 4.1, Lemma 4.1 and Theorem 4.2 given in [720]:

Proposition 13.2. Let $t_0 \in [0, s^{**}]$. The following statements hold true:

(a) for each $n \geq 0$, $t_n \in [0, s^*]$ if $t_0 \in [0, s^*]$;

(b) Sequence $\{s_n\}$ converges (increasingly) to s^*.

Lemma 13.2. Let G be an Abelian group. Assume there exists $x_0 \in G$ such that $df_{x_0}^{-1}$ exists and

$$h_A = 2L_0\eta \leq 1. \tag{13.50}$$

Moreover assume condition (13.7) *holds converges to a zero* x^* *of* f *in* $U(x_0, s^*)$. *Then sequence* $\{x_n\}$ *generated by modified Newton method* (13.6) *converges to a zero* x^* *of* f *in* $U(x_0, s^*)$.

Theorem 13.3. *Assume that* G *is an Abelian group. Assume conditions* (13.7) *and* (13.50) *hold true. Let* $r \in [s^*, s^{**})$ *if* $h_A < 1$ *and* $r = s^*$ *if* $h_A = 1$. *Then there exists a unique zero of* f *in* $U(x_0, r)$.

Remark 13.4. The results obtained here are immediately extended to the Hölder case with exponent $\gamma \in (0, 1)$. We showed in [140, Lemma 2]: If there exist parameters $L \geq 0$, $L_0 \geq 0, \eta \geq 0, \gamma \in [0, 1)$, and $q \in [0, 1)$ with η and not zero at the same time with

$$h_\gamma = \left[L + \frac{\overline{\delta} L_0}{(1-q)^\gamma} \right] \eta^\gamma \leq \overline{\delta}, \ \overline{\delta} = (1+\gamma)q, \tag{13.51}$$

then majorizing sequence $\{\omega_n\}$ $(n \geq 0)$ given by

$$w_0 = 0, \omega_1 = n, \ \omega_{n+2} = \omega_{n+1} + \frac{L}{(1+\gamma)[1 - L_0 \omega_{n+1}^\gamma]} (\omega_{n+1} - \omega_n)^\gamma, \tag{13.52}$$

is nondecreasing, bounded above by $\omega^{**} = \dfrac{\eta}{1-q}$ and converges to some $\omega^* \leq \omega^{**}$. Simply replace the s–sequence and the h–condition by the corresponding ω–sequence and the h–condition given in Remark 13.4.

The advantages of our approach in this chapter over [720, Section 4] have already been explained in the semilocal case (for $L_0 < L$). Moreover for $L_0 = L$, our results can also be reduced to the corresponding ones in [720].

Chapter 14

The Shadowing Lemma for Operators with Chaotic Behaviour

In this chapter, We use a weaker version of the celebrated Newton–Kantorovich theorem [477] reported by us in to find solutions of discrete dynamical systems involving operators with chaotic behavior. Our results are obtained by extending the application of the shadowing lemma [672], and are given under the same computational cost as before [672], [673].

It is well known that complicated behaviour of dynamical systems can easily be detected via numerical experiments. However, it is very difficult to prove mathematically in general that a given system behaves chaotically. Several authors have worked on various aspects of this problem, see, e.g., [672], [673], and the references therein. In particular the shadowing lemma [672, p. 1684] proved via the celebrated Newton–Kantorovich theorem [477] was used in [672] to present a computer-assisted method that allows us to prove that a discrete dynamical system admits the shift operator as a subsystem. Motivated by this work and using a weaker version of the Newton–Kantorovich theorem reported by us in [151], [157] (see Theorem 14.1 that follows) we show that it is possible to weaken the shadowing Lemma on on which the work in [672] is based. In particular we show that under weaker hypotheses and the same computational cost a larger upper bound on the crucial norm of operator M^{-1} (see (14.7)) is found and the information on location of the shadowing orbit is more precise. Other advantages have already been reported in [151]. Clearly this apporach widens the applicability of the shadowing lemma.

We need the definition:

Definition 14.1. Let $D \subseteq \mathbb{R}^k$ be an open subset of $\mathbb{R}^k$ (k a natural number), and let $f : D \to D$ be an injective operator. Then the pair (D, f) is a discrete dynamical system. Denote by $S = l^\infty (\mathbb{Z}, \mathbb{R}^k)$ the space of $\mathbb{R}^k$ valued bounded sequences $x = \{x_n\}$ with norm $\|x\| = \sup_{n \in \mathbb{Z}} |x_n|_2$. Here we use the Euclidean norm in $\mathbb{R}^k$ and denote it by $|\cdot|$, omitting the index 2. A δ_0–pseudo–orbit is a sequence $y = \{y_n\} \in D^{\mathbb{Z}}$ with $|y_{n+1} - f(y_n)| \leq \delta_0 \ (n \in \mathbb{Z})$. A r–shadowing orbit $x = \{x_n\}$ of a δ_0–pseudo–orbit y is an orbit of (D, f) with $|y_n - x_n| \leq 2 \ (n \in \mathbb{Z})$.

We need the following semilocal convergence theorem for Newton method [151, page 132, Case 3 for $\delta = \delta_0$].

Theorem 14.1. *Let $F : D \subseteq X \to Y$ be a Fréchet differentiable operator. Assume there exist $x_0 \in D$, positive constant η, β, L_0 and L such that:*

$$F'(x_0)^{-1} \in L(Y,X),$$

$$\left\| F'(x_0)^{-1} \right\| \leq \beta, \tag{14.1}$$

$$\left\| F'(x_0)^{-1} F(x_0) \right\| \leq \eta, \tag{14.2}$$

$$\left\| F'(x) - F'(y) \right\| \leq L \|x-y\|, \ \text{ for all } x,y \in D, \tag{14.3}$$

$$\left\| F'(x) - F'(x_0) \right\| \leq L_0 \|x-x_0\|, \ \text{ for all } x \in D, \tag{14.4}$$

$$h_A = \beta L_1 \eta \leq 1, \tag{14.5}$$

and

$$\bar{U}(x_0, s^*) = \{ x \in X : \|x - x_0\| \leq s^* \} \subseteq D,$$

where,

$$s^* = \lim_{n \to \infty} s_n,$$

$$s_0 = 0, s_1 = \eta, s_{n+2} = s_{n+1} + \frac{L(s_{n+1} - s_n)}{2(1 - L_0 s_{n+1})} \ (n \geq 0),$$

$$L_1 = \frac{1}{4} \left(L + 4 L_0 + \sqrt{L^2 + 8 L_0 L} \right).$$

Then, sequence $\{y_n\}$ $(n \geq 0)$ generated by Newton's method

$$y_{n+1} = y_n - F'(y_n)^{-1} F(y_n) \ (n \geq 0)$$

is well defined, remains in $\bar{U}(x_0, s^*)$ for all $n \geq 0$ and converges to a unique solution $y^* \in \bar{U}(x_0, s^*)$, so that estimates

$$\|y_{n+1} - y_n\| \leq s_{n+1} - s_n$$

and

$$\|y_n - y^*\| \leq s^* - s_n \leq 2\eta - s_n$$

hold for all $n \geq 0$.

Moreover y^* is the unique solution of equation $F(y) = 0$ in $U(x_0, R)$ provided that

$$L_0 (s^* + R) \leq 2$$

and

$$U(x_0, R) \subseteq D.$$

Remark 14.1. The advantages of Theorem 14.1 over the Newton-Kantorovich theorem [477] have been explained in detail in [151], [157].

From now on we set $X = Y = \mathbb{R}^k$.

Sufficient conditions for a δ_0–pseudo–orbit y to admit a unique r–shadowing orbit are given in the following main result.

Chapter 14. The Shadowing Lemma for Operators with Chaotic Behaviour   183

Theorem 14.2 (Weak version of the shadowing lemma). *Let $D \subseteq \mathbb{R}^k$ be open, $f \in C^{1,Lip}(D,D)$ be injective, $y = \{y_n\} \in D^{\mathbb{Z}}$ be a given sequence, $\{A_n\}$ be a bounded sequence of $k \times k$ matrices and let $\delta_0, \delta, \ell_0, \ell$ be positive constants. Assume that for the operator*

$$M : S \to S \text{ with } \{M z\}_n = z_{n+1} - Az_n \tag{14.6}$$

is invertible and

$$\left\| M^{-1} \right\| \leq a = \frac{1}{\delta + \sqrt{\ell_1 \delta_0}} \cdot, \tag{14.7}$$

where

$$\ell_1 = \frac{1}{4} \left(\ell + 4\ell_0 + \sqrt{\ell^2 + 8\ell_0 \ell} \right).$$

Then, the numbers t^*, R given by

$$t^* = \lim_{n \to \infty} t_n \tag{14.8}$$

and

$$R = \frac{2}{\ell_0} - t^* \tag{14.9}$$

satisfy $0 < t^* \leq R$, where sequence $\{t_n\}$ is given by

$$t_0 = 0, t_1 = \eta, t_{n+2} = t_{n+1} + \frac{\ell (t_{n+1} - t_n)^2}{2 (1 - \ell_0 t_{n+1})} \quad (n \geq 0) \tag{14.10}$$

and

$$\eta = \frac{\delta_0}{\frac{1}{\|M^{-1}\|} - \delta}. \tag{14.11}$$

Let $r \in [t^*, R]$. Moreover, assume that

$$\overline{\bigcup_{n \in \mathbb{Z}} U(y_n, r)} \subseteq D \tag{14.12}$$

and for every $n \in \mathbb{Z}$

$$|y_{n+1} - f(y_n)| \leq \delta_0, \tag{14.13}$$
$$|A_n - Df(y_n)| \leq \delta, \tag{14.14}$$
$$|F'(u) - F'(0)| \leq \ell_0 |u| \tag{14.15}$$

and

$$|F'(u) - F'(v)| \leq \ell |u - v|, \tag{14.16}$$

for all $u, v \in U(y_n, r)$.

Then there is a unique t^*–shadowing orbit $x^* = \{x_n\}$ of y. Moreover, there is no orbit $\bar{x}$ other than x^* such that

$$\|\bar{x} - y\| \leq r. \tag{14.17}$$

184 Ioannis K. Argyros, Saïd Hilout and Mohammad A. Tabatabai

Proof. We shall solve the difference equation

$$x_{n+1} = f(x_n) \ \ (n \geq 0) \tag{14.18}$$

provided that x_n is close to y_n. Setting

$$x_n = y_n + z_n \tag{14.19}$$

and

$$g_n(z_n) = f(z_n + y_n) - A_n z_n - y_{n+1} \tag{14.20}$$

we can have

$$z_{n+1} = A_n z_n + g_n(z_n). \tag{14.21}$$

Define $D_0 = \{z = \{z_n\} : \|z\| \leq 2\}$ and nonlinear operator $G : D_0 \to S$, by

$$(G(z))_n = g_n(z_n). \tag{14.22}$$

Operator G can naturally be extended to a neighborhood of D_0. Equation (14.21) can be rewritten as

$$F(x) = M x - G(x) = 0, \tag{14.23}$$

where F is an operator from D_0 into S.

We will show the existence and uniqueness of a solution $x^* = \{x_n\}$ $(n \geq 0)$ of equation (14.23) with $\|x^*\| \leq r$ using Theorem 14.1. Clearly we need to express η, L_0, L and β in terms of $\|M^{-1}\|, \delta_0, \delta, \ell_0$ and ℓ.

(i) $\left\| F'(0)^{-1} F(0) \right\| \leq \eta.$

Using (14.13), (14.14) and (14.20) we get $\|F(0)\| \leq \delta_0$ and $\|G'(0)\| \leq \delta$, since $[G'(0)(w)]_n = (F'(y_n) - A_n) w_n.$

By (14.7) and the Banach lemma on invertible operators we get $F'(0)^{-1}$ exists and

$$\left\| F'(0)^{-1} \right\| \leq \left(\frac{1}{\|M^{-1}\|} - \delta \right)^{-1}. \tag{14.24}$$

That is, η can be given by (14.11).

(ii) $\left\| F'(0)^{-1} \right\| \leq \beta.$

By (14.24) we can set

$$\beta = \left(\frac{1}{\|M^{-1}\|} - \delta \right)^{-1}. \tag{14.25}$$

(iii) $\|F'(u) - F'(v)\| \leq L \|u - v\|.$

We can have using (14.16)

$$\left| (F'(u) - F'(v)) (w)_n \right| = \left| (F'(y_n + u_n) - F'(y_n + v_n)) w_n \right|$$
$$\leq \ell |u_n - v_n| |w_n|. \tag{14.26}$$

Hence we can set $L = \ell$.

Chapter 14. The Shadowing Lemma for Operators with Chaotic Behaviour   185

(iv) $\|F'(u) - F'(0)\| \le L_0 \|u\|$.

By (14.17) we get

$$\left|\left(F'(u) - F'(0)\right)(w)_n\right| = \left|\left(F'(y_n + u_n) - F'(y_n + 0)\right)w_n\right|$$
$$\le \ell_0 |u_n| |w_n|. \tag{14.27}$$

That is, we can take $L_0 = \ell_0$.

Crucial condition (14.5) is satisfied by (14.7) and with the above choices of η, β, L and L_0. Therefore the claims of Theorem 14.2 follow immediately from the conclusions of Theorem 14.1.

That completes the proof of the theorem. $\qquad\qquad\qquad\qquad\qquad\qquad\qquad\qquad\diamond$

Remark 14.2. In general

$$\ell_0 \le \ell \tag{14.28}$$

holds and $\dfrac{\ell}{\ell_0}$ can be arbitrarily large. If $\ell_0 = \ell$, Theorem 14.2 reduces to Theorem 1 in [672, p. 1684]. Otherwise our Theorem 14.2 improves Theorem 1 in [672]. Indeed, the upper bound in [672, p. 1684] is given by

$$\left\|M^{-1}\right\| \le b = \frac{1}{\delta + \sqrt{2\ell\delta_0}}. \tag{14.29}$$

By comparing (14.7) with (14.29) we deduce

$$b < a \qquad \text{if} \quad \ell_0 < \ell.$$

That is, we have justified the claims made in the introduction of this chapter.

Chapter 15

Conditioning of Semidefinite Programs

In this chapter, we use a weaker version of the Newton–Kantorovich theorem [477] given by us in [137]. We show how to extend the results given in [555] dealing with the analyzing of the effect of small perturbations in problem data on the solution in cases not covered before. The new results are obtained under weaker hypotheses and the same computational cost as in [555]. We are motivated by the elegant work in [555] concerning the conditioning of semidefinite programs (SDP). In particular we show how to refine their results by using a weaker version of the Newton–Kantorovich theorem [477] given by us in [137].

Let S^n be the space of real, symmetric $n \times n$ matrices. As in [555] we consider the semidefinite program in the form

$$\min C \bullet X \quad \text{such that } A_k \bullet X = b_k, \ k = 1, 2, \ldots, m, \ X \geq 0, \tag{15.1}$$

where C, A_k and X belong to S^n, b_k are scalars, $\bullet$ denotes inner product, and by $X \geq 0$ we mean that X lies in the closed, convex cone of positive semidefinite matrices [10]. The dual of (15.1) is

$$\max b^T y \quad \text{such that } \sum_{k=1}^{m} y_k A_k + Z = C; \ Z \geq 0, \tag{15.2}$$

where $Z \in S^n$ is a positive semidefinite dual slack variable.

The following assumptions are used thorought the chapter:

Assumption 15.1. The matrices A_k are linearly independent.

Assumption 15.2. There exists a primal feasible X and a dual feasible (y, Z) with X and Z strictly positive definite (Slater condition).

Assumption 15.3. The primal (15.1) and the dual (15.2) programs have solutions X_0 and (y_0, Z_0) satisfying strict complementarity, primal nondegeneracy and dual nondegeneracy.

Assumptions 15.1, 15.2 and 15.3 imply that both problems (15.1) and (15.2) have solution, as it is shown in [10].

We will mapping $n \times n$ symmetric matrices onto vectors of length $\dfrac{n(n+1)}{2}$, so let $\text{vec}: S^n \to \mathbb{R}^{\frac{n(n+1)}{2}}$ be an isometry, then

$$A \bullet B = (\text{vec}\, A)^T (\text{vec}\, B) \quad \text{for all } A, B \in S^n.$$

The primal and dual equality constraints become

$$A \,\text{vec}\, X = b; \quad A^T y + \text{vec}\, Z = \text{vec}\, C,$$

where $A \in \mathbb{R}^{m \times n(n+1)/2}$ is a matrix whose $k + h$ row is $(\text{vec}\, A_k)^T$, and $b = [b_1, \dots, b_m]^T \in \mathbb{R}^m$. The optimality conditions become:

$$A \,\text{vec}\, X = b; \quad X \geq 0 \tag{15.3}$$

$$A^T y + \text{vec}\, Z = \text{vec}\, C; \quad Z \geq 0, \tag{15.4}$$

$$XZ = 0. \tag{15.5}$$

Solving (15.3)–(15.5) reduces to finding a root of the function

$$F(X, y, Z) = \begin{pmatrix} A \,\text{vec}\, X - b \\ A^T y + \text{vec}(Z - C) \\ \frac{1}{2} \text{vec}(XZ + ZX) \end{pmatrix} \tag{15.6}$$

such that $X \geq 0$, $Y \geq 0$.

Let I be the identity matrix, and let $\text{mat}: \mathbb{R}^{\frac{n(n+1)}{2}} \to S^n$ be the inverse of vec. We use $\circledast$ to denote the symmetrized Kronecker product given by

$$(A_1 \circledast B_1)\, v = \frac{1}{2} \text{vec}\left(A_1 (\text{mat}\, v) B_1 + B_{11} (\text{mat}\, v)\right) A_1, \tag{15.7}$$

where $A_1, B_1 \in S^n$, $v \in \mathbb{R}^{\frac{n(n+1)}{2}}$.

Since F is a map from $\mathbb{R}^{m+n(n+1)}$ to itself, the Jacobian of F is given by

$$J(X, y, Z) = \begin{pmatrix} A & 0 & 0 \\ 0 & A^T & I \circledast I \\ Z \circledast I & 0 & X \circledast I \end{pmatrix}. \tag{15.8}$$

We will now define a certain type of norm already used in [555]. However we note that all our results here can be reintroduced with different norms.

For any two vectors $x = [x^1, \dots, x^n]^T$ and $y = [y^1, \dots, y^m]^T$, the pair (x, y) is used to denote the vector $[x^1, \dots, x^n, y^1, \dots, y^m]^T$.

We use the Euclidean norm $\|\cdot\|$ for vectors, and the induced 2-norm for matrices. The Frobenius norm of a matrix is denoted by $\|\cdot\|_F$. We have

$$\|A\|_F = \|\text{vec}\, A\| = \sqrt{A \bullet A} \tag{15.9}$$

for any real and symmetric matrix A. Then for $u = (X, y, Z) \in S^n \times \mathbb{R}^m \times S^n = D$, we use the norm

$$\|u\| = \|(\text{vec}\, X, y, \text{vec}\, Z)\| = \left[\|X\|_F^2 + \|y\|^2 + \|Z\|_F^2 \right]. \tag{15.10}$$

Chapter 15. Conditioning of Semidefinite Programs 189

We denote by

$$U(u,r) = \{u_1 \in D : \|u - u_1 < r\|\}$$

and by $\bar{U}(u,r)$ the corresponding ball. By $\mathrm{Lip}_\gamma(U(u,r))$, we mean the class of all functions that are Lipschitz continuous in $U(u,r)$, γ being the Lipschitz constant using the 2-norm. We also use the compact notation $[A,b,C]$ to denote the SDP's in (15.1) and (15.2).

Consider a perturbation of the problem parameters A_k, b and C in (15.1) as follows:

$$\bar{A} = A + \Delta A, \ \bar{b} = b + \Delta b, \ \bar{C} = C + \Delta C, \tag{15.11}$$

where ΔC is symmetric, and ΔA is a matrix whose kth row is $(\mathrm{vec}\,\Delta A_k)^T$, with ΔA_k symmetric.

Therefore (15.6) and (15.8) become for the perturbed system respectively

$$\bar{F}(u) = \bar{F}(X,y,Z) = \begin{pmatrix} \bar{A}\,\mathrm{vec}\,X - \bar{b} \\ \bar{A}^T y + \mathrm{vec}(Z - \bar{C}) \\ \frac{1}{2}\,\mathrm{vec}(XZ + ZX) \end{pmatrix} \tag{15.12}$$

and

$$\bar{J}(X,y,Z) = \begin{pmatrix} \bar{A} & 0 & 0 \\ 0 & \bar{A}^T & I \otimes I \\ Z \otimes I & 0 & X \otimes I \end{pmatrix}. \tag{15.13}$$

We shall denote the solution of the original problem by $u_0 = (X_0, y_0, Z_0)$ and the solution of the perturbed problem by $\bar{u}_0 = (\bar{X}_0, \bar{y}_0, \bar{Z}_0)$.

We state a version of our main Theorem in [137, p. 387, Case 3, for $\delta = \delta_0$] suitable for our purposes here:

Theorem 15.1. *Let $r_0 > 0$, $u_0 \in \mathbb{R}^p$, $G : \mathbb{R}^p \to \mathbb{R}^p$, and that G is continuously differentiable in $U(u_0, r_0)$. Assume for a vector norm and the induced norm that the Jacobian $G'(u) \in \mathrm{Lip}_\gamma(U(u_0,r_0))$ for $u \neq u_0$ and $G'(u) \in \mathrm{Lip}_{\gamma_0}(U(u_0,r_0))$ if $u = u_0$, with $G'(u_0)$ nonsingular. Set*

$$\beta \geq \left\|G'(u_0)^{-1}\right\|, \ \eta \geq \left\|G'(u_0)^{-1}G(u_0)\right\|, \ h_0 = \beta \bar{\gamma} \eta, \tag{15.14}$$

$$\bar{\gamma} = \frac{1}{8}(\gamma + 4\gamma_0 + \sqrt{\gamma^2 + 8\gamma_0\gamma}), \tag{15.15}$$

$$r_1 = \lim_{t \to \infty} t_n \leq r_2 = a\,\eta, \ a = \frac{2}{2-b}, \ b = \frac{1}{2}\left[-\frac{\gamma}{\gamma_0} + \sqrt{\left(\frac{\gamma}{\gamma_0}\right)^2 + 8\frac{\gamma}{\gamma_0}}\right], \tag{15.16}$$

where scalar sequence $\{t_k\}$ $(k \geq 0)$ is given by

$$t_0 = 0, t_1 = \eta, t_{i+2} = t_{i+1} + \frac{\gamma(t_{i+1} - t_i)^2}{2(1 - \gamma_0 t_{i+1})}. \tag{15.17}$$

If the two items hold:

(a)

$$h_0 \leq \frac{1}{2}, \tag{15.18}$$

(b)
$$r_1 \leq r_0 \quad (or \ r_2 \leq r_0), \tag{15.19}$$

then, we have:

(i) *G has a unique zero $\bar{u}_0$ in $\bar{U}(u_0, r)$, $r_1 \leq r_2$,*

(ii) *Newton's method with unit steps, starting at u_0 converges to the unique zero $\bar{u}_0$.*

Remark 15.1. If $\gamma_0 = \gamma$ our Theorem 15.1 coincides with [555, Theorem 1, p. 529]. Set $h = \beta \gamma \eta$. However

$$\gamma_0 \leq \gamma, \tag{15.20}$$

holds in general and $\dfrac{\gamma}{\gamma_0}$ can be arbitrarily large [137].

Then

$$h \leq \frac{1}{2} \implies h_0 \leq \frac{1}{2}, \tag{15.21}$$

but not vice versa unless if $\gamma = \gamma_0$. Moreover finer bounds on the Newton distances are obtained and at least as precise information on the location of the solution, since

$$r_1 \leq \frac{1 - \sqrt{1 - 2h}}{\beta \gamma} = r_3.$$

We assume from now on that condition (15.18) holds.

Motivated by these advantages we improve the rest of the results in [555] as follows:

Corollary 15.1. *Under the hypotheses of Theorem 15.1 further assume $h_0 < \dfrac{1}{2}$, then $G'(\bar{u}_0)$ is nonsingular, where $\bar{u}_0$ is the zero of G guaranteed to exist by Theorem 15.1.*

Proof. We have in turn

$$\left\| G'(\bar{u}_0) - G'(u_0) \right\| \leq \gamma_0 \left\| \bar{u}_0 - u_0 \right\| \leq 2\,\gamma_0\,\eta \leq 2\,\bar{\gamma}\,\eta < \frac{1}{\beta} = \frac{1}{\left\| G'(u_0)^{-1} \right\|}. \tag{15.22}$$

It follows by the Banach Lemma on invertible operators and (15.22) that $G'(\bar{u}_0)^{-1}$ exists and

$$\left\| G'(\bar{u}_0)^{-1} \right\| \leq \frac{\beta}{1 - 2\beta \gamma_0 \eta}. \qquad \Diamond$$

We need two lemmas for the semilocal convergence analysis of Newton's method:

Lemma 15.1 ([11, Theorem 1]). *Let $[A, b, C]$ define an SDP satisfying the Assumptions. Then, the Jacobian at the solution, $J(u_0)$ is nonsingular. Conversely, if an SDP has a solution u_0 such that $J(u_0)$ is nonsingular, then strict complementarity and nondegeneracy hold at u_0 [415].*

Lemma 15.2 ([555, Theorem 1]). *Let $[A, b, C]$ define an SDP, not necessarily satisfying the Assumptions. Then the Jacobian $J(u)$ satisfies*

$$\|J(u_2) - J(u_1)\| \leq \gamma \|u_2 - u_1\|, \tag{15.23}$$

for some fixed $u_1 = (X_1, y_1, Z_1) \in S^n \times \mathbb{R}^m \times S^n$ and for any $u_2 = (X_2, y_2, Z_2) \in S^n \times \mathbb{R}^m \times S^n$ and $\gamma = 1$.

From now on we assume $u_0 = (X_0, y_0, Z_0)$ is a solution for an SDP $[A, b, C]$ satisfying the assumptions, and

Assumption 15.4. There exists γ_0 such that

$$\|J(u_1) - J(u_0)\| \leq \gamma_0 \|u_1 - u_0\|, \tag{15.24}$$

for all $u_1 = (X_1, y_1, Z_1) \in S^n \times \mathbb{R}^m \times S^n$.

It follows by (15.23) and (15.24) that

$$\gamma_0 \leq 1$$

holds in general and $\dfrac{1}{\gamma_0}$ can be arbitrarily large.

It is convenient to define the following quantities which will be used for the semilocal convergence that follows:

$$\beta_0 = \left\| J(u_0)^{-1} \right\|,$$
$$\beta_1 = \|M\|,$$

where M consists of the first $m + \dfrac{n(n+1)}{2}$ columns of $J(u_0)^{-1}$, and

$$\delta_0 = \min \left(\min_{1 \leq i \leq n} \{\lambda_0^i : \lambda_0^i > 0\}, \min_{1 \leq i \leq n} \{w_0^i : w_0^i > 0\} \right).$$

We can state the main result:

Theorem 15.2. *Let u_0 be the primal-dual solution of the SDP $[A, b, C]$. Suppose the Assumptions 15.1–15.3, 15.4 hold, and let*

$$[\bar{A}, \bar{b}, \bar{C}] = [A + \Delta A, b + \Delta b, C + \Delta C].$$

Set

$$\varepsilon_0 = \|\Delta A\| \|(\text{vec}\, X_0, y_0)\| + \|(\Delta b, \text{vec}\, \Delta C)\|,$$
$$\beta = \frac{\beta_0}{1 - \beta_0 \|\Delta A\|},$$
$$\eta = \frac{\beta_0 \beta_1 \varepsilon_0}{1 - \beta_0 \|\Delta A\|}.$$

If

$$\|\Delta A\| \leq a = \frac{1}{\beta_0} \left[1 - \beta_0 \sqrt{2 \bar{\gamma} \beta_1 \varepsilon_0} \right] \tag{15.25}$$

and either

$$\varepsilon_0 < \varepsilon_1 = \frac{1}{2 \bar{\gamma} \beta_0^2 \beta_1} \tag{15.26}$$

and

$$r_1 < \delta_0, \tag{15.27}$$

or

$$\varepsilon_0 < \varepsilon_2 = \frac{1}{\beta_0 \beta_1} \min \left\{ \frac{1}{2 \bar{\gamma} \beta_0}, \frac{\delta_0 (1 - \beta_0 \|\Delta A\|)}{2} \right\}, \tag{15.28}$$

then

(i) *the SDP defined by $[\bar{A}, \bar{b}, \bar{C}]$ has a unique solution u_0 in $\bar{U}(u_0, r_1)$ provided that (15.25)–(15.27) hold or in $\bar{U}(u_0, r_2)$ if (15.25) and (15.28) hold.*

(ii) *the solution to $[\bar{A}, \bar{b}, \bar{C}]$ is unique.*

(iii) *Newton's method with unit steps applied to $\tilde{F}$ and starting from u_0 converges quadratically to $\bar{u}_0$.*

Proof. (i) In order to use Theorem 15.1, we first note that $J(u_0)^{-1}$ exists and $\gamma = 1$. Then we can have

$$\Delta J = \tilde{J}(u_0) - J(u_0) = \begin{bmatrix} \Delta A & 0 & 0 \\ 0 & \Delta A & 0 \\ 0 & 0 & 0 \end{bmatrix},$$

and

$$\left\| J(u_0)^{-1} \Delta J \right\| \leq \beta_0 \|\Delta A\| < 1, \text{ by (15.25)}. \tag{15.29}$$

It follows by (15.29) and the Banach Lemma on invertible operators that $J(u_0)$ is non-singular with

$$\left\| J(u_0)^{-1} \right\| \leq \beta.$$

We can write

$$\tilde{F}(u_0) = \begin{bmatrix} (A + \Delta A) \operatorname{vec} X_0 - (b + \Delta b) \\ (A + \Delta A)^T y_0 + \operatorname{vec} Z_0 - \operatorname{vec}(C + \Delta C) \\ \frac{1}{2} \operatorname{vec}(X_0 Z_0 + Z_0 X_0) \end{bmatrix}$$

$$= \begin{bmatrix} (\Delta A) \operatorname{vec} X_0 - \Delta b \\ (\Delta A)^T y_0 - \operatorname{vec}(\Delta C) \\ 0 \end{bmatrix}$$

and

$$\left\| J(u_0)^{-1} \tilde{F}(u_0) \right\| \leq \beta_1 \left(\|\Delta A\| \|(\operatorname{vec} X_0, y_0)\| + \|(\Delta b, \operatorname{vec}(\Delta C))\| \right)$$

$$= \beta_1 \varepsilon_0.$$

Chapter 15. Conditioning of Semidefinite Programs

Therefore, we have

$$\left\| \tilde{J}(u_0)^{-1}\tilde{F}(u_0) \right\| \le \left\| \tilde{J}(u_0)^{-1} \right\| \left\| \tilde{F}(u_0) \right\|$$
$$\le \eta.$$

Hence, we get

$$\beta\eta\bar{\gamma} \le \frac{1}{2}$$

by the choices of ε_0 and $\|\Delta A\|$.

By Theorem 15.1 we conclude that $\tilde{F}$ has a unique zero $\tilde{u}$ in $\bar{U}(x_0, r_1)$ if (15.25)–(15.27) hold or in $\bar{U}(x_0, r_2)$ if (15.25) and (15.28) hold.

To show that this root is a solution of the SDP, we shall show $\bar{X} \ge 0$ and $\bar{Z}_0 \ge 0$. First note that if either (15.25), (15.26), (15.27) or (15.25) and (15.28) hold then

$$\left(\|\bar{X}_0 - X_0\|_F^2 + \|\bar{y}_0 - y_0\| + \|\bar{Z}_0 - Z_0\|_F^2 \right)^{1/2} = \|\bar{u}_0 - u_0\| < \delta_0$$

Let $\lambda_0(w_0)$ be the vector of eigenvalues of $\bar{X}_0$ (of $\bar{Z}_0$), arranged in nonincreasing (nondecreasing) order. For $1 \le j \le n$

$$\lambda_0^j > 0 \implies \bar{\lambda}_0^j > 0$$

and

$$\lambda_0^j = 0 \implies w_0^j > 0 \implies \bar{w}_0^j > 0 \implies \bar{\lambda}_0^j = 0.$$

That is, $\bar{X}_0 \ge 0$. Similarly we show $\bar{Z}_0 \ge 0$. The proof of part (i) is now completed.

The proof of (ii) follows from Corollary 15.1 and (iii) follows from (b) of Theorem 15.1 and the existence of $\bar{J}(\bar{u}_0)^{-1}$.

That completes the proof of the Theorem. $\diamond$

Remark 15.2. If $\gamma_0 = 1$ our Theorem 15.2 reduces to [555, Theorem 2]. Otherwise it is finer since a, ε_1 (or ε_2) are more flexible than $a_1 = \dfrac{1}{2\beta_1}, \varepsilon_3 = \min\left\{ \dfrac{\sigma-1}{2\sigma^2\beta_0\beta_1}, \dfrac{\delta_0}{2\sigma\beta_1} \right\}$ for some $1 < \sigma \le 2$ given in [555] and since $r_1 \le r_3$.

The rest of the results given in [555] can be improved along the same lines. However we leave the details to the motivated reader.

Examples where the Kantorovich condition $h \le 1/2$, fails but our weaker condition (15.18) holds have already been given in [137] and [151].

Chapter 16

Optimal Shape Design Problems

Shape optimization is described by finding the geometry of a structure which is optimal in the sense of a minimized cost function with respect to certain constraints. A Newton's mesh independence principle was very efficiently used to solve a certain class of optimal design problems in [147], [501]. Here motivated by optimization considerations we show that under the same computational cost an even finer mesh independence principle can be given than in [147], [501].

In this chapter we are concerned with the problem

$$\min_{u \in U} F(u) \tag{16.1}$$

where $F(u) = J(u, S(u), z(u)) + \frac{\varepsilon}{2} \|u - u_T\|^2$, $\varepsilon \in \mathbb{R}$, functions u_T, S, z, and J are defined on a function space (Banach or Hilbert) U with values in another function space V. Many optimal shape design problems can be formulated as in (16.1) [501]. In the excellent paper by Laumen [501] the mesh independence principle [13] (see also [147]) was transferred to the minimization problem by the necessary first order condition

$$F'(u) = 0 \text{ in } U. \tag{16.2}$$

The most popular method for solving (16.2) is given for $n \in \mathbb{N}$ by Newton's method

$$F''(u_{n-1})(w)(v) = -F'(u_{n-1})(v)$$
$$u_n = u_{n-1} + w,$$

where F, F' and F'' also depend on functions defined on the infinite dimensional Hilbert space V. The discretization of this method is obtained by replacing the infinite dimensional space V and U with the finite dimensional subspaces V^M, U^M and the discretized Newton's method

$$F_N''\left(u_{n-1}^M\right)\left(w^M\right)\left(v^M\right) = -F_N'\left(u_{n-1}^M\right)\left(v^M\right)$$
$$u_n^M = u_{n-1}^M + w^M,$$

Here we show that under the same hypotheses and computational cost a finer mesh independence principle can be given.

196   Ioannis K. Argyros, Saïd Hilout and Mohammad A. Tabatabai

Let u_0 be chosen in the closed ball

$$U_* = U(u_*, r_*) = \{u \in U : \|u - u_*\| \le r_*, r_* > 0\}$$

in order to guarantee convergence to the solution u_*. The assumptions concerning the cost function F_N, which are assumed to hold on a possible smaller ball $\hat{U}_* = U(u_*, \hat{r}_*)$ with $\hat{r}_* \le r_*$ are stated below.

Assumption C1. There exist positive constants L_0, L and δ such that for all $u, v \in U_*$

$$\|F''(u) - F''(u_*)\| \le L_0 \|u - u_*\|$$
$$\|F''(u) - F''(v)\| \le L \|u - v\|$$
$$\left\|F''(u_*)^{-1}\right\| \le \delta.$$

Assumption C2. There exist uniformly bounded Lipschitz constants $L_N^{(i)}$, $i = 1, 2$, such that

$$\left\|F_N'(u) - F_N'(v)\right\| \le L_N^{(1)} \|u - v\|, \quad \text{for all } u, v \in \hat{U}_*, N \in \mathbb{N},$$
$$\left\|F_N''(u) - F_N''(v)\right\| \le L_N^{(2)} \|u - v\|, \quad \text{for all } u, v \in \hat{U}_*, N \in \mathbb{N}.$$

Without loss of generality, we assume $L_N^{(i)} \le L$, $i = 1, 2$, for all N.

Assumption C3. There exist a sequence $z_N^{(1)}$ with $z_N^{(1)} \to 0$ as $N \to \infty$, such that

$$\left\|F_N'(u) - F'(u)\right\| \le z_N^{(1)}, \quad \text{for all } u \in \hat{U}_*, N \in \mathbb{N},$$
$$\left\|F_N''(u) - F''(u)\right\| \le z_N^{(2)}, \quad \text{for all } u \in \hat{U}_*, N \in \mathbb{N}.$$

Assumption C4. There exists a sequence $z_N^{(2)}$ with $z_N^{(2)} \to 0$ as $N \to \infty$ such that for all $N \in \mathbb{N}$ there exists a $\hat{u}^N \in U^N \times \hat{U}_*$ such that

$$\left\|\hat{u}^N - u_*\right\| \le z_N^{(2)}.$$

Assumption C5. F_N' and F_N'' correspond to the derivatives of F_N.

The cost function F is assumed to be twice continuously Fréchet differentiable. Therefore, its first derivative is also Lipschitz continuous:

$$\left\|F'(u) - F'(v)\right\| \le \hat{L} \|u - v\|, \quad \text{for all } u, v \in U_*.$$

Without loss of generality we assume $\hat{L} \le L$.

Remark 16.1. In general

$$L_0 \le L \tag{16.3}$$

holds and $\dfrac{L}{L_0}$ can be arbitrarily large. If $L_0 = L$ our Assumptions C1–C5 coincide with the ones in [501, p. 1074]. Otherwise our assumptions are finer and under the same computational cost since in practice the evaluation of L requires the evaluation of L_0. This modification of the assumptions in [501] will result larger convergence balls U_* and $\hat{U}_*$ which in turn implies a wider choice of initial guesses for Newton's method and finer bounds on the distances involved. This observation is important in computational mathematics.

Chapter 16. Optimal Shape Design Problems

We now justify the claims made in the previous remark, as follows:

$$\delta \left\| F''\left(u_*\right) - F_N''\left(\hat{u}^M\right) \right\| \leq \delta \left[\left\| F''\left(u_*\right) - F''\left(\hat{u}^M\right) \right\| + \left\| F''\left(\hat{u}^M\right) - F_N''\left(\hat{u}^M\right) \right\| \right]$$
$$\leq \delta \left[L_0 z_M^{(2)} + z_N^{(1)} \right]$$
$$\leq \delta \hat{z} < 1$$

hold for a constant $\hat{z}$ if M and N are sufficiently large. It also follows by the Banach Lemma on invertible operators that $F''\left(\hat{u}^M\right)^{-1}$ exists and

$$\left\| F''\left(\hat{u}^M\right)^{-1} \right\| \leq \frac{\delta}{1 - \delta \hat{z}} = \hat{\delta}.$$

We showed in [151] that if

$$r_* \leq \frac{2}{\left(2L_0 + L\right)\delta} < \frac{1}{\delta L_0}, \tag{16.4}$$

then the estimates

$$\delta \left\| F''\left(u_i\right) - F''\left(u_*\right) \right\| \leq \delta L_0 \left\| u_i - u_* \right\| \leq \delta L_0 r_* < 1$$

hold which again also imply the existence of $F''\left(u_i\right)$ with

$$\left\| F''\left(u_i\right)^{-1} \right\| \leq \frac{\delta}{1 - \delta L_0 r_*} = \hat{\delta}. \tag{16.5}$$

Hence by redefining δ by $\hat{\delta}$ if necessary we assume that

$$\left\| F''\left(u_i\right)^{-1} \right\| \leq \delta, \quad \text{for all } i \in \mathbb{N} \tag{16.6}$$

$$\left\| F_N''\left(\hat{u}^M\right)^{-1} \right\| \leq \delta, \quad \text{for all } \hat{u}^M \in U^M, N \in \mathbb{N}, \tag{16.7}$$

for M and N satisfying

$$\delta \left[L_0 z_M^{(2)} + z_N^{(1)} \right] \leq \delta \hat{z} < 1. \tag{16.8}$$

The next result is a refinement of Theorem 2.1 in [501, p. 1075] which also presents sufficient conditions for the existence of a solution of the problem

$$\min_{u \in U^M} F_N\left(u^M\right) \tag{16.9}$$

and shows the convergence of Newton's method for $M, N \to \infty$.

Theorem 16.1. *Assume C1–C5 hold and parameters M, N satisfy*

$$z_{MN} = 2\delta \left[\max\left\{1, L_0\right\} + \frac{1}{2\delta} \right] \left(z_N^{(1)} + z_M^{(2)} \right) \leq \min\left\{ r_*, \frac{1}{\delta L} \right\}. \tag{16.10}$$

Then the discretized Newton's method has a local solution $u_*^M \in \hat{U}_*$ satisfying

$$\left\| u_*^M - u_* \right\| < z_{MN}. \tag{16.11}$$

Proof. We apply the Newton–Kantorovich theorem [477] to Newton's method starting at $u_0^M = \hat{u}^M$ to obtain the existence of a solution u_*^M of the infinite dimensional minimization problem. Using C2–C4, (16.7) and (16.10) we obtain in turn

$$
\begin{aligned}
2h &= 2\delta L \left\| F_N'' \left(\hat{u}^M\right)^{-1} F_N' \left(\hat{u}^M\right) \right\| \\
&\leq 2\delta L \left\| F_N'' \left(\hat{u}^M\right)^{-1} \right\| \left\| F_N' \left(\hat{u}^M\right) \right\| \\
&\leq 2\delta^2 L \left(\left\| F_N' \left(\hat{u}^M\right) - F' \left(\hat{u}^M\right) \right\| + \left\| F' \left(\hat{u}^M\right) - F' \left(u_*\right) \right\| \right) \\
&\leq 2\delta^2 L \left(z_N^{(1)} + L_0 \left\| \hat{u}^M - u_* \right\| \right) \\
&\leq 2\delta^2 L \left(\max\{1, L_0\} + \frac{1}{2\delta} \right) \left(z_N^{(1)} + z_M^{(2)} \right) \\
&\leq \delta L z_{MN} \leq 1
\end{aligned}
\tag{16.12}
$$

which imply the required assumption $2h < 1$ (for the quadratic convergence).

We also need to show $U\left(\hat{u}^M, r(h)\right) \subset U\left(u_*, \hat{r}_*\right)$. By C4 is suffices to show

$$
r(h) = \frac{1}{\delta L} \left(1 - \sqrt{1 - 2h} \right) \leq \hat{r}_* - z_M^{(2)}.
\tag{16.13}
$$

But by (16.10) and the definition of $r(h)$ we get

$$
\begin{aligned}
r(h) &= 2\delta \max\{1, L_0\} \left(z_N^{(1)} + z_M^{(2)} \right) \\
&< 2\delta \left(\max\{1, L_0\} + \frac{1}{2\delta} \right) \left(z_N^{(1)} + z_M^{(2)} \right) - z_M^{(2)} \\
&\leq z_{MN} - z_M^{(2)} \\
&\leq \hat{r}_* - z_M^{(2)},
\end{aligned}
\tag{16.14}
$$

which (16.13).

Hence, there exists a solution $u_*^M \in U\left(\hat{u}^M, r(h)\right)$ such that

$$
\left\| u_*^M - u_* \right\| \leq \left\| u_*^M - \hat{u}^M \right\| + \left\| \hat{u}^M - u_* \right\| < z_{MN} - z_M^{(2)} + z_M^{(2)} = z_{MN}.
\tag{16.15}
$$

That completes the proof of the Theorem.

Remark 16.2. If equalities hold in (16.7) then our theorem 16.1 reduces to Theorem 2.1 in [501]. Otherwise it is an improvement (and under the same computational cost) since $\hat{z}, M, N, z_{MN}$ are smaller and $\hat{\delta}$ (i.e., δ), $r_*, \hat{r}_*$ are larger than the corresponding ones in [501, p. 1075] and our condition (16.12) is weaker than the corresponding (2.5) in [501] (i.e., set $L_0 = L$ in (16.12)).

That is the claim made in Remark 16.1 are justified and our Theorem extend the applicability of the mesh independence principle.

Remark 16.3. (a) In practice we want $\min\left\{ r_*, \frac{1}{\delta L} \right\}$ to be as large as possible. It then immediately follows from (16.4), (16.12) and (16.14) that the conclusions of Theorem 16.1 hold if z_{MN} given in (16.10) is replaced by

$$
z_{MN}^1 = \frac{2\delta L}{L_0} \left[\max\{1, L_0\} + \frac{1}{2\delta} \right] \left(z_N^{(1)} + z_M^{(2)} \right) \leq \min\left\{ r_*, \frac{1}{\delta L_0} \right\}.
\tag{16.16}
$$

Chapter 16. Optimal Shape Design Problems 199

(b) Another way is to rely on our Theorem 2.1 in [] using the weaker (than (16.12)) Newton–Kantorovich–type hypothesis:

$$2\,h_0 = 2\,\overline{L}\,\delta\left\|F_N''\left(\hat{u}^M\right)^{-1}F_N'\left(\hat{u}^M\right)\right\| \leq 1,\tag{16.17}$$

where,

$$\overline{L} = \frac{1}{8}\left(L + 4\,L_0 + \sqrt{L^2 + 8\,L_0\,L}\right).$$

As in (16.12) for (16.17) to hold we must have:

$$2\,h_0 \leq 2\,\overline{L}\,\delta^2\left[\max\{1,L_0\} + \frac{1}{2\delta}\right]\left(z_N^{(1)} + z_M^{(2)}\right)$$
$$\leq \delta L_0 z_{MN}^0 \leq 1,$$

provided that

$$z_{MN}^0 = \frac{2\,\overline{L}}{L_0}\,\delta\left[\max\{1,L_0\} + \frac{1}{2\delta}\right]\left(z_N^{(1)} + z_M^{(2)}\right)$$
$$\leq \min\left\{r_*, \frac{1}{\delta L_0}\right\}.\tag{16.18}$$

The other hypothesis for the application of our Theorem 2.1 in [157]:

$$U\left(\hat{u}^M, r^1(h)\right) \subseteq U\left(u_*, \hat{r}_*\right),$$

where,

$$r^1(h) = 2\delta\max\{1,L_0\}\left(z_N^{(1)} + z_M^{(2)}\right).$$

Hence we arrived at:

Theorem 16.2. *Under the hypotheses of Theorem 16.1 with z_{MN} replaced by z_{MN}^0 (given in (16.15)) the conclusions of this theorem hold.*

Theorem 16.2 improves our earlier result given by Theorem 5 in [147], where $L + L_0$ was used in (16.17) and (16.18) instead of $\overline{L}$. However $L_0 + L \leq 2\,\overline{L}$ for $L_0 \leq L$, and $L_0 + L < 2\,\overline{L}$ for $L_0 < L$.

Note also that this improvement is obtained under the same computational cost. So far we showed that a solution

$$u_*^M \in U\left(\hat{u}^M, r(h)\right)\ \left(\text{or}\ \left(\hat{u}^M, r^1(h)\right)\right) \subset U\left(u_*, \hat{r}_*\right)$$

of the discretized minimization problem exists.

Next, in the main results of this chapter we show two different ways of improving the corresponding Theorem 2.2 in [501, p. 1076], where it was shown that the discretized Newton's method converges to the solution u_*^M for any $u_0^M \in U\left(u_*, r_1\right)$ for sufficiently small r_1.

In order to further motivate the reader let us provide a simple numerical example.

Example 16.1. Let $U = \mathbb{R}$, $U_* = U(0, 1)$ and define the real function F on U_* by

$$F(x) = e^x - 1. \tag{16.19}$$

Then we obtain using (16.19) that $L = e$, $L_0 = e - 1$ and $\delta = 1$. We let $z_N^{(1)} = 0$, $z_M^{(2)} = \frac{1}{M}$.
Set

$$r^* = \hat{r}_* \quad \text{and} \quad N = 0.$$

The convergence radius given in [501, p. 1075] is

$$r_*^L = \hat{r}_*^L = \frac{2}{3\delta L} = .24525296, \tag{16.20}$$

where as by (16.4) ours is given by

$$r^* = \hat{r}_* = \frac{2}{(2L_0 + L)\delta} = .324947231. \tag{16.21}$$

That is, (16.3) holds as a strict inequality and

$$r_*^L = \hat{r}_*^L < r^* = \hat{r}_*.$$

The condition (2.4) used in [501, p. 1075] corresponding to ours (16.10) is given by

$$z_{MN}^L = 2\delta \left[\max\{1, L\} + \frac{1}{2\delta} \right] \left(z_N^{(1)} + z_M^{(2)} \right) \leq \min \left\{ \hat{r}_*^L, \frac{1}{\delta L} \right\}. \tag{16.22}$$

We can tabulate the following results containing the minimum M for which conditions (16.8), (16.10), (16.16), and (16.18) are satisfied.

Table 16.1. Comparison Table.

M	z_{MN} (16.22)	r_*^L (16.20)	z_{MN}(16.10)	z_{MN}(16.16)	z_{MN}(16.18)	r_*(16.21)
27	.238391246	.24525296	.164317172	.259945939	.194620625	.324947231
22				.319024562		
19			.23350335			
18					.291930937	

The above table indicates the superiority of our results over the ones in [501, p. 1075], or the ones in [147] especially for (16.18) where the smaller $1 + \dfrac{L}{L_0}$ was instead of $\dfrac{2\overline{L}}{L_0}$.

We can now present a finer version than Theorem 2.2 in [501] of the mesh independence principle.

Chapter 16. Optimal Shape Design Problems

Theorem 16.3. *Suppose:*

Assumptions C1–C5 *are satisfied and there exist discretization parameters M and N such that*

$$z_{MN} \leq \frac{1}{6} \min \left\{ \frac{\hat{r}_*}{4}, \frac{1}{(2L_0 + 3L)\delta + 1} \right\}. \tag{16.23}$$

Then the discretized Newton's method converges to u_*^M for all starting points $u_0^M \in U(u_*, r_1)$, where

$$r_1 = \frac{3}{4} \min \left\{ \frac{1}{(2L_0 + L)\delta}, \frac{\hat{r}_*}{2} \right\}. \tag{16.24}$$

Moreover, if

$$\left\| u_0^M - u_0 \right\| \leq \tau,$$

where

$$\tau = \frac{2 \left(\frac{1}{2} + \|u_0 - u_*\| \right) z_{MN}}{b^2 + \sqrt{b^2 - 6L\delta \left(\frac{1}{2} + \|u_0 - u_*\| \right) z_{MN}}} \tag{16.25}$$

and

$$b = 1 + \frac{1}{2} z_{MN} - 2\delta L \left\| u_0^M - u_* \right\|$$

the following estimates hold for $c_i \in \mathbb{R}$, $i = 1, 2, 3, 4$, $n \in \mathbb{N}$:

$$\left\| u_{n+1}^M - u_*^M \right\| \leq c_1 \left\| u_n^M - u_*^M \right\|^2, \tag{16.26}$$

$$\left\| u_n^M - u_n \right\| \leq c_2 z_{MN}$$

$$\left\| F_N' \left(u_n^M \right) - F'(u_n) \right\| \leq c_3 z_{MN}$$

and

$$\left\| u_n^M - u_*^M \right\| \leq \|u_n - u_*\| + c_4 z_{MN}.$$

Proof. We first show the convergence of the discretized Newton's method for all u_0^M in a suitable ball around u_*. Since the assumptions of Theorem 16.1 are satisfied, the existence of a solution $u_*^M \in \hat{U}_*$ is guaranteed. We shall show that the discretized Newton method converges to u_*^M if $u_0^M \in U(u_*, r_2)$, where

$$r_2 = \min \left\{ \frac{1}{(2L_0 + L)\delta}, \frac{\hat{r}_*}{2} \right\}.$$

The estimates

$$\left\| u_*^M - u_* \right\| + \left\| u_0^M - u_*^M \right\| \leq 2 \left\| u_*^M - u_* \right\| + \left\| u_0^M - u_* \right\|$$
$$\leq 2 z_{MN} + r_2 \leq \hat{r}_*$$

imply $U \left(u_*^M, \left\| u_0^M - u_*^M \right\| \right) \subset \hat{U}_*$. Hence, Assumptions C1–C5 hold in $U \left(u_*^M, \left\| u_0^M - u_*^M \right\| \right)$.

We can also have

$$\left\| F''(u_*)^{-1} \right\| \left\| F_N''(u_*^M) - F''(u_*) \right\|$$
$$\leq \delta \left[\left\| F_N''(u_*^M) - F_N''(u_*) \right\| + \left\| F_N''(u_*) - F''(u_*) \right\| \right]$$
$$\leq \delta \left[L \left\| u_*^M - u_* \right\| + z_{MN} \right] \qquad (16.27)$$
$$\leq (\delta L + 1) z_{MN}$$
$$\leq \frac{\delta L + \frac{1}{2}}{(2_0 + 3L)\delta + 1} < 1,$$

where we used

$$z_n^{(1)} \leq \frac{1}{2\delta} z_{MN} \qquad \text{(by (16.10))}.$$

It follows by (16.27) and the Banach Lemma on invertible operators that $F_N''(u_*^M)^{-1}$ exists, and

$$\left\| F_N''(u_*^M)^{-1} \right\| \leq \frac{\delta}{1 - \left(\delta L + \frac{1}{2}\right) z_{MN}}.$$

By the theorem on quadratic convergence of Newton's method and since all assumptions hold, the convergence to u_*^M has been established.

Using a refined formulation of this theorem given by us in [151], the convergence is guaranteed for all $u_0^M \in U(u_*^M, r_3)$, where

$$r_3 = \frac{2}{(2L_0 + L) \left\| F_N''(u_*^M)^{-1} \right\|}.$$

Therefore we should show

$$U(u_*, r_2) \subset U(u_*^M, r_3)$$

or equivalently,

$$\left\| u_0^M - u_*^M \right\| \leq \left\| u_0^M - u_* \right\| + \left\| u_* - u_*^M \right\|$$
$$\leq r_2 + z_{MN}$$
$$\leq \frac{1}{(2L_0 + L)\delta} + z_{MN}$$
$$= \frac{1 + (2L_0 + L)\delta}{(2L_0 + L)\delta}$$
$$\leq \frac{2\left(1 - L\delta z_{MN} - \frac{1}{2} z_{MN}\right)}{(2L_0 + L)\delta}$$
$$\leq \frac{2}{(2L_0 + L) \left\| F_N''(u_*^M)^{-1} \right\|} = r_3.$$

Hence, the discretized Newton's method converges to u_*^M for all $u_0^M \in U(u_*, r_2)$ such that (16.26) holds for $c_1 = \delta L$.

Next a proof by induction is used to show

$$\left\| u_n^M - u_n \right\| \leq \tau \leq c_2 z_{MN} \tag{16.28}$$

for all $u_0^M \in U(u_*, r_1)$, $r_1 = \frac{3}{4} r_2$, where τ is given by

$$\tau = \frac{2 \left(\frac{1}{2} + \| u_0 - u_* \| \right) z_{MN}}{b^2 + \sqrt{b^2 - 6L\delta \left(\frac{1}{2} + \| u_0 - u_* \| \right) z_{MN}}}$$

$$\leq \frac{\left(\frac{1}{2} + \| u_0 - u_* \| \right) z_{MN}}{b^2} =: c_2 z_{MN}$$

with $b = 1 + \frac{1}{2} z_{MN} - 2\delta L \left\| u_0^M - u_* \right\|$. The constant τ is well defined, since the inequalities

$$6L\delta \left(\frac{1}{2} + \| u_0 - u_* \| \right) z_{MN} \leq \frac{2L\delta + 1}{4 \left((2L_0 + 3L) \delta + 1 \right)} < \frac{1}{4} \text{ and } b \geq 1 - 2\delta L \rho_1 \geq \frac{1}{2}$$

imply

$$b^2 \geq \frac{1}{4} \geq 6 L \delta \left(\frac{1}{2} + \| u_0 - u_* \| \right) z_{MN}.$$

While the assertion (16.28) is fulfilled by assumption for $n = 0$, the induction step is based on the simple decomposition

$$\begin{aligned}
u_{i+1}^M - u_{i+1} = F_N'' \left(u_i^M \right)^{-1} \Big\{ &F_N'' \left(u_i^M \right) \left(u_i^M - u_i \right) - F_N' \left(u_i^M \right) + F_N' (u_i) \\
&+ \left(F_N'' \left(u_i^M \right) - F_N'' (u_i) \right) F'' (u_i)^{-1} F' (u_i) \\
&+ F_N'' (u_i) F'' (u_i)^{-1} F' (u_i) - F' (u_i) \\
&+ F' (u_i) - F_N' (u_i) \Big\}.
\end{aligned} \tag{16.29}$$

Assumptions C1–C4, equation (16.28), and the definition of z_{MN} imply

$$\begin{aligned}
\delta \left\| F_N'' \left(u_i^M \right) - F'' (u_i) \right\| &\leq \delta \left\| F_N'' \left(u_i^M \right) - F'' (u_i) \right\| + \delta \left\| F_N'' (u_i) - F'' (u_i) \right\| \\
&\leq \delta \left(L\tau + z_N^{(1)} \right) \\
&\leq \delta L\tau + \frac{1}{2} z_{MN} \\
&\leq \frac{\delta L z_{MN} + 2 \| u_0 - u_* \| \delta L z_{MN}}{1 - 2\delta L \| u_0 - u_* \|} + \frac{1}{2} z_{MN} \\
&\leq \frac{\frac{1}{3} \delta L + \frac{1}{4}}{(2L_0 + 3L) \delta + 1} < 1
\end{aligned}$$

resulting in the inequality

$$\left\| F_N'' \left(u_i^M \right)^{-1} \right\| \leq \frac{\delta}{1 - \left(L\delta\tau + \frac{1}{2} z_{MN} \right)}.$$

We obtain

$$\left\| F_N'' \left(u_i^M \right) \left(u_i^M - u_i \right) - F_N' \left(u_i^M \right) + F_N' (u_i) \right\| \leq \frac{1}{2} L \left\| u_i^M - u_i \right\|^2 \leq \frac{1}{2} L\tau^2,$$

and the convergence assertion $\|u_i - u_*\| \leq \|u_0 - u_*\|$ yields

$$\left\|\left(F_N''\left(u_i^M\right) - F_N''\left(u_i\right)\right) F''\left(u_i\right)^{-1} F'\left(u_i\right)\right\| \leq L\left\|u_i^M - u_i\right\|\left\|u_i - u_{i+1}\right\|$$

$$\leq 2L\tau\left\|u_0 - u_*\right\|.$$

The assumptions of the Theorem lead to

$$\left\|F_N''\left(u_i\right) F''\left(u_i\right)^{-1} F'\left(u_i\right) - F'\left(u_i\right)\right\| \leq$$

$$\leq \left\|-F_N''\left(u_i\right)\left(u_{i+1} - u_i\right) + F''\left(u_i\right)\left(u_{i+1} - u_i\right)\right\|$$

$$\leq \left\|F_N''\left(u_i\right) - F''\left(u_i\right)\right\|\left\|u_{i+1} - u_i\right\|$$

$$\leq z_n^{(1)} 2\left\|u_0 - u_*\right\|$$

$$\leq \frac{1}{8} z_{MN}\left\|u_0 - u_*\right\|$$

and $\|F'(u_i) - F_N'(u_i)\| \leq z_n^{(1)} \leq \frac{1}{28} z_{MN}$. Using the decomposition (16.29) the last inequalities complete the induction proof by

$$\left\|u_{i+1}^M - u_{i+1}\right\| \leq$$

$$\leq \frac{\delta}{1 - \left(L\delta\tau + \frac{1}{2}z_{MN}\right)}\left\{\frac{1}{2}L\tau^2 + 2L\left\|u_0 - u_*\right\|\tau + \left(\frac{1}{2} + \left\|u_0 - u_*\right\|\right)\frac{z_{MN}}{\delta}\right\}$$

$$= \tau.$$

The last equality is based on the fact that τ is equal to the smallest solution of the quadratic equation $3L\delta\tau^2 - 2b\tau + 2z_{MN}\left(\frac{1}{2} + \|u_0 - u_*\|\right) = 0$.

Finally, inequality (16.28) is shown by

$$\left\|F_N'\left(u_n^M\right) - F'\left(u_n\right)\right\| \leq \left\|F_N'\left(u_n^M\right) - F_N'\left(u_n\right)\right\| + \left\|F_N'\left(u_n\right) - F'\left(u_n\right)\right\|$$

$$\leq L\left\|u_n^N - u_n\right\| + z_{MN}$$

$$\leq \left(Lc_2 + 1\right) z_{MN} =: c_3 z_{MN}$$

and inequality (16.25) results from

$$\left\|\left(u_n^M - u_*^M\right) - \left(u_n - u_*\right)\right\| \leq \left\|u_n^M - u_n\right\| + \left\|u_*^M - u_*\right\|$$

$$\leq c_2 z_{MN} + z_{MN}$$

$$\leq \left(c_2 + 1\right) z_{MN} =: c_4 z_{MN}$$

The proof is completed. $\diamond$

Remark 16.4. The upper bounds on z_{MN} and r_1 were defined in [501] by

$$\frac{1}{6}\min\left\{\frac{\hat{r}_*}{4}, \frac{1}{6L\delta + 1}\right\} \tag{16.30}$$

and

$$\frac{3}{4}\min\left\{\frac{1}{3L\delta}, \frac{\hat{r}_*}{2}\right\} \tag{16.31}$$

respectively. By comparing (16.23), (16.24) with (16.30) and (16.31) respectively we conclude that our choices of z_{MN} and r_1 (or r_2 or r_3) are finer than the ones in [501]. However we leave the details to the motivated reader.

Chapter 17

Variational Inequalities

In this chapter, we present a Kantorovich–type semilocal convergence analysis of the Newton–Josephy method for solving a certain class of variational inequalities. By using a combination of Lipschitz and center–Lipschitz conditions, and our new idea of recurrent functions, we provide an analysis with the following advantages over the earlier works [726], [727] (under the same or less computational cost): weaker sufficient convergence conditions, larger convergence domain, finer error bounds on the distances involved, and an at least as precise information on the location of the solution.

We are concerned with the problem of approximating a solution $x^\star \in \mathcal{D}$ of variational inequalities $VI(\mathcal{D}, F)$, such that:

$$(y - x^\star)^T F(x^\star) \geq 0, \quad \text{for all } y \in \mathcal{D}, \tag{17.1}$$

where F is a Fréchet–differentiable operator defined on a non–empty, convex, and closed subset $\mathcal{D}$ of $\mathbb{R}^n$, with values in $\mathbb{R}^n$.

Moreover, the nonlinear complementarity problem is a special case of (17.1). In this case, we seek: $x^\star \in \mathcal{D} = \mathbb{R}^n_+$, such that

$$x^{\star T} F(x^\star) = 0, \quad F(x^\star) \in \mathcal{D}.$$

If $F(x) = \nabla G(x)$ $(x \in \mathcal{D})$ in (17.1), where $G : \mathcal{D} \longrightarrow \mathbb{R}$ is a real differentiable function, then, (17.1) corresponds to the first order necessary optimality condition of the following optimization–problem

$$\min_{x \in \mathcal{D}} G(x).$$

A particular case of the feasible set $\mathcal{D}$ in (17.1) is studied in [294], where,

$$\mathcal{D} = \{x \in \mathbb{R}^n : g_i(x) \geq 0, \ i = 1, \ldots, m; \ h_j(x) = 0, \ j = 1, \ldots, l\},$$

and g_i $(1 \leq i \leq m)$, h_j $(1 \leq j \leq l)$ are concave, affine real functions (see, [294], [738], [739]). An elegant extensive study on the theory, applications of variational inequalities, and their complete connections with other well–known problems in mathematical programming in finite dimensional spaces is presented in [420].

Newton–Josephy method (NJM) generates a sequence $\{x_k\}$, so that x_{k+1} solves the F_k k–th linearized subproblem $VI(\mathcal{D}, F_k)$, where

$$F_k(x) = F(x_k) + F'(x_k)(x - x_k), \quad (k \geq 0), \tag{17.2}$$

and

$$(F'(x))_{ij} = \left(\frac{\partial F_i(x)}{\partial x_j} \right) \quad [155], [477], [568]. \tag{17.3}$$

(NJM) is used in [664] for solving monotone variational inequalities. Moreover, the superlinear rate of convergence is established under standard conditions. A perturbed version of (NJM) is presented in [465]. Using D–gap function, the constrained variational inequality problem is solved by (17.2) in [580]. There is an extensive literature on the local as well as the semilocal convergence of Newton–type methods under various Lipschitz–type hypotheses on F'. A survey of such results can be found in [155], [189], [477], [568].

In this chapter, we are motivated by the works of Wang–Shen in [727], Wang [724]–[726], and optimization considerations. Wang and Shen provided a Newton–Kantorovich–type semilocal convergence result requiring computations only at the initial guess x_0. Their studies improve earlier works with more complicated hypotheses, which are difficult, or impossible to test [420], [470], [625], [625].

Using automatic differentiation, and interval computation [9], techniques of hypotheses in [727] can be tested. As far as we know the studies in [726], [727] provide so far the weakest sufficient convergence conditions for (NJM) under certain hypotheses (see Theorem 17.2).

In this chapter, we expand the applicability of (NJM). First, we introduce Lipschitz and center–Lipschitz conditions on F'. Using a combination of the two, and our new idea of recurrent functions, we provide under less or the same computational cost a new semilocal convergence analysis of (NJM) with the following advantages over the works in [727], [726]:

(a) Weaker sufficient convergence conditions;

(b) Larger convergence domain;

(c) Finer error bounds on the distances $\| x_{k+1} - x_k \|$, $\| x_k - x^\star \|$ $(k \geq 0)$;

(d) An at least as precise information on the location of the solution $x^\star$.

Finally, we provide a numerical example, where we compare our error bounds with the corresponding ones in [727], and [726].

The Euclidean norm is used in this chapter. We shall use three lemmas:

Lemma 17.1 ([727]). *Let $B \in \mathbb{R}^{n \times n}$ be symmetric. Then, B is positively definite if and only if there exists a symmetric positive definite matrix A such that:*

$$\| A - B \| \, \| A^{-1} \| < 1.$$

If the above inequality holds, then

$$\| B^{-1} \| \leq \frac{\| A^{-1} \|}{1 - \| I - A^{-1} B \|} \leq \frac{\| A^{-1} \|}{1 - \| A^{-1} \| \, \| A - B \|}.$$

Lemma 17.2 ([271]). *Let $\mathcal{D} \subseteq \mathbb{R}^n$ be nonempty, closed and convex. If $F : \mathcal{D} \subseteq \mathbb{R}^n \longrightarrow \mathbb{R}^n$ is strictly monotone in $\mathcal{D}$, then, $VI(\mathcal{D}, F)$ has at most one solution.*
Moreover, if it is continuous and strongly monotone, then $VI(\mathcal{D}, F)$ is uniquely solvable.

We finally need a lemma on majorizing sequences for (NJM).

Lemma 17.3. *Assume there exist constants $L_0 > 0$, $L > 0$, with $L_0 \leq L$, and $\eta > 0$, such that:*

$$h = \alpha \, \eta \leq \frac{1}{2}, \tag{17.4}$$

where,

$$\alpha = \frac{L_0 \, (2 \, L_0 + L + \sqrt{(2 \, L_0 + L)^2 + 24 \, L^2})}{6 \, (2 \, L_0 - 5 \, L + \sqrt{(2 \, L_0 + L)^2 + 24 \, L^2})}. \tag{17.5}$$

Then, sequence $\{t_k\}$ $(k \geq 0)$ given by

$$t_0 = 0, \quad t_1 = \eta,$$
$$t_{k+1} = t_k + \frac{L \, (t_k - t_{k-1})^2}{2 \, (1 - \frac{1}{3} \, (L_0 \, t_{k-1} + L \, (t_k - t_{k-1})))} \quad (k \geq 1), \tag{17.6}$$

is well defined, nondecreasing, bounded from above by $t^{\star\star}$, and converges to its unique least upper bound $t^\star \in [0, t^{\star\star}]$, where,

$$t^{\star\star} = \frac{2 \, \eta}{2 - \delta}, \tag{17.7}$$

$$\delta = \frac{12 \, L}{2 \, L_0 + L + \sqrt{(2 \, L_0 + L)^2 + 24 \, L^2}}. \tag{17.8}$$

Proof. The denominator in (17.5) is positive, since $L_0 > 0$, and $L > 0$.
We shall first show

$$0 \leq t_{k+1} - t_k \leq \frac{\delta}{2} \, (t_k - t_{k-1}) \leq \cdots \leq \left(\frac{\delta}{2}\right)^k \eta, \quad (k \geq 1), \tag{17.9}$$

and

$$L_0 \, t_{k-1} + L \, (t_k - t_{k-1}) < 3 \tag{17.10}$$

hold, using induction on k.
For $k = 1$, (17.9), and (17.10) become:

$$0 \leq t_2 - t_1 \;=\; \frac{L \, \eta^2}{2 \, (1 - \frac{1}{3} \, L \, \eta)} \leq \frac{\delta}{2} \, \eta, \tag{17.11}$$

$$L \, \eta < 1, \tag{17.12}$$

respectively. It follows from (17.4)–(17.6), and (17.8) that estimates (17.9), and (17.10) hold for $k = 1$.

Let us assume (17.9), and (17.10) hold for all $m \leq k$. Then in view of (17.9), we have:

$$t_{k+1} - t_k \leq \left(\frac{\delta}{2}\right)^k \eta, \qquad (17.13)$$

and

$$t_k \leq \frac{1 - \left(\frac{\delta}{2}\right)^k}{1 - \frac{\delta}{2}} \eta < t^{**}. \qquad (17.14)$$

By the induction hypotheses to show (17.9), and (17.10), it suffices:

$$L\,(t_{k+1} - t_k) + \frac{\delta}{3}\,(L_0\,t_k + L\,(t_{k+1} - t_k)) \leq \delta, \qquad (17.15)$$

or

$$\left\{ L\left(\frac{\delta}{2}\right)^k + \frac{\delta}{3}\left(L_0 \frac{1 - \left(\frac{\delta}{2}\right)^k}{1 - \frac{\delta}{2}} + L\left(\frac{\delta}{2}\right)^k \right) \right\} \eta \leq \delta. \qquad (17.16)$$

Estimate (17.16) motivates us (after dividing (17.16) by $\frac{\delta}{2}$) to set $s = \frac{\delta}{2}$, and introduce functions f_k on $[0, 1)$:

$$f_k(s) = L\,s^{k-1}\,\eta + \frac{2}{3}\left(L_0\,(1 + s + s^2 + \cdots + s^{k-1}) + L\,s^k \right)\eta - 2, \quad (k \geq 1). \qquad (17.17)$$

We shall show instead of (17.16):

$$f_k(\frac{\delta}{2}) \leq 0 \qquad (k \geq 1). \qquad (17.18)$$

We need a relationship between two consecutive functions f_k:

$$
\begin{aligned}
f_{k+1}(s) \\
=\ & L\,s^k\,\eta + \frac{2}{3}\left(L_0\,(1 + s + s^2 + \cdots + s^{k-1} + s^k) + L\,s^{k+1} \right)\eta - 2 \\
=\ & f_k(s) + L\,s^k\,\eta + \frac{2}{3}\left(L_0\,(1 + s + \cdots + s^k) + L\,s^{k+1} \right)\eta - 2 - \\
& L\,s^{k-1}\,\eta - \frac{2}{3}\left(L_0\,(1 + s + \cdots + s^{k-1}) + L\,s^k \right)\eta + 2 \\
=\ & f_k(s) + g(s)\,s^{k-1}\,\eta,
\end{aligned}
\qquad (17.19)
$$

where,

$$g(s) = \frac{1}{3}\,(2\,L\,s^2 + (2\,L_0 + L)\,s - 3\,L). \qquad (17.20)$$

Note that $\frac{\delta}{2}$, given by (17.8) is the unique positive root of polynomial g.

Define function f_∞ on $[0,1)$ by

$$f_\infty(s) = \lim_{k \to \infty} f_k(s). \tag{17.21}$$

Then, we have by (17.17), and (17.21):

$$f_\infty\left(\frac{\delta}{2}\right) = \lim_{k \to \infty} f_k\left(\frac{\delta}{2}\right) = 2\left\{\frac{L_0\,\eta}{3\left(1-\frac{\delta}{2}\right)} - 1\right\}. \tag{17.22}$$

Using (17.19), and (17.20), we get:

$$f_\infty\left(\frac{\delta}{2}\right) = f_k\left(\frac{\delta}{2}\right) = f_{k-1}\left(\frac{\delta}{2}\right) = \cdots = f_1\left(\frac{\delta}{2}\right). \tag{17.23}$$

It follows from (17.18), and (17.23), that we only need to show

$$f_\infty\left(\frac{\delta}{2}\right) \le 0, \tag{17.24}$$

which is true by (17.22), and (17.4).

Hence, estimate (17.18) holds.

It follows that sequence $\{t_n\}$ is nondecreasing, bounded from above by $t^{\star\star}$, and as such it converges to its unique least upper bound $t^\star \in [0, t^{\star\star}]$.

That completes the proof of Lemma 17.3. $\diamond$

We can provide the main semilocal convergence result for (NJM):

Theorem 17.1. *Let $F : \mathcal{D} \subseteq \mathbb{R}^n \longrightarrow \mathbb{R}^n$ be a Fréchet–differentiable operator.*

Assume:

there exist $x_0 \in \mathcal{D}$, constants $\beta > 0$, $\gamma_0 > 0$, $\gamma > 0$, $\eta > 0$, with $\gamma_0 \le \gamma$, such that for all $x, y \in \mathcal{D}$:

$F'(x_0)$ is positive definite, x_1 is the unique solution of subproblem $VI(\mathcal{D}, F_0)$

$$\|\widetilde{F'(x_0)}^{-1}\| \le \beta, \qquad \|x_1 - x_0\| \le \eta, \tag{17.25}$$

$$\|F'(x) - F'(x_0)\| \le \gamma_0 \,\|x - x_0\|, \tag{17.26}$$

$$\|F'(x) - F'(y)\| \le \gamma \,\|x - y\|, \tag{17.27}$$

$$h_{AH} = \overline{\alpha}\,\eta \le \frac{1}{2}, \tag{17.28}$$

$$\overline{U}(x_0, r^\star) = \{x \in \mathbb{R}^n,\, \|x - x_0\| \le r^\star\} \subseteq \mathcal{D}, \tag{17.29}$$

where,

$$\widetilde{F'(x)} = \frac{1}{2}\left(F'(x) + F'(x)^T\right) \qquad x \in \mathcal{D}, \tag{17.30}$$

$$\overline{\alpha} = \frac{\beta\,\gamma_0}{2}\,\frac{2\,\gamma_0 + \gamma + \sqrt{(2\,\gamma_0 + \gamma)^2 + 24\,\gamma^2}}{2\,\gamma_0 - 5\,\gamma + \sqrt{(2\,\gamma_0 + \gamma)^2 + 24\,\gamma^2}}, \tag{17.31}$$

$$r^\star = \lim_{k \to \infty} r_k, \tag{17.32}$$

$$r_0 = 0, \quad r_1 = \eta,$$

$$r_{k+1} = r_k + \frac{3\,\beta\,\gamma\,(r_k - r_{k-1})^2}{2\,(1 - (\gamma_0\,\beta\,r_{k-1} + \gamma\,\beta\,(r_k - r_{k-1})))} \qquad (k \geq 1). \tag{17.33}$$

Then,

(a) *Scalar sequence $\{r_k\}$ is nondecreasing, bounded from above by $r^{\star\star}$, which is given by:*

$$r^{\star\star} = \frac{2\,\eta}{2 - \delta}, \tag{17.34}$$

and converges to its unique least upper bound $r^\star \in [0, r^{\star\star}]$. Moreover, estimate (17.9) holds, with, L_0, L, $t^\star$, $t^{\star\star}$, h given by

$$L_0 = 3\,\gamma_0\,\beta, \ \ L = 3\,\gamma\,\beta, \ \ t^\star = r^\star, \ \ t^{\star\star} = r^{\star\star}, \ \ h = h_{AH}. \tag{17.35}$$

(b) *Sequence $\{x_k\}$ generated by (NJM) is well defined, remains in $\overline{U}(x_0, r^\star)$ for all $k \geq 0$, and converges to a unique solution $x^\star \in \overline{U}(x_0, r^\star)$ of $VI(\mathcal{D}, F)$.*

Moreover the following estimates hold for all $k \geq 0$:

$$\| x_{k+1} - x_k \| \leq r_{k+1} - r_k, \tag{17.36}$$

and

$$\| x_k - x^\star \| \leq r^\star - r_k. \tag{17.37}$$

Proof.

(a) This part of the proof follows immediately from Lemma 17.3 by simply using (17.35).

(b) By using induction on k, we shall show that sequence $\{x_k\}$ is well defined,

$$\| x_{k+1} - x_k \| \leq r_{k+1} - r_k, \tag{17.38}$$

and

$$\overline{U}(x_{k+1}, r^\star - r_{k+1}) \subseteq \overline{U}(x_k, r^\star - r_k) \tag{17.39}$$

hold for all k.

Note that in view of (17.33):

$$r^\star \geq r_1. \tag{17.40}$$

Let $z \in \overline{U}(x_1, r^\star - r_1)$. Then, estimate

$$\begin{aligned} \| z - x_0 \| \ &\leq \ \| z - x_1 \| + \| x_1 - x_0 \| \\ &\leq \ r^\star - r_1 + r_1 - r_0 = r^\star - r_0, \end{aligned} \tag{17.41}$$

implies $z \in \overline{U}(x_0, r^\star - r_0)$.

We also have

$$\| x_1 - x_0 \| \leq \eta = r_1 - r_0, \tag{17.42}$$

which together with (17.41) show (17.38), and (17.39) for $k = 0$. Let us assume $\{x_m\}$ is well defined, and (17.38), (17.39) hold for all $m \leq k$.

Let $x \in \overline{U}(x_0, r^\star)$. Using (17.26), and (17.30), we get:

$$
\begin{aligned}
& \beta \parallel \widetilde{F}'(x) - \widetilde{F}'(x_0) \parallel \\
= \ & \beta \left\| \frac{1}{2}(F'(x) + F'(x)^T) - \frac{1}{2}(F'(x_0) + F'(x_0)^T) \right\| \\
\leq \ & \beta \parallel F'(x) - F'(x_0) \parallel \\
\leq \ & \beta \gamma_0 \parallel x - x_0 \parallel \leq \beta \gamma_0 \, r^\star = L_0 \, t^\star < 1.
\end{aligned}
\tag{17.43}
$$

By Lemma 17.1, we have $\widetilde{F}'(x)$, $F'(x)$ are positively definite, and

$$
\parallel \widetilde{F}'(x)^{-1} \parallel \leq \frac{\beta}{1 - \beta \gamma_0 \parallel x - x_0 \parallel}.
\tag{17.44}
$$

In particular, for $x = x_1$, Lemma 17.1 implies $VI(\mathcal{D}, F_1)$ is uniquely solvable, and consequently x_2 is well defined.

We have by the induction hypotheses

$$
\begin{aligned}
\parallel x_k - x_0 \parallel \ & \leq \ \sum_{i=1}^{k} \parallel x_i - x_{i-1} \parallel \\
& \leq \ \sum_{i=1}^{k} (r_i - r_{i-1}) = r_k - r_0 = r_k,
\end{aligned}
\tag{17.45}
$$

and

$$
\begin{aligned}
\parallel x_k + \theta(x_k - x_{k-1}) - x_0 \parallel \ & \leq \ r_{k-1} + \theta(r_k - r_{k-1}) \\
& \leq \ r^\star,
\end{aligned}
\tag{17.46}
$$

for all $\theta \in (0, 1)$.

Using (17.26), (17.27), and (17.38), we obtain the estimate:

$$
\begin{aligned}
& \parallel F(x_{k-1}) - F(x_k) - F'(x_k)(x_{k-1} - x_k) \parallel \\
& \left\| \int_0^1 (F'(x_k + \theta(x_{k-1} - x_k)) - F'(x_k))(x_{k-1} - x_k) \, d\theta \right\| \\
& \leq \gamma \int_0^1 \parallel x_k + \theta(x_{k-1} - x_k) - x_k \parallel d\theta \parallel x_{k-1} - x_k \parallel \\
& = \frac{\gamma}{2} \parallel x_{k-1} - x_k \parallel^2 \leq \frac{\gamma}{2}(r_{k-1} - r_k)^2.
\end{aligned}
\tag{17.47}
$$

The induction hypotheses are:

$$
\begin{aligned}
& x_{k-1}, x_k \in \mathcal{D}, \\
& x_k, x_{k+1} \text{ solve } VI(\mathcal{D}, F_{k-1}), \ VI(\mathcal{D}, F_k), \ respectively,
\end{aligned}
\tag{17.48}
$$

$$
\begin{aligned}
& (x_{k+1} - x_k)^T F_{k-1}(x_k) \\
& = (x_{k+1} - x_k)^T (F(x_{k-1}) + F'(x_{k-1})(x_k - x_{k-1})) \geq 0,
\end{aligned}
\tag{17.49}
$$

and

$$(x_k - x_{k+1})^T \, F_k(x_{k+1})$$
$$= (x_k - x_{k+1})^T \, (F(x_k) + F'(x_k) \, (x_{k+1} - x_k)) \geq 0. \tag{17.50}$$

It then follows that estimates (17.49), and (17.50) give in turn:

$$(x_k - x_{k+1})^T \, F'(x_{k-1}) \, (x_k - x_{k+1})$$
$$\leq (x_k - x_{k+1})^T \, \Big(F(x_k) - F(x_{k-1}) + F'(x_k) \, (x_{k+1} - x_k) -$$
$$F'(x_{k-1}) \, (x_{k+1} - x_{k-1}) \Big)$$
$$= (x_k - x_{k+1})^T \, \Big((F'(x_k) - F'(x_{k-1})) \, (x_{k+1} - x_{k-1}) -$$
$$(F(x_{k-1}) - F(x_k) - F'(x_k) \, (x_{k-1} - x_k)) \Big) \tag{17.51}$$
$$\leq \| x_k - x_{k+1} \| \, \Big(\| F'(x_k) - F'(x_{k-1}) \| \, \| x_{k+1} - x_{k-1} \| +$$
$$\| F(x_{k-1}) - F(x_k) - F'(x_k) \, (x_{k-1} - x_k) \| \Big)$$
$$\leq \Big(\gamma \, (\| x_{k+1} - x_k \| + \| x_k - x_{k-1} \|) \, \| x_k - x_{k-1} \| +$$
$$\frac{\gamma}{2} \, \| x_k - x_{k-1} \|^2 \Big) \, \| x_{k+1} - x_k \| .$$

We also have:

$$(x_k - x_{k+1})^T \, F'(x_{k-1}) \, (x_k - x_{k+1})$$
$$= (x_k - x_{k+1})^T \, \widetilde{F}'(x_{k-1}) \, (x_k - x_{k+1}) \tag{17.52}$$
$$\geq \frac{\| x_{k+1} - x_k \|^2}{\| F'(x_{k-1})^{-1} \|} \geq \frac{\| x_{k+1} - x_k \|^2}{\beta_{k-1}},$$

where,

$$\| F'(x_{k-1})^{-1} \| \leq \beta_{k-1} \leq \frac{\beta}{1 - \beta \, \gamma_0 \, \| x_{k-1} - x_0 \|}, \quad (k > 1), \tag{17.53}$$

and

$$\beta_0 = \beta. \tag{17.54}$$

In view of (17.51), and (17.52), we obtain:

$$\| x_{k+1} - x_k \| \leq \beta_{k-1} \Big(\gamma \, \| x_k - x_{k-1} \| \, (\| x_{k+1} - x_k \| +$$
$$\| x_k - x_{k-1} \|) + \frac{\gamma}{2} \, \| x_k - x_{k-1} \|^2 \Big), \tag{17.55}$$

or

$$\| x_{k+1} - x_k \| \leq \frac{3 \, \gamma \, \beta_{k-1} \, \| x_k - x_{k-1} \|^2}{2 \, (1 - \beta_{k-1} \, \gamma \, \| x_k - x_{k-1} \|)}$$
$$\leq \frac{3 \, \gamma \, \beta \, (r_k - r_{k-1})^2}{2 \, (1 - \beta \, (\gamma_0 \, (r_{k-1} - r_0) + \gamma \, (r_k - r_{k-1})))} \tag{17.56}$$
$$= r_{k+1} - r_k,$$

Chapter 17. Variational Inequalities　213

which implies (17.38) holds for all k.

Let $w \in \overline{U}(x_{k+1}, r^\star - r_{k+1})$, then we have:

$$
\begin{aligned}
\| w - x_k \| &\leq \| w - x_{k+1} \| + \| x_{k+1} - x_k \| \\
&\leq r^\star - r_{k+1} + r_{k+1} - r_k = r^\star - r_k.
\end{aligned}
\tag{17.57}
$$

Hence, we deduce:

$$
w \in \overline{U}(x_k, r^\star - r_k).
\tag{17.58}
$$

The induction for (17.38), and (17.39) is completed. Lemma 17.3 implies that $\{r_k\}$ is a complete sequence. In view of (17.38), and (17.39), $\{x_k\}$ $(k \geq 0)$ is also a complete sequence, and as such it converges to some $x^\star \in \overline{U}(x_0, r^\star)$ (since $\overline{U}(x_0, r^\star)$ is a closed set), with

$$
(v - x_{k+1})^T \left(F(x_k) + F'(x_k)(x_{k+1} - x_k) \right) \geq 0, \quad \text{for all } v \in \mathcal{D}.
\tag{17.59}
$$

Estimate (17.39) follows from (17.38) by using standard majorization techniques.

Finally, by letting $k \longrightarrow \infty$ in (17.59), and the continuity of F, F', we get:

$$
(x - x^\star)^T F(x^\star) \geq 0, \quad \text{for all } x \in \mathcal{D}.
\tag{17.60}
$$

Hence, we deduce $x^\star$ is a solution of $VI(\mathcal{D}, F)$.

That completes the proof of Theorem 17.1.　　　　　　　　　　　　　　　　$\diamondsuit$

Note that the case $\gamma = 0$ is not included, since it is already covered in [727].

Remark 17.1. Note that

$$
\gamma_0 \leq \gamma
\tag{17.61}
$$

holds in general, and $\dfrac{\gamma}{\gamma_0}$ can be arbitrarily large.

We state the corresponding Theorem 1 given by Wang and Shen [727], so we can compare it with our Theorem 17.1.

Theorem 17.2. *Let $F : \mathcal{D} \subseteq \mathbb{R}^n \longrightarrow \mathbb{R}^n$ be a twice Fréchet–differentiable operator. Assume:*
there exist $x_0 \in \mathcal{D}$, constants $\beta > 0$, $\gamma > 0$, and $\eta > 0$, such that for all $x, y \in \mathcal{D}$: $F'(x_0)$ is positive definite, x_1 is the unique solution of subproblem $VI(\mathcal{D}, F_0)$

$$
\| \widetilde{F'(x_0)}^{-1} \| \leq \beta, \qquad \| x_1 - x_0 \| \leq \eta,
\tag{17.62}
$$

$$
\| F''(x) \| \leq \gamma \quad \text{for all } x \in \mathcal{D},
\tag{17.63}
$$

$$
h_{WS} = \overline{\overline{\alpha}}\, \eta \leq \frac{1}{2},
\tag{17.64}
$$

and

$$
\overline{U}(x_0, s^\star) \subseteq \mathcal{D},
\tag{17.65}
$$

where, $\widetilde{F'(x)}$ is defined by (17.30),

$$\overline{\overline{\alpha}} = \frac{2\beta\gamma}{7-\sqrt{33}}, \tag{17.66}$$

$$s^\star = \lim_{k\to\infty} s_k, \tag{17.67}$$

$$s_0 = 0, \quad s_1 = \eta,$$

$$s_{k+1} = s_k + \frac{3\beta\gamma(s_k - s_{k-1})^2}{2(1-\gamma\beta s_{k-1})} \qquad (k \geq 1). \tag{17.68}$$

Then, sequence $\{x_k\}$ generated by (NJM) is well defined, remains in $\overline{U}(x_0,s^\star)$ for all $k \geq 0$, and converges to a unique solution $x^\star \in \overline{U}(x_0,s^\star)$ of $VI(\mathcal{D},F)$.

Moreover, the following estimates hold for all $k \geq 0$:

$$\| x_{k+1} - x_k \| \leq s_{k+1} - s_k, \tag{17.69}$$

and

$$\| x_k - x^\star \| \leq s^\star - s_k \leq \frac{1}{a}\left(\frac{3a}{2(1-a)}\right)^k \left(\frac{h_0}{a}\right)^{2^k - 1} \eta, \tag{17.70}$$

where,

$$a = \frac{7-\sqrt{33}}{4}, \quad \text{and} \quad h_0 = \beta\gamma\eta.$$

We can now compare Theorem 17.1 to Theorem 17.2

Remark 17.2. (a) **Improvement 1: Case $\gamma_0 = \gamma$.** Theorem 17.2 requires that operator F is twice–Fréchet differentiable. However, according to Theorem 17.1, condition (17.63) can be replaced by the weaker (17.27).

(b) **Improvement 2: Case $\gamma_0 < \gamma$.** It follows from (17.28), (17.31), (17.64), and (17.66) that

$$\overline{\alpha} < \overline{\overline{\alpha}}. \tag{17.71}$$

Proof of estimate (17.71). In view of (17.31), and (17.66), we must have:

$$\frac{\gamma_0}{2} \frac{2\gamma_0 + \gamma + \sqrt{(2\gamma_0 + \gamma)^2 + 24\gamma^2}}{2\gamma_0 - 5\gamma + \sqrt{(2\gamma_0 + \gamma)^2 + 24\gamma^2}} < \frac{2\gamma}{7 - \sqrt{33}}. \tag{17.72}$$

Set $\gamma_0 = \gamma t$, $(t \in (0,1))$. Then estimate (17.72) holds, if

$$f(t) < 0, \qquad t \in (0,1), \tag{17.73}$$

where,

$$f(t) = t\frac{2t + 1 + \sqrt{(2t+1)^2 + 24}}{2t - 5 + \sqrt{(2t+1)^2 + 24}} - \frac{4}{7 - \sqrt{33}}.$$

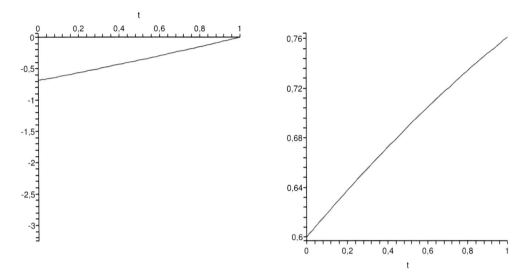

Figure 17.1. Left: Function f on intervall $[0,1]$. Right: Function f' on intervall $[0,1]$.

The set of solutions for inequality (17.73) is $(0,\infty)$. For validating this result, see Figure 17.1.

Using Maple to further validate (17.71), we derive f. Then, the derivative f' is strictly positive (see Figure 17.1), and f is strictly increasing.

Hence, we have
$$h_{WS} \leq \frac{1}{2} \Longrightarrow h_{AH} \leq \frac{1}{2}. \tag{17.74}$$

But not necessarily vice versa, unless if $\gamma_0 = \gamma$. Indeed, we then have:
$$\overline{\alpha} = \overline{\overline{\alpha}},$$

and
$$h_{WS} = h_{AH}.$$

Note that in this case, (17.4) or (17.74) coincide with the Kantorovich hypothesis (see also (17.68)).

Indeed, let us define scalar iteration $\{w_n\}$ by:
$$w_0 = 0, \quad w_1 = \eta, \quad w_{k+1} = w_k + \frac{\lambda(w_k - w_{k-1})^2}{2(1 - \lambda w_k)}, \quad (n \geq 1).$$

Then, the Kantorovich hypothesis for the convergence of sequence $\{w_k\}$ is given by
$$h_K = \lambda \eta \leq \frac{1}{2}.$$

Note that, iteration $\{w_k\}$ is not exactly the same as $\{s_k\}$ given by (17.68). That is why (17.4) or (17.64) do not look exactly like the Kantorovich hypothesis. However, according to the proofs of Lemma 17.3, Theorem 1 in [727], (17.4), and (17.64), respectively are the correct analogs of the Kantorovich hypothesis.

Moreover, using (17.33), (17.68), and a simple induction argument, we conclude:

$$r_k < s_k, \qquad (k > 1), \tag{17.75}$$

$$r_{k+1} - r_k < s_{k+1} - s_k, \qquad (k > 1), \tag{17.76}$$

$$r^\star - r_k \leq s^\star - s_k, \qquad (k \geq 1), \tag{17.77}$$

and

$$r^\star \leq s^\star. \tag{17.78}$$

Note that under condition (17.28), the quadratic order of convergence for sequences $\{x_n\}$, and $\{s_n\}$ was established in [727] (see (17.70)). In view of (17.74), (17.76), and (17.77), our estimates (17.35), (17.36) are even tighter than (17.69), and (17.70), respectively.

(c) The limit point $r^\star$ in (17.29) can be replaced by $r^{\star\star}$ given in closed form by (17.34). Hence, we do know the location of the solution of each subprogram.

(d) It follows from the proof of Theorem 17.1 that sequence $\{v_k\}$ given by

$$v_0 = 0, \quad v_1 = \eta,$$

$$v_{k+1} = v_k + \frac{\beta\, \gamma_1\, (v_k - v_{k-1})^2}{2\, (1 - (\gamma_0\, \beta\, v_{k-1} + \overline{\gamma}_1\, \beta\, (v_k - v_{k-1})))} \qquad (k \geq 1) \tag{17.79}$$

is also majorizing sequence for $\{x_k\}$.

If $\gamma_0 < \gamma$, a simple inductive argument shows:

$$v_k < r_k, \qquad (k > 1), \tag{17.80}$$

$$v_{k+1} - v_k < r_{k+1} - r_k, \qquad (k \geq 1), \tag{17.81}$$

$$v^\star - v_k \leq r^\star - s_k, \qquad (k \geq 1), \tag{17.82}$$

and

$$v^\star = \lim_{k \to \infty} v_k \leq r^\star. \tag{17.83}$$

where,

$$\gamma_1 = \begin{cases} 3\,\gamma_0, & if \quad k = 1 \\ 3\,\gamma, & if \quad k > 1 \end{cases}, \qquad \overline{\gamma}_1 = \begin{cases} \gamma_0, & if \quad k = 1 \\ \gamma, & if \quad k > 1 \end{cases}. \tag{17.84}$$

The motivation that leads to the definition of sequence $\{v_k\}$ is the following: Simply notice that in (17.51), we used (17.27) for the derivation of the upper bound $\| x_2 - x_1 \|$ (given by (17.56)). However, the more precise than (17.27), condition (17.26) is actually needed.

Chapter 17. Variational Inequalities

All the above (see, Remarks 17.1 and 17.2) justify the claims made in the Introduction of this chapter. Hence the applicability of (NJM) has been extended under less, or the same computational cost. In practice, the computation of γ requires that of γ_0. Hence, (17.26) is not an additional hypothesis.

The proofs of the remaining results are omitted, since they can be obtained from the corresponding ones in [727] by simply replacing $\{s_k\}$, s^*, h_{WS} by $\{r_k\}$, r^*, h_{AH}, respectively.

Theorem 17.3. *Assume hypotheses of Theorem 17.1 hold. Then, for any $y_0 \in \overline{U}(x_0, \eta) \cap \mathcal{D}$, (NJM) $\{y_k\}$ is well defined, remains in $\overline{U}(x_0, r^*)$ for all $k \geq 0$, and converges to a solution y^* of $VI(\mathcal{D}, F)$, which is contained in $\overline{U}(x_0, \eta) \cap \mathcal{D}$.*

Theorem 17.4. *Assume hypotheses of Theorem 17.1 hold, and the limit x^* of (NJM) $\{x_k\}$ is included in $U(x_0, \eta)$, then x^* is the unique solution of $VI(\mathcal{D}, F)$ in $U(x_0, \eta)$.*

Theorem 17.5. *Assume hypotheses of Theorem 17.1 hold, and $h_{AH} < \dfrac{1}{2}$, then x^* is the unique solution of $VI(\mathcal{D}, F)$ in $U(x_0, \eta)$.*

Theorem 17.6. *Assume hypotheses of Theorem 17.1 hold, and $h_{AH} < \dfrac{1}{2}$, then starting from any $y_0 \in \overline{U}(x_0, \eta)$, (NJM) $\{y_k\}$ is quadratically convergent.*

Finally, the results obtained here can also be provided in affine invariant form, if we simply replace F by $\widetilde{F'(x_0)}^{-1} F$ [155], [477], [568]. The advantages of this approach have already been explained in detail [364].

We further extend the applicability of (NJM). In particular, we show Theorems 1.2, and 1.3 given by Wang [726] can always be replaced by the following corresponding results.

The following two results correspond to Theorems 1.2, and 1.3 in [726], respectively. The proofs are obtained by simply using (17.26) instead of (17.27) to obtain the upper bounds on the inverses of the operators involved, and majorizing sequence $\{r_k\}$ instead of $\{s_k\}$ (see also (17.44)). The error bounds are obtained by letting $L_0 = \gamma_0 \beta$, and $L = \gamma \beta$ in Lemma 3.1.

Theorem 17.7. *Let $F : \mathcal{D} \subseteq \mathbb{R}^n \longrightarrow \mathbb{R}^n$ be a Fréchet–differentiable operator.*
Assume:
there exist $x_0 \in \mathcal{D}$, constants $\beta > 0$, $\gamma_0 > 0$, $\gamma > 0$, $\eta > 0$, with $\gamma_0 \leq \gamma$, such that for all $x, y \in \mathcal{D}$:
$F'(x_0)$ is positive definite, x_1 is the unique solution of subproblem $VI(\mathcal{D}, F_0)$

$$F_0(x) = F(x_0) + F'(x_0)(x - x_0),$$
$$\| \widetilde{F'(x_0)}^{-1} \|_2 \leq \beta, \qquad \| x_1 - x_0 \|_2 \leq \eta,$$
$$\| F'(x) - F'(x_0) \|_2 \leq \gamma_0 \| x - x_0 \|_2,$$
$$\| F'(x) - F'(y) \|_2 \leq \gamma \| x - y \|_2,$$
$$h_{AH} = \overline{\alpha} \, \eta \leq \frac{1}{2}, \tag{17.85}$$
$$\overline{U}(x_0, r^*) = \{x \in \mathbb{R}^n, \, \| x - x_0 \|_2 \leq r^*\} \subseteq \mathcal{D},$$

where,

$$\overline{\alpha} = \frac{\beta}{8}\left(\gamma + 4\gamma_0 + \sqrt{\gamma^2 + 8\gamma_0\gamma}\right),$$

$$r^\star = \lim_{k\to\infty} r_k,$$

$$r_0 = 0, \quad r_1 = \eta, \quad r_{k+1} = r_k + \frac{\beta\gamma(r_k - r_{k-1})^2}{2(1 - \beta\gamma_0 r_{k-1})} \qquad (k \geq 1). \qquad (17.86)$$

Then,

(a) *Scalar sequence $\{r_k\}$ is nondecreasing, bounded from above by $r^{\star\star}$, which is given by:*

$$r^{\star\star} = \frac{2\eta}{2 - \delta},$$

and converges to its unique least upper bound $r^\star \in [0, r^{\star\star}]$, where

$$\delta = \frac{4\gamma}{\gamma + \sqrt{\gamma^2 + 8\gamma_0\gamma}}.$$

(b) *Sequence $\{x_k\}$ generated by (NJM) is well defined, remains in $\overline{U}(x_0, r^\star)$ for all $k \geq 0$, and converges to a unique solution $x^\star \in \overline{U}(x_0, r^\star)$ of $VI(\mathcal{D}, F)$.*

Moreover the following estimates hold for all $k \geq 0$:

$$\| x_{k+1} - x_k \|_2 \leq r_{k+1} - r_k \leq \left(\frac{\delta}{2}\right)^k (2h_{AH})^{2^k - 1}\eta,$$

and

$$\| x_k - x^\star \|_2 \leq r^\star - r_k \leq \left(\frac{\delta}{2}\right)^k \frac{(2h_{AH})^{2^k - 1}\eta}{1 - (2h_{AH})^{2^k}}, \qquad (2h_{AH} < 1).$$

Theorem 17.8. *Let $\ell = \{\ell_i\} \in \{\mathbb{R} \cup \{-\infty\}\}^n$, $u = \{u_i\} \in \{\mathbb{R} \cup \{+\infty\}\}^n$ be given with $\ell_i < u_i$, $i = 1, 2, \ldots, n$.*

Denote:

$$[\ell, u] = \{x = \{x_i\} \in \mathbb{R}^n : \ell_i \leq x_i \leq u_i, \quad i = 1, 2, \cdots, n\}.$$

$F : \mathcal{D} \supseteq [\ell, u] \longrightarrow \mathbb{R}^n$ is a Fréchet–differentiable operator defined on a convex set $\mathcal{D}$.

Assume there exist $x_0 \in \mathcal{D}$, constants $\beta > 0$, $\gamma_0 > 0$, $\gamma > 0$, $\eta > 0$, with $\gamma_0 \leq \gamma$, such that for all $x, y \in \mathcal{D}$:

$F'(x_0)$ is an $\mathcal{H}$–matrix whose diagonal elements are all positive,

$$\| \overline{F'(x_0)^{-1}} \|_\infty \leq \beta, \qquad \| x_1 - x_0 \|_\infty \leq \eta,$$

$$\| F'(x) - F'(x_0) \|_\infty \leq \gamma_0 \| x - x_0 \|_\infty,$$

$$\| F'(x) - F'(y) \|_\infty \leq \gamma \| x - y \|_\infty,$$

$$h_{AH} = \overline{\alpha}\eta \leq \frac{1}{2},$$

$$\overline{U}(x_0, r^\star) = \{x \in \mathbb{R}^n, \| x - x_0 \|_\infty \leq r^\star\} \subseteq \mathcal{D},$$

where, h_{AH} is given in Theorem 17.7, $\overline{F'(x_0)^{-1}}$ is the comparison matrix of $F'(x_0)$, and x_1 is the unique solution of subproblem $VI([\ell, u], F_0)$, where F_0 is given Theorem 17.7.

Then, the conclusions of Theorem 17.7 hold with $\| . \|_2$ replaced by the $\| . \|_\infty$ norm.

Chapter 17. Variational Inequalities

Remark 17.3. The majorizing sequence $\{s_k\}$ used in [726] is given by:

$$s_0 = 0, \quad s_1 = \eta, \quad s_{k+1} = s_k + \frac{\beta\,\gamma\,(s_k - s_{k-1})^2}{2\,(1 - \beta\,\gamma\,s_{k-1})} \qquad (k \geq 1). \qquad (17.87)$$

If $\gamma_0 = \gamma$, then $r_k = s_k$, $(k \geq 0)$, and our results reduce to the ones in [726]. Otherwise, they constitute an improvement (see, Remark 17.2).

Note also that if $\gamma_0 < \gamma$, then we have:

$$h_W = \beta\,\gamma\,\eta \leq \frac{1}{2} \Longrightarrow h_{AH} \leq \frac{1}{2},$$

since $\overline{\alpha} < \gamma$ (but not vice verca).

Moreover, according to Lemma 3.1, the ratio of quadratic convergence "$2\,h_{AH}$" (for $L = \gamma\,\beta$, $L_0 = \gamma_0\,\beta$) is smaller than "$2\,h_W$" given by Wang [726].

Furthermore, as noted in Remark 17.2, sequence $\{v_k\}$ given by:

$$v_0 = 0, \quad v_1 = \eta, \quad v_{k+1} = v_k + \frac{\beta\,\gamma_1\,(v_k - v_{k-1})^2}{2\,(1 - \beta\,\gamma_0\,v_{k-1})} \qquad (k \geq 1) \qquad (17.88)$$

is also majorizing for $\{x_k\}$, where,

$$\gamma_1 = \begin{cases} \gamma_0, & if \quad k = 1 \\ \gamma, & if \quad k > 1 \end{cases}.$$

For example that $\gamma_0 < \gamma$, $h_{AH} \leq \frac{1}{2}$, but $h_{WS} > \frac{1}{2}$, the reader can refer to Chapter 1 (see also Chapter 4).

Remark 17.4. We shall compare majorizing sequence $\{r_k\}$, $\{s_k\}$, $\{v_k\}$, with each other. Let us choose $\alpha := .7$ in Example 4.1, then, we have

$$\eta = .1, \quad \gamma_0 = 6.9, \quad \gamma = 7.8, \quad \beta = \frac{1}{3}.$$

The hypotheses of our Theorem 17.8, and Theorem 1.3 in [726] are satisfied. Then, we can provide the comparison table.

Comparison table.

	(17.2)	(17.88)	(17.86)	(17.87)
k	$\|x_{k+1} - x_k\|$	$v_{k+1} - v_k$	$r_{k+1} - r_k$	$s_{k+1} - s_k$
0	.1414213562	.1	.1	.1
1	0.1107329219e-1	0.1150000000e-1	0.1300000000e-1	0.1300000000e-1
2	0.1423335877e-3	0.2232792208e-3	0.2853246753e-3	0.2968918919e-3
3	0.2590350763e-7	0.8716252243e-7	0.1429985426e-6	0.1622603145e-6
4	0.3716428396e-10	0.1329207578e-13	0.3595020992e-13	0.4851937919e-13
5	$\sim$	0.3091142050e-27	0.2272172307e-26	0.4338317565e-26
6	$\sim$	0.1671749770e-54	0.9076552540e-53	0.3468434894e-52
7	$\sim$	0.4889628797e-109	0.1448372442e-105	0.2216960533e-104

The table shows that our error bounds (17.88) and (17.86) are tighter than (17.87) proposed in [726].

Chapter 18

Directional Secant–Type Methods

In this chapter, a semilocal convergence analysis for directional Secant–type methods in n–variables is provided. Using weaker hypotheses than in the related work by An and Bai [27], and motivated by optimization considerations, we provide under the same computational cost a semilocal convergence analysis with the following advantages: weaker convergence conditions; larger convergence domain; finer error estimates on the distances involved, and an at least as precise information on the location of the zero of the mapping. A numerical example where our results apply to solve an equation but not the ones in [27] is also provided in this chapter.

We are concerned with the problem of approximating a solution x^* of equation

$$F(x) = 0, \tag{18.1}$$

where F is a differentiable mapping defined on a convex subset $\mathcal{D}$ of $\mathbb{R}^n$ (n a natural number) with values in $\mathbb{R}$.

A large number of problems in applied mathematics and also in engineering are solved by finding the solutions of certain equations. In computer graphics, we usually compute the intersection $C = \mathcal{A} \cap \mathcal{B}$ of two surfaces $\mathcal{A}$ and $\mathcal{B}$ in $\mathbb{R}^3$ [257], [512]. If the two surfaces are explicitly given by

$$\mathcal{A} = \{(u,v,w)^T : w = F_1(u,v)\}$$

and

$$\mathcal{B} = \{(u,v,w)^T : w = F_2(u,v)\},$$

then the solution $x^* = (u^*, v^*, w^*)^T \in C$ must satisfy the nonlinear equation

$$F_1(u^*, v^*) = F_2(u^*, v^*)$$

and

$$w^* = F_1(u^*, v^*).$$

Hence, we must solve a nonlinear equation in two variables $x = (u,v)^T$ of the form

$$F(x) = F_1(x) - F_2(x) = 0,$$

which is a special case of equation (18.1).

In mathematical programming [588], for an equality–constraint optimization problem, e.g.,

$$\min \; \psi(x)$$
$$s.t. \quad F(x) = 0$$

where, ψ, $F : \mathcal{D} \subseteq \mathbb{R}^n \longrightarrow \mathbb{R}$ are nonlinear mappings, we usually seek a feasible point to start a numerical algorithm, which again requires the determination of $x^\star$.

An and Bai [27] used the directional Secant method (DSM)

$$x_{k+1} = x_k + h_k$$
$$h_k = -\frac{\theta_k \, F(x_k)}{F(x_k + \theta_k \, d_k) - F(x_k)} \, d_k$$
$$(k \geq 0, \quad x_0 \in \mathbb{R}^n, \quad d_k \in \mathbb{R}^n, \quad \| \, d_k \, \| = 1, \quad \theta_k \geq 0)$$

to generate a sequence $\{x_k\}$ converging to $x^\star$.

(DSM) is a usefull alternative to directional Newton method (DNM) [170], [257], [504]:

$$x_{k+1} = x_k - \frac{F(x_k)}{\nabla F(x_k) \cdot d_k} \, d_k \quad (k \geq 0),$$

where,

$$\nabla F(x_k) = \left(\frac{\partial F(x_k)}{\partial x_1}, \frac{\partial F(x_k)}{\partial x_2}, \cdots, \frac{\partial F(x_k)}{\partial x_n} \right),$$

is the gradient of F and d_k is a direction at x_k.

(DNM) converges quadratically to $x^\star$, if x_0 is close enough to $x^\star$ [504]. However, as already noted in [504], the computation of the gradient $\nabla F(x_k)$ may be too expensive as it is the case when the number n of unknowns is large.

In some applications, the mapping F may not be differentiable, or the gradient is impossible to compute. The (DSM) avoids these obstacles.

Note that if $n = 1$, (DSM) reduces to the classical Secant method [137].

We consider two choices of the directions d_k for locating $x^\star$. The first is called the directional near–gradient Secant method (DNGSM), and the second is called the directional maximal–component–modulus Secant method (DMCMSM).

The quadratic convergence of (DSM) was established for directions d_k sufficiently close to the gradients $\nabla F(x_k)$, and under standard Newton–Kantorovich–type hypotheses [27]. We are motivated by the papers [27], [170], [504], and optimization considerations. By introducing the center–Lipschitz condition and using it, in combination with the Lipschitz condition (along the lines of our works in [136], [189]), we provide a semilocal convergence analysis with the following advantages over the work in [27]:

1. Weaker hypotheses;

2. Larger convergence domain for (DSM);

3. Finer error bounds on the distances $\| \, x_{k+1} - x_k \, \|$, $\| \, x_k - x^\star \, \|$ $(k \geq 0)$;

4. An at least as precise information on the location of the zero $x^\star$.

Chapter 18. Directional Secant–Type Methods

We use the Euclidean norms for both vector and matrix. The unit direction d_k is chosen such that

$$d_k \approx \frac{\nabla F(x_k)}{\|\nabla F(x_k)\|}.$$

As in [27], we need two lemmas to establish semilocal convergence of (DNGSM).

Lemma 18.1 ([568, 3.2.2]). *Let $C \subseteq \mathbb{R}^n$ be a convex set, and $F : C \longrightarrow \mathbb{R}$ a differentiable mapping.*

Then, for any $x, y \in C$, there exists a vector $\mu \in (x, y)$, such that:

$$F(y) - F(x) = \nabla F(\mu)(y - x), \tag{18.2}$$

where,

$$(x, y) = \{z : z = x + \theta y, \quad 0 < \theta < 1\}, \tag{18.3}$$

represents the open straight line between the points x and y.

Lemma 18.2 ([568, 3.3.10]). *Assume that $F : C \longrightarrow \mathbb{R}$ is a twice differentiable mapping.*
Then, for any $x, y \in C$, there exists a vector $\lambda \in (x, y)$, such that:

$$F(y) - F(x) - \nabla F(x)(y - x) = \frac{1}{2}(y - x)^T F''(\lambda)(y - x). \tag{18.4}$$

We can state the main semilocal convergence result for (DNGSM).

Theorem 18.1. *Let $\mathcal{D}_0 \subseteq \mathbb{R}^n$ be a convex set, $F : \mathcal{D}_0 \subseteq \mathbb{R}^n \longrightarrow \mathbb{R}$ be a differentiable mapping.*

Let $x_0 \in \mathcal{D}_0$, satisfy

$$F(x_0) \neq 0, \quad \text{and} \quad \nabla F(x_0) \neq 0, \tag{18.5}$$

$d_0 \in \mathbb{R}^n$ be a given unit vector, θ_0 be a given positive parameter, h_0, and x_1 be defined as

$$h_0 = \frac{\theta_0 F(x_0)}{F(x_0 + \theta_0 d_0) - F(x_0)} d_0, \tag{18.6}$$

$$x_1 = x_0 + h_0. \tag{18.7}$$

We suppose that for $U_0 \subseteq \mathcal{D}_0$, and $F \in C^2[U_0]$, there exist a constants M_0, and M, such that:

$$\|\nabla F(x) - \nabla F(x_0)\| \leq M_0 \|x - x_0\| \qquad x \in U_0, \tag{18.8}$$

and

$$M = \sup_{x \in U_0} \|F''(x)\|. \tag{18.9}$$

Moreover, let the sequence $\{x_k\}$ be generated by (DNGSM), and further assume that the unit direction d_k, and positive parameters α, β, θ_k, satisfy for all $k \geq 0$:

$$\frac{|\nabla F(x_k) \cdot d_k|}{\|\nabla F(x_k)\|} \geq \frac{|\nabla F(x_0) \cdot d_0|}{\|\nabla F(x_0)\|}, \tag{18.10}$$

$$\theta_k \leq \alpha \|h_k\|, \tag{18.11}$$

$$|\nabla F(x_k) \cdot d_k| \leq \beta |\nabla F(x) \cdot d_k|, \qquad x \in (x_k, x_k + \theta_k d_k), \tag{18.12}$$

and

$$p = \overline{L} \parallel h_0 \parallel \leq \frac{1}{2}, \tag{18.13}$$

where,

$$L_0 = M_0 \parallel \nabla F(x_0) \parallel^{-1}, \tag{18.14}$$

$$L = \beta (1 + \alpha) M |\nabla F(x_0) \cdot d_0|^{-1}, \tag{18.15}$$

$$U_0 = \{x \in \mathbb{R}^n : \parallel x - x_1 \parallel \leq t^\star\} \subseteq \mathcal{D}_0, \tag{18.16}$$

$$t_1 = \parallel h_0 \parallel, \tag{18.17}$$

and $\delta, \overline{L}$ are given in Lemma 3.1.

Then, sequence $\{x_k\}$ generated by (DNGSM) is well defined, remains in U_0 for all $k \geq 0$, and converges to a zero $x^\star \in U_0$ of mapping F.

Moreover, the following hold

$$\begin{aligned} \nabla F(x) \neq 0 \quad &\text{for all} \quad x \in U_0, \quad \text{and} \\ \nabla F(x^\star) \neq 0 \quad &\text{unless} \quad \parallel x^\star - x_0 \parallel = t^\star \end{aligned} \tag{18.18}$$

Furthemore, the following estimates hold for all $k \geq 0$:

$$\parallel x_{k+1} - x_k \parallel \leq t_{k+1} - t_k \leq \left(\frac{\delta}{2}\right)^k (2p)^{2^k - 1} \parallel h_0 \parallel, \tag{18.19}$$

$$\parallel x_k - x^\star \parallel \leq t^\star - t_k \leq \left(\frac{\delta}{2}\right)^k \frac{(2p)^{2^k - 1} \parallel h_0 \parallel}{1 - (2p)^{2^k}}, \quad (2p < 1), \tag{18.20}$$

where, $t^\star, \{t_k\}$ are given in Lemma 3.1,

$$\eta = \parallel h_0 \parallel, \quad \text{and} \quad q = p. \tag{18.21}$$

Proof. We shall show using mathematical induction on $k \geq 0$:

$$\parallel x_{k+1} - x_k \parallel \leq t_{k+1} - t_k, \tag{18.22}$$

and

$$\overline{U}(x_{k+1}, t^\star - t_{k+1}) \subseteq \overline{U}(x_k, t^\star - t_k). \tag{18.23}$$

For every $z \in \overline{U}(x_1, t^\star - t_1)$

$$\begin{aligned} \parallel z - x_0 \parallel &\leq \parallel z - x_1 \parallel + \parallel x_1 - x_0 \parallel \\ &\leq t^\star - t_1 + t_1 - t_0 = t^\star - t_0, \end{aligned}$$

shows $z \in \overline{U}(x_0, t^\star - t_0)$.

Since, also

$$\parallel x_1 - x_0 \parallel = \parallel h_0 \parallel \leq \eta = t_1 - t_0,$$

estimates (18.22), and (18.23) hold for $k = 0$.

Chapter 18. Directional Secant–Type Methods 225

Assume (18.22), and (18.23) hold for all $i \leq k$. Then we have:

$$
\begin{aligned}
\| x_{i+1} - x_0 \| & \leq \| x_{i+1} - x_i \| + \| x_i - x_{i-1} \| + \cdots + \| x_1 - x_0 \| \\
& \leq (t_{i+1} - t_i) + (t_i - t_{i-1}) + \cdots + (t_1 - t_0) = t_{i+1},
\end{aligned}
$$

and

$$
\| x_i + t\,(x_{i+1} - x_i) - x_0 \| \leq t_i + t\,(t_{i+1} - t_i) \leq t^\star, \quad t \in [0,1].
$$

Using condition (18.8), and (18.14) for $x = x_i$, we get in turn:

$$
\begin{aligned}
\| \nabla F(x_i) \| & \geq \| \nabla F(x_0) \| - \| \nabla F(x_i) - \nabla F(x_0) \| \\
& \geq \| \nabla F(x_0) \| - M_0 \| x_i - x_0 \| \\
& \geq \| \nabla F(x_0) \| - M_0\,(t_i - t_0) \\
& \geq \| \nabla F(x_0) \|\,(1 - L_0\,t_i) > 0 \\
& \quad \text{(by Lemma 3.1 and (18.5)).}
\end{aligned}
\tag{18.24}
$$

We need the approximation

$$
F(x_i) = \int_{x_{i-1}}^{x_i} (x_i - x)^T\,F''(x)\,dx + r_{i-1},
\tag{18.25}
$$

where,

$$
r_i = F(x_i) + \nabla F(x_i)\,h_i \qquad (i \geq 0).
\tag{18.26}
$$

Using (18.9), and (18.11), and the induction hypotheses, we have:

$$
\begin{aligned}
& \left| \int_{x_{i-1}}^{x_i} (x_i - x)^T\,F''(x)\,dx \right| \\
= {}& \left| \int_0^1 (1-t)\,(h_{i-1})^T\,F''(x_{i-1} + t\,h_{i-1})\,h_{i-1}\,dt \right| \\
\leq {}& \frac{1}{2}\,M\,\| h_{i-1} \|^2,
\end{aligned}
\tag{18.27}
$$

and,

$$
\begin{aligned}
|r_{i-1}| & = \left| F(x_{i-1}) - \frac{\theta_{i-1}\,F(x_{i-1})\,\nabla F(x_{i-1}) \cdot d_{i-1}}{F(x_{i-1} + \theta_{i-1}\,d_{i-1}) - F(x_{i-1})} \right| \\
& = \left| \frac{F(x_{i-1})}{2}\,\frac{(\theta_{i-1}\,d_{i-1})^T\,F''(\lambda)\,(\theta_{i-1}\,d_{i-1})}{F(x_{i-1} + \theta_{i-1}\,d_{i-1}) - F(x_{i-1})} \right|, \\
& \quad \lambda \in (x_{i-1}, x_{i-1} + \theta_{i-1}\,d_{i-1}) \\
& \leq \frac{1}{2}\,M\,\theta_{i-1} \left| \frac{\theta_{i-1}\,F(x_{i-1})}{F(x_{i-1} + \theta_{i-1}\,d_{i-1}) - F(x_{i-1})} \right| \\
& \leq \frac{1}{2}\,M\,\alpha\,\| h_{i-1} \|^2.
\end{aligned}
\tag{18.28}
$$

In view of (18.12), (18.24)–(18.28), we get

$$
\begin{aligned}
|F(x_i)| & \leq \frac{1}{2}\,M\,\| h_{i-1} \|^2 + \frac{1}{2}\,\alpha\,M\,\| h_{i-1} \|^2 \\
& = \frac{1}{2}\,(1 + \alpha)\,M\,\| h_{i-1} \|^2,
\end{aligned}
\tag{18.29}
$$

so,

$$
\begin{aligned}
\| h_i \| &\leq \left| \frac{\theta_i \, F(x_i)}{F(x_i + \theta_i \, d_i) - F(x_i)} \right| \\
&\leq \frac{1}{2} \, (1 + \alpha) \, M \, \| h_{i-1} \|^2 \left| \frac{\theta_i}{F(x_i + \theta_i \, d_i) - F(x_i)} \right| \\
&\leq \frac{1}{2} \, (1 + \alpha) \, M \, \| h_{i-1} \|^2 \, \frac{1}{|\nabla F(\mu_i) \cdot d_i|}, \\
&\quad (\mu_i \in (x_i, x_i + \theta_i \, d_i)) \\
&\leq \frac{\beta}{2} \, \frac{(1 + \alpha) \, M \, \| h_{i-1} \|^2}{|\nabla F(x_i) \cdot d_i|} \\
&\leq \frac{\beta}{2} \, \frac{(1 + \alpha) \, M \, \| h_{i-1} \|^2}{|\nabla F(x_0) \cdot d_0|} \, \frac{\| \nabla F(x_0) \|}{\| \nabla F(x_i) \|} \\
&\leq \frac{\beta \, (1 + \alpha) \, M \, \| h_{i-1} \|^2}{2 \, |\nabla F(x_0) \cdot d_0|} \, \frac{\| \nabla F(x_0) \|}{\| \nabla F(x_0) \| \, (1 - L_0 \, t_i)} \\
&\leq \frac{L}{2} \, \frac{\| h_{i-1} \|^2}{1 - L_0 \, t_i} \leq \frac{L}{2} \, \frac{(t_i - t_{i-1})^2}{1 - L_0 \, t_i} = t_{i+1} - t_i,
\end{aligned}
\tag{18.30}
$$

which shows (18.22) for all i.

Moreover, for every $w \in \overline{U}(x_{i+2}, t^\star - t_{i+2})$, we obtain:

$$
\begin{aligned}
\| w - x_{i+1} \| &\leq \| w - x_{i+2} \| + \| x_{i+2} - x_{i+1} \| \\
&\leq t^\star - t_{i+2} + t_{i+2} - t_{i+1} = t^\star - t_{i+1},
\end{aligned}
$$

showing (18.23) for all $k \geq 0$.

Lemma 3.1 implies that $\{t_n\}$ is a Cauchy sequence. It then follows from (18.22), and (18.23), that $\{x_n\}$ is a Cauchy sequence too, and as such it converges to some $x^\star \in U_0$ (since U_0 is a closed set).

The vector $x^\star$ is a zero of F, since by (18.29):

$$
0 \leq |F(x_i)| \leq \frac{1}{2} \, (1 + \alpha) \, M \, (t_i - t_{i-1})^2 \longrightarrow 0 \quad \text{as} \quad i \to \infty.
\tag{18.31}
$$

Furthermore, we shall show (18.18).

Using (18.8) for $x \in U_0$, Lemma 3.1, and the definition of L_0, we get

$$
\begin{aligned}
\| \nabla F(x) - \nabla F(x_0) \| &\leq M_0 \, \| x - x_0 \| \\
&\leq M_0 \, t^\star \leq \| \nabla F(x_0) \|
\end{aligned}
$$

If $\| x - x_0 \| < t^\star$, then we have:

$$
\| \nabla F(x) - \nabla F(x_0) \| \leq M_0 \, \| x - x_0 \| < M_0 \, t^\star \leq \| \nabla F(x_0) \|,
$$

or

$$
\| \nabla F(x_0) \| > \| \nabla F(x) - \nabla F(x_0) \|,
$$

which shows $\nabla F(x) \neq 0$.

Finally, the left hand side inequality in (18.20) follows from (18.19) by using standard majorization techniques [155], [568], [719].

That completes the proof of Theorem 18.1. $\quad\diamond$

We state Theorems 3.1 and 3.2 in a condensed form [27, p. 294], so we can compare them with our Theorem 18.1.

Chapter 18. Directional Secant–Type Methods

Theorem 18.2. *Let $\mathcal{D}_0 \subseteq \mathbb{R}^n$ be a convex set, $F : \mathcal{D}_0 \subseteq \mathbb{R}^n \longrightarrow \mathbb{R}$ be a differentiable mapping.*

Let $x_0 \in \mathcal{D}_0$, satisfy

$$F(x_0) \neq 0, \quad \text{and} \quad \nabla F(x_0) \neq 0,$$

$d_0 \in \mathbb{R}^n$ be a given unit vector, θ_0 be a given positive parameter, h_0, and x_1 be defined by (18.6), and (18.7) respectively.

Let $U_0 \subseteq \mathcal{D}_0$, $F \in C^2[U_0]$, and

$$M = \sup_{x \in U_0} \| F''(x) \|$$

be such that

$$r_0 = \ell_0 \parallel h_0 \parallel \leq \frac{1}{2}, \tag{18.32}$$

where,

$$\ell_0 = 2\,M\,|\nabla F(x_0) \cdot d_0|^{-1}, \tag{18.33}$$

and

$$U_0 = \{ x \in \mathbb{R}^n : \parallel x - x_1 \parallel \leq \parallel h_0 \parallel \}.$$

In addition, let the sequence $\{x_k\}$ be generated by (DNGSM), and assume that the unit direction d_k, and positive parameter θ_k, satisfy for all $k \geq 0$:

$$\frac{|\nabla F(x_{k+1}) \cdot d_{k+1}|}{\| \nabla F(x_{k+1}) \|} \geq \frac{|\nabla F(x_k) \cdot d_k|}{\| \nabla F(x_k) \|}, \tag{18.34}$$

$$\theta_k \leq \frac{1}{2} \parallel h_k \parallel,$$

and

$$|\nabla F(x_k) \cdot d_k| \leq 2\,|\nabla F(x) \cdot d_k|, \qquad x \in (x_k, x_k + \theta_k d_k).$$

Then, the following hold:

(a) *$\{x_k\} \subseteq U_0$, and there exists $x^\star \in U_0$, such that $x_k \longrightarrow x^\star$, and $F(x^\star) = 0$;*

(b)

$$\nabla F(x) \neq 0 \qquad (x \in U_0);$$

(c) *For $k \geq 0$,*

$$\nabla F(x_k) \cdot d_k \neq 0, \quad \parallel h_k \parallel \leq \frac{1}{2} \parallel h_{k-1} \parallel, \tag{18.35}$$

$$\begin{aligned} \parallel x_{k+1} - x_k \parallel \ &\leq \ \frac{3\,M}{2\,|\nabla F(x_k) \cdot d_k|} \parallel x_k - x_{k-1} \parallel^2 \\ &\leq \ \frac{3\,M\,\| \nabla F(x_0) \|}{2\,a\,|\nabla F(x_0) \cdot d_0|} \parallel x_k - x_{k-1} \parallel^2, \end{aligned} \tag{18.36}$$

and

$$\begin{aligned} \parallel x_{k+1} - x^\star \parallel \ &\leq \ \frac{3\,M}{2\,|\nabla F(x_k) \cdot d_k|} \parallel x_k - x_{k-1} \parallel^2 \\ &\leq \ \frac{3\,M\,\| \nabla F(x_0) \|}{2\,a\,|\nabla F(x_0) \cdot d_0|} \parallel x_k - x_{k-1} \parallel^2, \end{aligned} \tag{18.37}$$

where,

$$a = \min_{x \in U_0} \| \nabla F(x) \| > 0; \qquad (18.38)$$

(d) *If for*

$$\gamma = \frac{\| \nabla F(x_0) \| \ |\nabla F(x_0) \cdot d_0|^{-1}}{a}, \qquad (18.39)$$

$$\ell = \frac{3\gamma M}{4}, \qquad (18.40)$$

$$r = \ell \| h_0 \| < \frac{1}{2}, \qquad (18.41)$$

then,

$$\| x_{k+1} - x_k \| \le \frac{(2q)^{2^k-1}}{1 - (2q)^{2^k-1}} \| h_0 \|, \qquad (18.42)$$

and

$$\| x_k - x^\star \| \le \frac{(2q)^{2^k-1}}{1 - (2q)^{2^k-1}} \| h_0 \| . \qquad (18.43)$$

We need the definition of Lipschitz continuity.

Definition 18.1. Let $F : U_0 \subseteq \mathbb{R}^n \longrightarrow \mathbb{R}$ be a differentiable mapping. If there exists a constant $\overline{M} \ge 0$, such that:

$$\| \nabla F(x) - \nabla F(y) \| \le \overline{M} \| x - y \| \quad \text{for all} \quad x, y \in U_0 \qquad (18.44)$$

then, we say ∇F is Lipschitz continuous.

Note that in view of (18.44), there exists center–Lipschitz constant $M_0 \ge 0$, such that (18.8) is satisfied.

Clearly,

$$M_0 \le \overline{M}, \qquad (18.45)$$

$$M_0 \le M \qquad (18.46)$$

hold in general, and $\dfrac{M}{M_0}, \dfrac{\overline{M}}{M_0}$ can be arbitrarly large.

Note that if F is twice differentiable, and (18.44) holds, then $\overline{M}$ can replace M in hypothesis (18.9).

Next, we provide a comparison between Theorems 18.1 and 18.2.

Remark 18.1. (a) We compare sufficient convergence condition (18.13) to (18.32) for $\alpha = \dfrac{1}{2}$, and $\beta = 2$.

We have

$$r_0 \le \frac{1}{2} \implies p \le \frac{1}{2} \qquad (18.47)$$

but not necessarily vice verca, unless if $M_0 = M$.

Hence, condition (18.32) can always be replaced by (18.13).

Chapter 18. Directional Secant–Type Methods

That is, the applicability of (DNGSM) has been extended.

Note also that (18.8) is not an additional hypothesis, since in practice the computation of M requires that of constant M_0.

According to (18.35), only linear convergence is shown provided that hypothesis (18.32) holds.

However, in Theorem 18.1, we showed quadratic convergence (for $p \leq \frac{1}{2}$). Theorem 18.1 avoids the computation of a given (18.38), which is expensive in general, and also uses (18.10) instead of the more difficult to verify (18.34).

Finally, note that if $p \in [0, \frac{1}{2})$, and $\beta \in [0, 2)$, then condition (18.13) is even weaker than (18.32).

(b) Condition (18.10) is equivalent to $c_{k+1} \leq c_0$, since $\| d_k \| = 1$, where,

$$c_k = \angle(\nabla F(x_k), d_k),$$

with $\angle(.,.)$ being the angle between two vectors u, v given by:

$$\angle(u, v) = \arccos \frac{u \cdot v}{\| u \| \cdot \| v \|}, \quad u \neq 0, \quad v \neq 0.$$

(c) Condition (18.11) is equivalent to

$$|F(x_k + \theta_k d_k) - F(x_k)| \leq \alpha |F(x_k)|.$$

(d) If $M\theta_k \leq b |\nabla F(x_k) \cdot d_k|$, $b \in [0, 1)$, then, condition (18.12) holds for $\beta = 1 - b$. Indeed, we have

$$|\nabla F(x) \cdot d_k - \nabla F(x_k) \cdot d_k| \leq M \theta_k,$$

so,

$$\begin{aligned} |\nabla F(x) \cdot d_k| &\geq |\nabla F(x_k) \cdot d_k| - |\nabla F(x) \cdot d_k - \nabla F(x_k) \cdot d_k| \\ &\geq |\nabla F(x_k) \cdot d_k| - b |\nabla F(x_k) \cdot d_k| = \beta |\nabla F(x_k) \cdot d_k|. \end{aligned}$$

We use in the following ∞–norm for both vector and matrix. By simply replacing the Euclidean norm by ∞–norm, and $|\nabla F(x_k) \cdot d_k|$ by $\| \nabla F(x_k) \|_\infty$ in the proof of Theorem 18.1, we arrive at the following corresponding semilocal convergence result for (DMCMCM):

Theorem 18.3. *Let $\mathcal{D}_0 \subseteq \mathbb{R}^n$ be a convex set, $F : \mathcal{D}_0 \subseteq \mathbb{R}^n \longrightarrow \mathbb{R}$ be a differentiable mapping.*

Let $x_0 \in \mathcal{D}_0$, satisfy

$$F(x_0) \neq 0, \quad \text{and} \quad \nabla F(x_0) \neq 0, \tag{18.48}$$

θ_0 be a given positive parameter, h_0, and x_1 be defined as

$$h_0 = \frac{\theta_0 F(x_0)}{F(x_0 + \theta_0 d_0) - F(x_0)} d_0,$$

$$x_1 = x_0 + h_0,$$

where,

$$d_0 = e_{m(0)}$$

is the $m(0)$th unit vector in $\mathbb{R}^n$.

We suppose that for $U_0 \subseteq \mathcal{D}_0$, and $F \in C^2[U_0]$, there exist constants M_0, and M, such that:

$$\| \nabla F(x) - \nabla F(x_0) \|_\infty \leq M_0 \| x - x_0 \|_\infty \qquad x \in U_0,$$

and

$$M = \sup_{x \in U_0} \| F''(x) \|_\infty.$$

Moreover, let the sequence $\{x_k\}$ be generated by (DMCMSM), and further assume direction d_k, and positive parameters α, β, θ_k, satisfy for all $k \geq 0$:

$$\| \nabla F(x_k) \|_\infty \leq \beta \| \nabla F(x) \cdot e_{m(k)} \|_\infty, \quad x \in (x_k, x_k + \theta_k d_k),$$

$$\theta_k \leq \alpha \| h_k \|_\infty,$$

and

$$p = \overline{L} \| h_0 \|_\infty \leq \frac{1}{2},$$

where,

$$L_0 = M_0 \| \nabla F(x_0) \|_\infty^{-1},$$

$$L = \beta (1 + \alpha) M \| \nabla F(x_0) \|_\infty^{-1},$$

$$U_0 = \{ x \in \mathbb{R}^n : \| x - x_1 \|_\infty \leq t^\star \} \subseteq \mathcal{D}_0,$$

with δ, and $\overline{L}$ are given in Lemma 3.1.

Then, sequence $\{x_k\}$ generated by (DMCMSM) is well defined, remains in U_0 for all $k \geq 0$, and converges to a zero $x^\star \in U_0$ of mapping F.

Moreover, the following hold

$$\nabla F(x) \neq 0 \quad \text{for all} \quad x \in U_0, \quad \text{and}$$
$$\nabla F(x^\star) \neq 0 \quad \text{unless} \quad \| x^\star - x_0 \|_\infty = t^\star.$$

Furthemore, the following estimates hold for all $k \geq 0$:

$$\| x_{k+1} - x_k \|_\infty \leq t_{k+1} - t_k \leq \left(\frac{\delta}{2} \right)^k (2p)^{2^k - 1} \| h_0 \|_\infty,$$

and

$$\| x_k - x^\star \|_\infty \leq t^\star - t_k \leq \left(\frac{\delta}{2} \right)^k \frac{(2p)^{2^k - 1} \| h_0 \|_\infty}{1 - (2p)^{2^k}}, \quad (2p < 1),$$

where, $t^\star$, $\{t_k\}$ are given in Lemma 3.1, and

$$\eta = \| h_0 \|_\infty, \quad q = p.$$

As in Remark 18.1, Theorem 18.3 compares favorably to Theorem 3.3 in [27, p. 300].
We provide now a numerical example to show that hypotheses of Theorem Thorem 18.1 are satisfied, but not corresponding hypotheses of Theorem 18.2.

Chapter 18. Directional Secant–Type Methods   231

Example 18.1. Let $n = 2$. Here, we use the Euclidean inner product, and the corresponding norm for both vector and matrix.

Choose:

$$x_0 = (1,1)^T, \quad \mathcal{D}_0 = \{x \in \mathbb{R}^2 : \| x - x_0 \| \leq 1 - \xi\},$$

for $\xi \in [0,1)$, and define mapping F on $\mathcal{D}_0$ by:

$$F(x) = \frac{\lambda_1^3 + \lambda_2^3}{2} - 2\,\xi, \quad x = (\lambda_1, \lambda_2)^T. \tag{18.49}$$

Then, the gradient ∇F of mapping F is given by

$$\nabla F(x) = \frac{3}{2} \, (\lambda_1^2, \lambda_2^2)^T. \tag{18.50}$$

Using (18.6), (18.8), (18.9), (18.14), (18.15), we obtain for $\alpha = \frac{1}{2}, \beta = 2$:

$$M_0 = \frac{3\sqrt{2}}{2} \, (3 - \xi), \quad M = 3\sqrt{2}\,(2 - \xi),$$

$$L_0 = 3 - \xi, \quad L = 3\,(2 - \xi), \quad \| \nabla F(x_0) \| = \frac{3\sqrt{2}}{2},$$

$$F(x_0) = 1 - \sqrt{2}\,\xi, \quad \text{and} \quad \eta = \frac{\sqrt{2}}{3} \, (1 - 2\,\xi).$$

We can choose the directions d_k by

$$d_k = \frac{\nabla F(x_k)}{\| \nabla F(x_k) \|},$$

so that condition (18.10), and (18.34) are satisfied as equalities.

Let $\xi = .4162$. Then, condition (18.32) is violated, since, we have $\eta = .079007397$, $M = 6.71949432$, and

$$\frac{4\sqrt{2}}{3} \, (2 - \xi) \, (1 - 2\,\xi) = .500527666 > .5.$$

Hence, there is no guarantee that (DNGSM) starting at x_0 converges to the solution $x^\star$ of equation $F(x) = 0$.

Moreover, we obtain:

$$L_0 = 2.5838, \quad L = 4.7514, \quad \overline{L} = 3.259626413,$$

$$\delta = 1.207332475, \quad t^{\star\star} = .199345613 < 1 - \xi = .5838, \quad \text{and} \quad q = .257534598 < .5.$$

That is condition (18.13) is satisfied, and $U_0 \subseteq \mathcal{D}$. The conclusions of Theorem 18.1 apply for solving equation $F(x) = 0$. We found $x^\star = (.940684577, .940684577)$.

Remark 18.2. The results extend in a Hilbert space setting. Indeed, let F be a differentiable operator defined on a convex subset $\mathcal{D}$ of a Hilbert space $\mathcal{H}$ with values in $\mathbb{R}$. Here $x \cdot y$ denotes the inner product of elements x, and y in $\mathcal{H}$, and $\| x \| = (x \cdot x)^{1/2}$.

Moreover, instead of condition (18.10), assume:

$$\| d_k \| = 1,$$

and, there exists $\kappa \in [0,1]$, such that:

$$|F'(x_k) \cdot d_k| \geq \kappa \, \| F'(x_k) \| \, .$$

Note that in the case of (18.10), we can set:

$$\kappa = \frac{|F'(x_k) \cdot d_k|}{\| F'(x_0) \|} \leq 1.$$

Define

$$L_0 = \frac{M_0}{\kappa \, \| F'(x_0) \|}, \quad L = \frac{\beta \, (1 + \alpha) \, M}{\kappa \, \| F'(x_0) \|}, \quad \eta = \| h_0 \|$$

in the case of Theorem 18.1, and

$$L_0 = \frac{M_0}{\kappa \, \| F'(x_0) \|_\infty}, \quad L = \frac{\beta \, (1 + \alpha) \, M}{\kappa \, \| F'(x_0) \|_\infty}, \quad \eta = \| h_0 \|_\infty$$

in the case of Theorem 18.3.

Then, due to the proofs of Theorems 18.1 and 18.3, the results hold in this more general setting.

Chapter 19

Directional Newton–Type Methods

In this chapter, we provide a semilocal convergence analysis for directional Newton–like methods in n–variables containing non–differentiable terms using our new idea of recurrent functions. In the special case of Newton's method, our convergence analysis has the following advantages over earlier the work in [504]: weaker convergence conditions; larger convergence domain; finer error estimates on the distances involved, and an at least as precise information on the location of the zero of the function. A numerical example where our results apply but others fail is also provided in this chapter.

We are concerned with the problem of approximating a zero $x^\star$ of equation

$$F(x) + G(x) = 0, \tag{19.1}$$

where F is a differentiable function defined on a convex subset $\mathcal{D}$ of $\mathbb{R}^n$ (n in $\mathbb{N}^\star$) with values in $\mathbb{R}$, and $G : \mathcal{D} \longrightarrow \mathbb{R}$ is a continuous function. Throughout this chapter $\| B \|$ denotes the matrix norm corresponding to the Euclidean norm $\| x \|$.

We introduce the directional Newton–like method (DNLM) defined recursively by

$$x_{k+1} = x_k + h_k$$
$$h_k = -\frac{F(x_k) + G(x_k)}{A(x_k) \cdot d_k} d_k$$
$$(k \geq 0, \quad x_0 \in \mathbb{R}^n, \quad d_k \in \mathbb{R}^n),$$

where, $A(x) \in L(\mathbb{R}^n, \mathbb{R})$ the space of bounded linear functions from $\mathbb{R}^n$ into $\mathbb{R}$.

We usually choose $A(x)$ to be an approximation to the gradient $\nabla F(x)$ of function F [151], [170], [257], [504]. Another possible choice is $A(x) \approx \nabla F(x) + G(x + \theta_k d_k) - G(x)$, where, $\{\theta_k\}$ is a scalar sequence satisfying $\theta_k \leq c \| h_k \|$, $(k \geq 0)$, $c \in [0, 1)$ [137], [190], [372], [429].

If, $G(x) = 0$, and $A(x) = \nabla F(x)$ $(x \in \mathcal{D})$, we obtain the directional Newton's method (DNM)

$$x_{k+1} = x_k + h_k$$
$$h_k = -\frac{F(x_k)}{\nabla F(x_k) \cdot d_k} d_k$$
$$(k \geq 0, \quad x_0 \in \mathbb{R}^n, \quad d_k \in \mathbb{R}^n)$$

studied in [170], [257], [504], where, ∇F denotes the gradient of F.

We shall also study the directional residual Newton–like method (DRNLM)

$$x_{k+1} = x_k + h_k$$
$$h_k = -\frac{F(x_k) + G(x_k) + s_k}{A(x_k) \cdot d_k} d_k$$
$$(k \geq 0, \quad x_0 \in \mathbb{R}^n, \quad d_k \in \mathbb{R}^n),$$

where, residual sequence $\{s_k\}$ satisfies

$$\left| \frac{s_k}{A(x_k) \cdot d_k} \right| \leq \lambda \, \| F(x_k) + G(x_k) \|, \qquad (\lambda \geq 0), \quad [70], \, [190].$$

We provide a semilocal convergence analysis for all these methods. The advantages of our approach in the case of Newton's method over the work in [504] have already been stated in the abstract of this chapter.

We need the following result on majorizing sequences for (DNLM) and (DNM), respectively.

Lemma 19.1 ([170], [190]). *Assume:*

there exist constants $L \geq 0$, $K \geq 0$, with $K \leq L$, and $\eta \geq 0$, such that:

$$q_0 = \beta \, \eta \begin{cases} \leq \dfrac{1}{2}, & if \quad L_0 \neq 0 \\[2mm] < \dfrac{1}{2}, & if \quad L_0 = 0 \end{cases},$$

where,

$$\beta = \frac{1}{8} \left(K + 4 L + \sqrt{K^2 + 8 L K} \right).$$

Then, sequence $\{s_k\}$ $(k \geq 0)$ given by

$$s_0 = 0, \quad s_1 = \eta, \quad s_{k+1} = s_k + \frac{K \, (s_k - s_{k-1})^2}{2 \, (1 - L \, s_k)} \qquad (k \geq 1),$$

is well defined, nondecreasing, bounded from above by $s^{\star\star}$, and converges to its unique least upper bound $s^\star \in [0, s^{\star\star}]$, where

$$s^{\star\star} = \frac{2 \, \eta}{2 - \overline{\delta}},$$

$$1 \leq \overline{\delta} = \frac{4 \, K}{K + \sqrt{K^2 + 8 L K}} < 2 \quad for \, L \neq 0.$$

Moreover, the following estimates hold:

$$0 \leq s_{k+1} - s_k \leq \frac{\overline{\delta}}{2} \, (s_k - s_{k-1}) \leq \cdots \leq \left(\frac{\overline{\delta}}{2} \right)^k \eta, \quad (k \geq 1),$$

$$s_{k+1} - s_k \leq \left(\frac{\overline{\delta}}{2} \right)^k (2 \, q_0)^{2^k - 1} \eta, \quad (k \geq 0),$$

$$0 \leq s^\star - s_k \leq \left(\frac{\overline{\delta}}{2} \right)^k \frac{(2 \, q_0)^{2^k - 1} \eta}{1 - (2 \, q_0)^{2^k}}, \quad (2 \, q_0 < 1), \quad (k \geq 0).$$

Chapter 19. Directional Newton–Type Methods 235

We can show the following semilocal convergence result for (DNLM).

Theorem 19.1. *Let $F : \mathcal{D} \subseteq \mathbb{R}^n \longrightarrow \mathbb{R}$ be a differentiable function, $G : \mathcal{D} \longrightarrow \mathbb{R}$ be a continuous function, and let $A(x) \in L(\mathbb{R}^n, \mathbb{R})$ be an approximation of $F'(x)$, and $x_0 \in \mathcal{D}$.*
 Assume:

$$F(x_0) + G(x_0) \neq 0, \qquad A(x_0) \neq 0; \tag{19.2}$$

there exist an open convex subset $\mathcal{D}$ of $\mathbb{R}^n$, and constants $\overline{K} > 0$, $\overline{M} > 0$, $\overline{\ell} \geq 0$, $\eta > 0$, $\mu_0 \geq 0$, $\mu_1 \geq 0$, $L > 0$, such that for all $x, y \in \mathcal{D}$:

$$0 < \left\| \frac{F(x_0) + G(x_0)}{A(x_0) \cdot d_0} d_0 \right\| < \eta, \tag{19.3}$$

$$\| F'(x) - F'(y) \| \leq \overline{K} \, \| x - y \|, \tag{19.4}$$

$$\| F'(x) - A(x) \| \leq \overline{M} \, \| x - x_0 \| + \mu_0, \tag{19.5}$$

$$\| A(x) - A(x_0) \| \leq \overline{L} \, \| x - x_0 \| + \overline{\ell}, \tag{19.6}$$

$$\| G(x) - G(y) \| \leq \mu_1 \, \| x - y \|, \tag{19.7}$$

$$U_0 = \overline{U}(x_0, t^\star) = \{x \in X, \, \| x - x_0 \| \leq t^\star\} \subseteq \mathcal{D}; \tag{19.8}$$

Sequence $\{x_k\}$ ($k \geq 0$) generated by (DNLM) satisfies

$$\angle(d_k, A(x_k)) \leq \angle(d_0, A(x_0)) \qquad k \geq 0, \tag{19.9}$$

and the hypotheses of Lemma 1.3 hold with $\ell = 0$,

$$\begin{aligned} \mu &= (\mu_0 + \mu_1) \, |A(x_0) \cdot d_0|^{-1}, \quad K = \overline{K} \, |A(x_0) \cdot d_0|^{-1}, \\ M &= \overline{M} \, |A(x_0) \cdot d_0|^{-1}, \quad L = \overline{L} \, |A(x_0) \cdot d_0|^{-1}, \quad \ell = \overline{\ell} \, |A(x_0) \cdot d_0|^{-1}. \end{aligned} \tag{19.10}$$

Then, sequence $\{x_k\}$ ($k \geq 0$) is well defined, remains in U_0 for all $k \geq 0$, and converges to a zero $x^\star \in U_0$ of function $F + G$.
 Moreover, $A(x^\star) \neq 0$, unless

$$\| x^\star - x_0 \| = t^\star. \tag{19.11}$$

Furthemore, the following estimates hold for all $k \geq 0$:

$$\| x_{k+1} - x_k \| \leq t_{k+1} - t_k, \tag{19.12}$$

and

$$\| x_k - x^\star \| \leq t^\star - t_k, \tag{19.13}$$

where, sequence $\{t_k\}$ ($k \geq 0$), and $t^\star$ are given in Lemma 1.3.
 Finally, the solution $x^\star$ of equation (19.1) is unique in $U^\circ(x_0, R)$, ($R \geq t^\star$) provided that:

$$\frac{K}{2} R + (M + L) \, t^\star + \mu + \ell \leq 1, \quad \text{and} \quad U^\circ(x_0, R) \subseteq \mathcal{D}. \tag{19.14}$$

Note that the condition (19.9) is equivalent to:

$$\frac{|A(x_k) \cdot d_k|}{\| A(x_k) \|} \geq \frac{|A(x_0) \cdot d_0|}{\| A(x_0) \|} \quad (k \geq 0). \tag{19.15}$$

Proof. We shall show using mathematical induction on $k \geq 0$:

$$\| x_{k+1} - x_k \| \leq t_{k+1} - t_k, \tag{19.16}$$

and

$$\overline{U}(x_{k+1}, t^\star - t_{k+1}) \subseteq \overline{U}(x_k, t^\star - t_k). \tag{19.17}$$

For every $z \in \overline{U}(x_1, t^\star - t_1)$

$$
\begin{aligned}
\| z - x_0 \| &\leq \| z - x_1 \| + \| x_1 - x_0 \| \\
&\leq t^\star - t_1 + t_1 - t_0 = t^\star - t_0,
\end{aligned}
$$

shows $z \in \overline{U}(x_0, t^\star - t_0)$.

Since, also

$$\| x_1 - x_0 \| = \| h_0 \| \leq \eta = t_1 - t_0,$$

estimates (19.16), and (19.17) hold for $k = 0$.

Assume (19.16), and (19.17) hold for all $i \leq k$. Then we have:

$$
\begin{aligned}
\| x_{i+1} - x_0 \| &\leq \| x_{i+1} - x_i \| + \| x_i - x_{i-1} \| + \cdots + \| x_1 - x_0 \| \\
&\leq (t_{i+1} - t_i) + (t_i - t_{i-1}) + \cdots + (t_1 - t_0) = t_{i+1},
\end{aligned}
\tag{19.18}
$$

and

$$\| x_i + t\,(x_{i+1} - x_i) - x_0 \| \leq t_i + t\,(t_{i+1} - t_i) \leq t^\star, \quad t \in [0,1].$$

Using condition (19.6) for $x = x_i$, Lemma 1.3, and (19.2), we get:

$$
\begin{aligned}
\| A(x_i) \| &\geq \| A(x_0) \| - \| A(x_i) - A(x_0) \| \\
&\geq \| A(x_0) \| - (\overline{L}\,\| x_i - x_0 \| + \overline{\ell}) \\
&\geq \| A(x_0) \| - (\overline{L}\,(t_i - t_0) + \overline{\ell}) \\
&= \| A(x_0) \| - (\overline{L}\,t_i + \overline{\ell}) \\
&= \| A(x_0) \| \left(1 - \| A(x_0) \|^{-1}\,(\overline{L}\,t_i + \overline{\ell})\right) \\
&= \| A(x_0) \| \left(1 - (L\,t_i + \ell)\right) > 0.
\end{aligned}
\tag{19.19}
$$

In view of (19.4), (19.5), (19.7), (19.10), (19.18), and the induction hypotheses, we obtain in turn:

$$
\begin{aligned}
&|F(x_i + G(x_i))| \\
&\leq \int_0^1 \| F'(x_i + \theta\,(x_{i-1} - x_i)) - F'(x_i) \| \| x_{i-1} - x_i \| \, d\theta + \\
&\quad \| F'(x_{i-1}) - A(x_{i-1}) \| \| x_i - x_{i-1} \| + \| G(x_i) - G(x_{i-1}) \| \\
&\leq \frac{K}{2}\,\| x_i - x_{i-1} \|^2 + (\overline{M}\,\| x_{i-1} - x_0 \| + \mu_0 + \mu_1)\,\| x_i - x_{i-1} \| \\
&\leq \frac{K}{2}\,(t_i - t_{i-1})^2 + (\overline{M}\,t_{i-1} + \mu_0 + \mu_1)\,(t_i - t_{i-1}).
\end{aligned}
\tag{19.20}
$$

Chapter 19. Directional Newton–Type Methods   237

By (DNLM), Lemma 1.3, (19.9), (19.10), (19.19), and (19.20), we have:

$$
\begin{aligned}
\| x_{i+1} - x_i \| = \| h_i \| &= \frac{|F(x_i) + G(x_i)|}{|A(x_i) \cdot d_i|} \\
&\leq \frac{|F(x_i) + G(x_i)|}{\| A(x_i) \|} \frac{\| A(x_0) \|}{|A(x_0) \cdot d_0|} \\
&\leq \frac{|F(x_i) + G(x_i)| \ \| A(x_0) \|}{\| A(x_0) \| \ (1 - (Lt_i + \ell)) \ |A(x_0) \cdot d_0|} \\
&\leq \frac{\dfrac{K}{2} (t_i - t_{i-1})^2 + (M \, t_{i-1} + \mu) \, (t_i - t_{i-1})}{1 - (L \, t_i + \ell)} \\
&= t_{i+1} - t_i,
\end{aligned}
\tag{19.21}
$$

which shows (19.16) for all $i \geq 0$.

Then, for every $w \in \overline{U}(x_{i+2}, t^\star - t_{i+2})$, we obtain:

$$
\begin{aligned}
\| w - x_{i+1} \| &\leq \| w - x_{i+2} \| + \| x_{i+2} - x_{i+1} \| \\
&\leq t^\star - t_{i+2} + t_{i+2} - t_{i+1} = t^\star - t_{i+1},
\end{aligned}
\tag{19.22}
$$

showing (19.17) for all $i \geq 0$.

Lemma 1.3 implies that $\{t_n\}$ is a Cauchy sequence. It then follows from (19.16), and (19.17) that $\{x_n\}$ is a Cauchy sequence too, and as such it converges to some $x^\star \in U_0$ (since, U_0 is a closed set).

The point $x^\star$ is a zero of $F + G$, since

$$
0 \leq |F(x_i) + G(x_i)| \leq \frac{\overline{K}}{2} (t_i - t_{i-1})^2 + (\overline{M} \, t^\star + \mu_0 + \mu_1) \, (t_i - t_{i-1}) \xrightarrow[i \to \infty]{} 0.
\tag{19.23}
$$

Moreover, we shall prove $A(x^\star) \neq 0$, unless, if $\| x^\star - x_0 \| = t^\star$.

Using (19.6) for $x \in U_0$, we get

$$
\begin{aligned}
\| A(x) - A(x_0) \| &\leq \overline{L} \, \| x - x_0 \| + \overline{\ell} \\
&\leq \overline{L} \, t^\star + \overline{\ell} \leq \| A(x_0) \| .
\end{aligned}
$$

If $\| x - x_0 \| < t^\star$, then:

$$
\| A(x) - A(x_0) \| \leq \overline{L} \, t^\star + \overline{\ell} = \| A(x_0) \|,
$$

or

$$
\| A(x_0) \| > \| A(x) - A(x_0) \|,
$$

which shows $A(x) \neq 0$.

In particular, we have $A(x^\star) \neq 0$.

Estimates (19.13) follows from (19.12) by using standard majorization techniques.

Finally to show uniqueness, let $y^\star \in U^\circ(x_0, R)$ be a zero of function $F + G$.

We have by (19.4), (19.5), (19.7), (19.10), (19.14), (DNLM), and (19.19):

$$\| y^\star - x_{i+1} \|$$

$$\leq \left((1 - (L t^\star + \ell)) \, |A(x_0) \cdot d_0| \right)^{-1} \times$$

$$\left(\int_0^1 \| F'(x_i + \theta (y^\star - x_i)) - F'(x_i) \| \, d\theta + \right.$$

$$\left. \| F'(x_i) - A(x_i) \| \right) \| y^\star - x_i \| + \| G(x_i) - G(y^\star) \| \Big)$$

$$\leq (1 - (L t^\star + \ell))^{-1} \left(\frac{K}{2} \| y^\star - x_i \| + \right. \tag{19.24}$$

$$\left. M \| x_i - x_0 \| + \mu \right) \| y^\star - x_i \|$$

$$\leq (1 - (L t^\star + \ell))^{-1} \left(\frac{K}{2} \| y^\star - x_0 \| + M t_i + \mu \right) \| y^\star - x_i \|$$

$$< (1 - (L t^\star + \ell))^{-1} \left(\frac{K}{2} R + M t^\star + \mu \right) \| y^\star - x_i \|$$

$$\leq \| y^\star - x_i \|,$$

showing $\lim_{i \longrightarrow \infty} x_i = y^\star$. But we have $\lim_{m \longrightarrow \infty} x_i = x^\star$. Hence, we deduce $x^\star = y^\star$.

That completes the proof of Theorem 19.1. $\diamondsuit$

Remark 19.1. Note that $t^\star$ can be replaced by $t^{\star\star}$ given in closed form by Lemma 1.3 for hypotheses (19.8), and (19.14).

In the special case of Newton's method, i.e., for $G(x) = 0$, and $A(x) = \nabla F(x)$, $(x \in \mathcal{D})$, simply using Lemma 19.1, instead of Lemma 1.3, we arrive at:

Proposition 19.1 ([170]). *Let $F : \mathcal{D} \subseteq \mathbb{R}^n \longrightarrow \mathbb{R}$ be a differentiable function.*
Assume:

$$F(x_0) \neq 0, \qquad \nabla F(x_0) \neq 0;$$

there exist an open convex subset $\mathcal{D}$ of $\mathbb{R}^n$, and constants $\overline{K}, \overline{L}, \eta > 0$, such that for all $x, y \in \mathcal{D}$:

$$0 < \left\| \frac{F(x_0)}{\nabla F(x_0) \cdot d_0} d_0 \right\| < \eta,$$

$$\| F'(x) - F'(y) \| \leq \overline{K} \, \| x - y \|,$$

$$\| F'(x) - F'(x_0) \| \leq \overline{L} \, \| x - x_0 \|,$$

$$U_0 \subseteq \mathcal{D},$$

sequence $\{x_k\}$ $(k \geq 0)$ generated by (DNM) satisfies

$$\angle(d_k, \nabla F(x_k)) \leq \angle(d_0, \nabla F(x_0)) \qquad k \geq 0,$$

where, each $d_k \in \mathbb{R}^n$ is such that $\| d_k \| = 1$. and
the hypotheses of Lemma 19.1 hold with

$$K = \overline{K} \, |\nabla F(x_0) \cdot d_0|^{-1}, \quad L = \overline{L} \, |\nabla F(x_0) \cdot d_0|^{-1}.$$

Chapter 19. Directional Newton–Type Methods 239

Then, sequence $\{x_k\}$ $(k \geq 0)$ generated by (DNM) is well defined, remains in U_0 for all $k \geq 0$, and converges to a zero $x^\star$ of function F.

Furthemore, the following estimates hold for all $k \geq 0$:

$$\| x_{k+1} - x_k \| \leq s_{k+1} - s_k \leq \left(\frac{\overline{\delta}}{2}\right)^k (2\, q_0)^{2^k - 1}\, \eta,$$

and

$$\| x_k - x^\star \| \leq s^\star - s_k \leq \left(\frac{\overline{\delta}}{2}\right)^k \frac{(2\, q_0)^{2^k - 1}\, \eta}{1 - (2\, q_0)^{2^k}}, \quad (2\, q_0 < 1),$$

where, sequence $\{s_k\}$ $(k \geq 0)$, $s^\star$, $\overline{\delta}$, and q_0 are given in Lemma 19.1.

We also provide the following semilocal convergence result for (DRNLM).

Proposition 19.2. *Assume:*
hypotheses (19.2)–(19.8) *hold;*
there exists sequence $\{s_k\}$ $(k \geq 0)$, $\lambda \geq 0$, satisfying

$$\left| \frac{s_k}{A(x_k) \cdot d_k} \right| \leq \lambda\, \| F(x_k) + G(x_k) \|, \quad (k \geq 0); \tag{19.25}$$

hypotheses of Lemma 1.3 *with $\ell = 0$ or Lemma* 19.2 *(or Lemma* 19.1 *if $G(x) = 0$ on $\mathcal{D}$) hold with*

$$\mu = (\mu_0 + \mu_1)\, (1 + \lambda)\, |A(x_0) \cdot d_0|^{-1},$$

$$K = \overline{K}\, (1 + \lambda)\, |A(x_0) \cdot d_0|^{-1},$$

$$M = \overline{M}\, (1 + \lambda)\, |A(x_0) \cdot d_0|^{-1}$$

and L, ℓ, as defined in Theorem 19.1.

Then, the conclusions of Theorem 19.1, *excluding the uniqueness part, hold for* (DRNLM).

Proof. The proof is similar to Theorem 19.1, with only addition due to (19.25) that estimate $|F(x_i) + G(x_i) + r_i|$ is bounded above by the right hand side of estimate (19.20) times $1 + \lambda$. That completes the proof of Proposition 19.2. $\quad\diamondsuit$

Remark 19.2. Another choice of d_k is the unit vector $e^{m(k)}$, where $m(k)$ is the index of component of $A(x_k)$ of maximal modulus

$$|A(x_k)[m(k)]| := \max_{j = 1, \cdots, n} |A(x_k)[j]|.$$

For this choice of d_k, (DNLM), (DNM), and (DRNLM) become

$$x_{k+1} = x_k - \frac{F(x_k) + G(x_k)}{A(x_k)[m(k)]}\, e^{m(k)},$$

$$x_{k+1} = x_k - \frac{F(x_k)}{\nabla F(x_k)[m(k)]}\, e^{m(k)},$$

and

$$x_{k+1} = x_k - \frac{F(x_k) + G(x_k) + s_k}{A(x_k)[m(k)]} \, e^{m(k)},$$

respectively.

In this setting, by simply replacing the Euclidean norm by the ∞–norm $\| \cdot \|_\infty$, we obtain the results of Theorem 19.1, Propositions 19.1, and 19.2.

Remark 19.3. The results obtained in this chapter extend in a Hilbert space setting. Indeed, let F be a differentiable operator defined on a convex subset $\mathcal{D}$ of a Hilbert space $\mathcal{H}$ with values in $\mathbb{R}$.

Moreover, instead of conditions (19.9) or (19.15), assume:

$$\| d_k \| = 1,$$

and, there exists $\gamma \in [0, 1]$, such that:

$$|A(x_k) \cdot d_k| \geq \gamma \, \| A(x_k) \| \, .$$

Note that in the case of (19.15), we can always set:

$$\gamma = \frac{|A(x_0) \cdot d_0|}{\| A(x_0) \|} \leq 1.$$

Here $x \cdot y$ denotes the inner product of elements x, and y in $\mathcal{H}$, and $\| x \| = (x \cdot x)^{1/2}$.
Let

$$\eta = \| \frac{F(x_0) + G(x_0)}{A(x_0) \cdot d_0} \, d_0 \| \, .$$

In the case of:
Theorem 19.1, define

$$K = \frac{\overline{K}}{\gamma \, \| A(x_0) \|}, \quad M = \frac{\overline{M}}{\gamma \, \| A(x_0) \|}, \quad \mu = \frac{\mu_0 + \mu_1}{\gamma \, \| A(x_0) \|},$$

$$L = \frac{\overline{L}}{\| A(x_0) \|}, \quad \ell = \frac{\overline{\ell}}{\| A(x_0) \|};$$

Proposition 19.1,

$$K = \frac{\overline{K}}{\gamma \, \| \nabla F(x_0) \|}, \quad L = \frac{\overline{L}}{\| \nabla F(x_0) \|};$$

Proposition 19.2 $(G \neq 0)$,

$$K = \frac{(1 + \lambda) \, \overline{K}}{\gamma \, \| A(x_0) \|}, \quad M = \frac{(1 + \lambda) \, \overline{M}}{\gamma \, \| A(x_0) \|}, \quad \mu = \frac{(1 + \lambda) \, (\mu_0 + \mu_1)}{\gamma \, \| A(x_0) \|},$$

$$L = \frac{\overline{L}}{\| A(x_0) \|}, \quad \ell = \frac{\overline{\ell}}{\| A(x_0) \|};$$

Proposition 19.2 $(G = 0)$,

$$K = \frac{(1 + \lambda) \, \overline{K}}{\gamma \, \| A(x_0) \|}, \quad L = \frac{\overline{L}}{\| A(x_0) \|}.$$

Then, using analogous proofs, the results of Theorem 19.1, Propositions 19.1, and 19.2 hold in this more general setting.

Chapter 19. Directional Newton–Type Methods

The advantages of our results over the corresponding ones in [504, Theorem 1] have already been stated in the abstract of this chapter.

We now refer the motivated reader to [504, Section 4] for further applications (see, also [429], [671]). Clearly, the applicability of the results listed in [504] has now been expanded in view of our results. More applications, and other relevant work can be found in [568], [719]). Maple programs to the methods mentioned here can be downloaded from [257].

We provide an example where our Theorem 19.1 can apply to solve an equation, but not corresponding Theorem 1 in [504].

Example 19.1. Let $n = 2$. Here, we use the Euclidean inner product, and the corresponding norm for both vector and matrix.

Case $G(x) = 0$, $A(x) = \nabla F(x)$, $(x \in \mathcal{D})$, and $\lambda = 0$.

Choose:

$$x_0 = (1,1)^T, \quad \mathcal{D} = \{x : \| x - x_0 \| \leq 1 - b\} \quad \text{for} \quad b \in [0,1),$$

and define function F on $\mathcal{D}$ by

$$F(x) = \frac{\theta_1^3 + \theta_2^3}{2} - b, \quad x = (\theta_1, \theta_2)^T. \tag{19.26}$$

Using hypotheses of Proposition 19.1, we obtain the parameters:

$$\overline{K} = 3\,(2-b)\,\sqrt{2}, \quad \overline{L} = \frac{3\,(3-b)\,\sqrt{2}}{2}, \quad \| \nabla F(x_0) \| = \frac{3\,\sqrt{2}}{2},$$

and

$$F(x_0) = 1 - b, \quad \text{and} \quad \eta = \frac{\sqrt{2}}{3} F(x_0).$$

We can choose the directions d_k by

$$d_k = \frac{\nabla F(x_k)}{\| \nabla F(x_k) \|},$$

so that condition (19.9) is satisfied as equality.

Then, condition

$$q = K\,\eta \leq \frac{1}{2}$$

used in [504] is violated for say $b = .6166$, since

$$K = 2.7668, \quad \eta = .180736493,$$

and

$$q = \frac{2\,\sqrt{2}}{3}\,(2-b)\,(1-b) = .500061729 > .5.$$

Hence, there is no guarantee that (DNM) starting at x_0 converges to a zero $x^\star$ of function F.

However, our conditions hold.

Indeed, we have:

$$L = 2.3834, \quad \beta = 2.50910088, \quad \overline{\delta} = 1.050097978.$$

Then, we have:

$$q_0 = .453486093 < .5, \quad and \quad s^{\star\star} = .38053712 < 1 - b = .3834.$$

Hence, the hypotheses of Proposition 19.1 are satisfied.

That is our Proposition 19.1 guarantees the existence of a zero $x^\star$ in $\overline{U}(x_0, s^\star)$ of function F, obtained as the limit of (DNM) starting at x_0. We found $x^\star = (.851140338, .851140338)^T$.

Case $G(x) = 0, A(x) = \nabla F(x), (x \in \mathcal{D})$, and $\lambda = .09$.

In this case, we similarly have:

$$K = 3.015812, \quad L = 2.597906, \quad \eta = .180736493,$$

$$\beta = 2.734919959, \quad \overline{\delta} = 1.050097978, \quad and \quad s^{\star\star} = .380537127.$$

Hence, we get:
$$q_0 = .494299842 < .5,$$

and
$$s^{\star\star} < 1 - b.$$

That is the conclusions of Proposition 19.2 hold again for equation $F(x) = 0$.

Remark 19.4. The hypotheses of Lemma 1.3 have been left as uncluttered as possible. Note that these hypotheses involve only computations only at the initial point x_0. Below, we shall provide some simpler but stronger hypotheses under which the hypotheses of Lemma 1.3 with $\ell = 0$ hold.

Lemma 19.2. *Let $K > 0$, $M > 0$, $\mu > 0$, with $L > 0$, and $\eta > 0$, be such that:*

$$\mu < \alpha, \quad 2M < K,$$

and
$$0 < h_A = a\eta \leq \frac{1}{2}, \tag{19.27}$$

where,
$$a = \frac{1}{4(\alpha - \mu)} \max\{2L\alpha^2 + 2L\alpha + K\alpha + 2M, K + 2\alpha L\}. \tag{19.28}$$

Then, the following hold:

f_1 has a positive root $\dfrac{\delta}{2}$,

$$\max\{\delta_0, \delta\} \leq 2\,\alpha,$$

and

the conclusions of Lemma 1.3 *hold, with* α *replacing* $\dfrac{\delta}{2}$.

Proof. It follows from Lemma 1.3, and (19.27) that

$$f_m(\alpha) = f_1(\alpha) \leq 0, \quad (m \geq 1), \tag{19.29}$$

which together with $f_1(0) = 2\,(M\,\eta + \mu) > 0$, imply that there exists a positive root $\dfrac{\delta}{2}$ of polynomial f_1, satisfying

$$\delta \leq 2\,\alpha. \tag{19.30}$$

It also follows from (1.97), and (19.27) that

$$\delta_0 \leq 2\,\alpha, \tag{19.31}$$

and (1.105) holds, with α replacing $\dfrac{\delta}{2}$ (by (19.29)). Note also that $\alpha \in (0,1)$, by (1.95).

That completes the proof of Lemma 19.2 $\qquad\qquad\qquad \diamond$

Chapter 20

Directional Two–Step Methods

In this chapter, a semilocal convergence analysis for directional two–step Newton methods in a Hilbert space setting is provided. Two different techniques are used to generate the sufficient convergence results, as well as the corresponding error bounds. The first technique uses our new idea of recurrent functions, whereas the second uses recurrent sequences. We also compare the results of the two techniques.

We are concerned with the problem of approximating a zero $x^\star$ of a differentiable function F defined on a convex subset $\mathcal{D}$ of a Hilbert space $\mathcal{H}$, with values in $\mathbb{R}$. The directional Newton method (DNM) [504] given by

$$x_{k+1} = x_k - \frac{F(x_k)}{\nabla F(x_k) \cdot d_k} d_k \quad (k \geq 0)$$

is used to generate a sequence $\{x_k\}$ converging to $x^\star$, when $\mathcal{H} = \mathbb{R}^i$ (i is a natural number).

Let us explain how (DNM) is conceived. We start with an initial guess $x_0 \in \mathcal{D}$, where F is differentiable and a direction vector d_0.

Then, we restrict F on the line $\mathcal{A} = \{x_0 + \theta\, d_0, \theta \in \mathbb{R}\}$, where it is a function of one variable $f(\theta) = F(x_0 + \theta\, d_0)$.

Set $\theta_0 = 0$ to obtain the Newton iteration for f, that is the next point:

$$v_1 = -\frac{f(0)}{f'(0)}.$$

The corresponding iteration for F is

$$x_1 = x_0 - \frac{F(x_0)}{\nabla F(x_0) \cdot d_0} d_0.$$

If $i = 1$, (DNM) reduces to the classical Newton method [70], [192], [568], [719]. A semilocal convergence analysis for the (DNM) was provided in the elegant work by Levin and Ben–Israel in [257], [504]. The convergence of the method was established for directions d_k sufficiently close to the gradients $\nabla F(x_k)$, and under standard Newton–Kantorovich–type hypotheses [70], [192], [568].

In particular the following conditions were used:

$(\mathcal{A}_1)$

$$|F(x_0)| \leq \lambda;$$

$(\mathcal{A}_2)$

$$|\nabla F(x_0) \cdot d_0| \geq a;$$

$(\mathcal{A}_3)$

$$\sup_{x \in \mathcal{D}} \| F''(x) \| = M;$$

$(\mathcal{A}_4)$

$$h_k = \lambda M a^{-2} \leq \frac{1}{2}.$$

In view of $(\mathcal{A}_3)$, there exist M_0, such that:

$(\mathcal{A}_5)$

$$\| \nabla F(x) - \nabla F(x_0) \| \leq M_0 \| x - x_0 \|, \quad x \in \mathcal{D}.$$

Clearly,

$$M_0 \leq M \tag{20.1}$$

hold in general, and $\dfrac{M}{M_0}$ can be arbitrarily large. Note that $(\mathcal{A}_5)$ is not an additional hypothesis, since in practice the computation of M, requires that of M_0.

Using hypotheses $(\mathcal{A}_1)$, $(\mathcal{A}_2)$, $(\mathcal{A}_3)$, $(\mathcal{A}_5)$, and replacing $(\mathcal{A}_4)$ by

$(\mathcal{A}_4')$

$$h_A = \lambda \overline{M} a^{-2} \leq \frac{1}{2},$$

where,

$$\overline{M} = \frac{1}{8} (M + 4 M_0 + \sqrt{M^2 + 8 M_0 M}),$$

Argyros [170] provided a semilocal convergence analysis with the following advantages over the work in [504]:

1. Weaker hypotheses;

2. Larger convergence domain for (DNM);

3. Finer error bounds on the distances $\| x_{k+1} - x_k \|$, $\| x_k - x^\star \|$ $(k \geq 0)$;

4. An at least as precise information on the location of the zero $x^\star$.

The main idea used the needed $(\mathcal{A}_5)$ instead of $(\mathcal{A}_3)$ for the computation of the bounds on $\| \nabla F(x_n) \|$. This modification leads to more precise estimates which in turn provide the advantages as mentioned above.

Note that

$$h_K \leq \frac{1}{2} \Longrightarrow h_A \leq \frac{1}{2} \tag{20.2}$$

but not necessarily vice verca (unless if $M = M_0$).

Hence, the applicability of (DNM) is extended under the same or less computational cost.

Motivated by these ideas, we provide a semilocal convergence analysis of the two step directional Newton method (TSDNM) for starting $x_0 \in \mathcal{D}$:

$$y_n = x_n - \frac{F(x_n)}{\nabla F(x_n) \cdot d_n} d_n$$

$$x_{n+1} = y_n - \frac{F(y_n)}{\nabla F(x_n) \cdot d_n} d_n.$$

We provide two different and competing techniques to generate the sufficient semilocal convergence conditions as well as the corresponding error bounds for the cubically convergent (TSDNM). The first technique uses the concept of recurrent sequences, which is well established by Ezquerro and Hernández [372], [378] for Chebyshev–type and other high order methods. The second technique uses our new idea of recurrent functions already employed in a Banach space setting [170], [192]. Moreover, we show that the R–order of convergence for the (TSDNM) is at least three. A priori error bounds are also provided.

Throughout this chapter, we shall also use condition:

$(\mathcal{A}_6)$ $\| d_n \| = 1$, and there exists $\gamma \in [0, 1]$, such that

$$|\nabla F(x_n) \cdot d_n| \geq \gamma \, \| \nabla F(x_n) \|,$$

or equivalently

$$\angle(\nabla F(x_n), d_n) \leq \arccos \gamma,$$

where,

$$\angle(u, v) = \arccos \frac{u \cdot v}{\| u \| \cdot \| v \|}, \quad u \neq 0, \quad v \neq 0,$$

$u \cdot v$ denotes the scalar product between two elements u, v of $\mathcal{H}$, and $\| u \| = (u \cdot u)^{1/2}$.

Furthermore, we use instead of $(\mathcal{A}_3)$ weaker condition:

$(\mathcal{A}_7)$

$$\| \nabla F(x) - \nabla F(y) \| \leq M \, \| x - y \|, \quad x, y \in \mathcal{D}.$$

Finally, we denote by (C) set of conditions $(\mathcal{A}_1)$, $(\mathcal{A}_2)$, $(\mathcal{A}_6)$, $(\mathcal{A}_7)$, whereas $(\mathcal{A}_1)$, $(\mathcal{A}_2)$, $(\mathcal{A}_5)$, and $(\mathcal{A}_7)$ are denoted by $(C^\star)$.

Definition 20.1. Let $\gamma \in [0, 1]$, $a > 0$, $\lambda \geq 0$, and $M > 0$ be given parameters. Following [372], it is convenient for us to define

Functions:

$$g_0(t) = 1 - \gamma t \left(1 + \frac{1}{2} t\right)$$

$$g_1(t) = \frac{1}{8} t^2 (t + 4),$$

$$g_2(t) = \frac{g_1(t)}{g_0^2(t)},$$

$$g(t) = g_2(t) - 1,$$

Parameters:

$$v = -1 + \sqrt{\frac{\gamma + 2}{\gamma}},$$

$$c_0 = \lambda,$$

$$\varsigma_0 = a,$$

$$e_0 = \frac{M\lambda}{(\gamma a)^2},$$

and Recurrent Sequences:

$$c_{n+1} = g_1(e_n)\, c_n,$$

$$\varsigma_{n+1} = g_0(e_n)\, \varsigma_n,$$

$$e_{n+1} = g_2(e_n)\, e_n.$$

In Lemmas 20.1–20.3, we study the properties of functions, parameters, and sequences given in Definition 20.1.

Lemma 20.1. *Let functions g_i $i = 0, 1, 2$, g, parameter v, and sequences $\{c_n\}$, $\{\varsigma_n\}$, $\{e_n\}$ be as in Definition 20.1.*

Then, the following hold:

(a) *v is the only positive zero of function g_0.*

   Function g has a minimal positive zero v_0, such that $v_0 < v$;

Let $t \in (0, v_0)$.

(b) *g_0 is decreasing, and $g_0(t) \in (0, 1)$;*

(c) *g_1, g_2 are increasing, $g_1(t) \in (0, 1)$, and $g_2(t) \in (0, 1)$;*

(d) *For $\mu \in (0, 1]$,*
$$g_0(\mu t) \ge g_0(t), \quad g_1(\mu t) \le \mu^2\, g_1(t).$$

   and
$$g_2(\mu t) \le \mu^2\, g_2(t).$$

(e) *Let $e_0 < v_0$. The following hold*
$$g_i(e_n) \in (0, 1), \quad i = 0, 1, 2 \quad (n \ge 0),$$

   and sequences $\{c_n\}$, $\{\varsigma_n\}$, $\{e_n\}$ are decreasing.

Proof.

(a) Using the quadratic formula, we obtain that v is the only positive zero of g_0. Moreover, for any fixed $\mu \in (0, 1]$, we have $g_2(0) = 0$, and $\lim_{t \to v^-} g_2(t) = +\infty$. It follows from the intermediate value theorem that function g has at least a real positive zero in $(0, v)$. Let us denote the minimal positive zero of g by v_0. Clearly, we have $v_0 < v$.

Chapter 20. Directional Two–Step Methods 249

(b) We have $g_0'(t) = -\gamma(1+t) < 0$ for $t \geq 0$. That is function g_0 is decreasing in $(0, v_0)$. Moreover, by $g_0(0) = 1$, and $g_0(v_0) > g_0(v) = 0$, we deduce $g_0(t) \in (0, 1)$.

(c) By the definition of function g_1, we conclude that is increasing. Moreover, we get

$$g_2'(t) = \frac{g_1'(t)\,g_0(t) - 2\,g_0'(t)\,g_1(t)}{g_0^3(t)} > 0,$$

which implies that function g_2 is increasing too. That it is, we have $g_2(t) \in (0, 1)$. Furthemore, $g_1(t) < g_0^2(t)$. Hence, we get $g_1(t) \in (0, 1)$.

(d) By (b), we have $g_0(\mu t) \geq g_0(t)$. Moreover, by the definition of function g_1, and g_2, we have:

$$g_1(\mu t) \leq \mu^2\, g_1(t) \quad \text{and} \quad g_2(\mu t) \leq \mu^2\, g_2(t).$$

(e) This part follows from (a)–(d).

That completes the proof of Lemma 20.1. $\diamondsuit$

Lemma 20.2. *Under hypotheses of Lemma* 20.1(e), *the following estimates hold*

$$g_2(e_n) \leq g_2(e_0)^{3^n}, \tag{20.3}$$

and

$$e_n \leq e_0 \left(g_2(e_0)\right)^{\frac{3^n-1}{2}} \tag{20.4}$$

for all $n \geq 0$.

Proof. We shall use induction. Estimates (20.3), and (20.4) hold for $n = 0$. By Lemma 20.1, we have

$$g_2(e_n) < 1,$$

$$g_2(e_n) = g_2(g_2(e_{n-1})\,e_{n-1}) \leq g_2(e_{n-1})^3 \leq \cdots \leq g_2(e_0)^{3^n},$$

and

$$e_{n+1} = g_2(e_n)\,e_n \leq g_2(e_0)^{3^n}\,e_n \leq \cdots \leq e_0 \left(g_2(e_0)\right)^{\frac{3^n-1}{2}}.$$

That completes the proof of Lemma 20.2. $\diamondsuit$

Lemma 20.3. *Under hypotheses of Lemma* 20.1(e), *the following estimates hold*

$$\frac{g_1(e_n)}{g_0(e_n)} \leq g_0(e_0)\,g_2(e_0)^{3^n}, \tag{20.5}$$

$$\prod_{i=0}^{n} \frac{g_1(e_i)}{g_0(e_i)} \leq g_0(e_0)^{n+1} \left(g_2(e_0)\right)^{\frac{3^{n+1}-1}{2}}, \tag{20.6}$$

$$c_n \leq c_0\, g_0(e_0)^{2^n} \left(g_2(e_0)\right)^{\frac{3^n-1}{2}}, \tag{20.7}$$

for all $n \geq 0$,

and

$$c_n \longrightarrow 0 \quad \text{as} \quad n \longrightarrow \infty. \tag{20.8}$$

Moreover, for any μ_1, $\mu_2 \in (0,1)$, we have:

$$\Xi = \sum_{i=n}^{n+k} \mu_1^i \mu_2^{3^i} \leq \mu_1^n \mu_2^{3^n} \frac{1 - \mu_1^{k+1} \mu_2^{3^n (3^k+1)}}{1 - \mu_1 \mu_2^{2 \times 3^n}} \tag{20.9}$$

for all $n \geq 0$, and $k \geq 1$.

Proof. Estimates (20.5), and (20.6) hold for $n = 0$. Define function g_3 by

$$g_3(t) = \frac{g_1(t)}{g_0(t)}.$$

By Lemma 20.1, we have:

$$g_3(e_n) = g_0(e_n) g_2(e_n) < 1.$$

Moreover, by (20.3) and (20.4), we get in turn:

$$g_3(e_n) = g_3\left(e_0 \left(g_2(e_0)\right)^{\frac{3^n-1}{2}}\right) \leq g_3(e_0) g_2(e_0)^{3^n-1} = g_0(e_0) g_2(e_0)^{3^n},$$

which shows (20.5).

Furthermore, we obtain:

$$\prod_{i=0}^{n} \frac{g_1(e_i)}{g_0(e_i)} \leq \prod_{i=0}^{n} \left(g_0(e_0) g_2(e_0)^{3^i}\right) \leq g_0(e_0)^{n+1} \left(g_2(e_0)\right)^{\frac{3^{n+1}-1}{2}},$$

which shows (20.6).

Estimate (20.7) holds for $n = 0$. As in (20.5) and (20.6), we get in turn:

$$g_1(e_n) = g_1\left(e_0 \left(g_2(e_0)\right)^{\frac{3^n-1}{2}}\right)$$
$$\leq g_1(e_0) g_2(e_0)^{3^n-1} = g_0(e_0)^2 g_2(e_0)^{3^n},$$

and consequently,

$$c_{n+1} = g_1(e_n) c_n \leq g_0(e_0)^2 g_2(e_0)^{3^n} c_n$$
$$\leq c_0 g_0(e_0)^{2(n+1)} \left(g_2(e_0)\right)^{\frac{3^{n+1}-1}{2}}$$

which shows (20.7).

In view of the estimates $g_2(e_0) < 1$, and $g_0(e_0) < 1$, we obtain (20.22). Finally, (20.9) follows from the estimate:

$$\Xi \leq \mu_1^n \mu_2^{3^n} + \mu_1 \mu_2^{2 \times 3^n} \left(\Xi - \mu_1^{n+k} \mu_2^{3^{n+k}}\right).$$

That completes the proof of Lemma 20.3. $\diamondsuit$

We also need the following Ostrowski–type identities for function F [151], [568], [610], [671].

Chapter 20. Directional Two–Step Methods 251

Lemma 20.4. *Assume:*

$F: \mathcal{D} \subseteq \mathcal{H} \longrightarrow \mathbb{R}$ be continuously differentiable, $\nabla F(x)$ $(x \in \mathcal{D})$ be the gradient of F, and iterates $\{x_n\}$, $\{y_n\}$ $(n \geq 0)$ generated by (TSDNM) are well defined.

Then, the following identities hold for all $n \geq 0$:

$$F(y_n) = \int_0^1 (\nabla F(x_n + \theta(y_n - x_n)) - \nabla F(x_n))(y_n - x_n)\, d\theta \qquad (20.10)$$

and

$$
\begin{aligned}
&F(x_{n+1}) \\
&= \int_0^1 (\nabla F(x_n + \theta(x_{n+1} - x_n)) - \nabla F(x_n + \theta(y_n - x_n)))(y_n - x_n)\, d\theta \qquad (20.11) \\
&\quad + \int_0^1 (\nabla F(x_n + \theta(x_{n+1} - x_n)) - \nabla F(x_n))(x_{n+1} - y_n)\, d\theta.
\end{aligned}
$$

Proof. We have in turn:

$$
\begin{aligned}
F(y_n) &= F(x_n) + \nabla F(x_n)(y_n - x_n) \\
&\quad + \int_0^1 (\nabla F(x_n + \theta(y_n - x_n)) - \nabla F(x_n))(y_n - x_n)\, d\theta \\
&= \int_0^1 (\nabla F(x_n + \theta(y_n - x_n)) - \nabla F(x_n))(y_n - x_n)\, d\theta,
\end{aligned}
$$

which shows (20.10).

Moreover, we get

$$
\begin{aligned}
F(x_{n+1}) &= F(x_n) + \nabla F(x_n)(x_{n+1} - x_n) \\
&\quad + \int_0^1 (\nabla F(x_n + \theta(x_{n+1} - x_n)) - \nabla F(x_n))(x_{n+1} - x_n)\, d\theta \\
&= -F(y_n) + \int_0^1 (\nabla F(x_n + \theta(x_{n+1} - x_n)) - \nabla F(x_n))(x_{n+1} - x_n)\, d\theta \\
&= -F(y_n) + \int_0^1 (\nabla F(x_n + \theta(x_{n+1} - x_n)) - \nabla F(x_n))(x_{n+1} - y_n)\, d\theta \\
&\quad + \int_0^1 (\nabla F(x_n + \theta(x_{n+1} - x_n)) - \nabla F(x_n))(y_n - x_n)\, d\theta,
\end{aligned}
$$

which shows (20.11) (by (20.10)).

That completes the proof of Lemma 20.4. $\diamondsuit$

Following [151], [372], [429], we also need relations between recurrent sequences.

Lemma 20.5. *Assume condition $(\mathcal{C})$, and hypotheses of Lemma 20.1(e) hold.*

Let us define:

$$r^\star = \frac{\lambda}{a\,\gamma}\left(\frac{1}{1 - g_0(e_0)\,g_2(e_0)} + \frac{e_0}{2\,(1 - g_0(e_0)\,g_2(e_0))^2}\right). \qquad (20.12)$$

Then, the following hold for all $n \geq 0$:

(a)

$$|F(x_n)| \leq c_n,$$

(b)
$$\| \nabla F(x_n) \| \geq \varsigma_n,$$

(c)
$$\| y_{n+1} - x_{n+1} \| \leq \frac{g_1(e_n)}{g_0(e_n)} \| y_n - x_n \|,$$

(d)
$$\frac{M \| y_n - x_n \|}{|\nabla F(x_n) \cdot d_n|} \leq e_n,$$

(e)
$$\| x_{n+1} - y_n \| \leq \frac{1}{2} e_n \| y_n - x_n \|,$$

(f)
$$\| x_{n+1} - x_n \| \leq \left(1 + \frac{1}{2} e_n\right) \| y_n - x_n \|,$$

and

(g)
$$x_n, y_n \in U(x_0, r^\star).$$

Proof. Let $n = 0$. Using (TSDNM), $(\mathcal{A}_1)$, $(\mathcal{A}_2)$, and $(\mathcal{A}_6)$, we get:

$$\| y_0 - x_0 \| = \frac{|F(x_0)|}{|\nabla F(x_0) \cdot d_0|} \leq \frac{|F(x_0)|}{\gamma \| \nabla F(x_0) \|} \leq \frac{c_0}{\gamma \varsigma_0} < r^\star,$$

which shows $y_0 \in U(x_0, r^\star)$.

In view of $(\mathcal{A}_7)$, and (20.10) for $n = 0$, we have:

$$|F(y_0)| \leq \frac{M}{2} \| y_0 - x_0 \|^2,$$

so,

$$
\begin{aligned}
\| x_1 - y_0 \| &\leq \frac{|F(y_0)|}{\gamma \| \nabla F(x_0) \|} \\
&\leq \frac{M}{2\gamma\varsigma_0} \| x_0 - y_0 \|^2 \leq \frac{1}{2} e_0 \| x_0 - y_0 \|,
\end{aligned}
$$

and

$$\| x_1 - x_0 \| \leq \| x_1 - y_0 \| + \| y_0 - x_0 \| \leq (1 + \frac{1}{2} e_0) \| y_0 - x_0 \| < r^\star,$$

which shows $x_1 \in U(x_0, r^\star)$.

In view of condition $(\mathcal{A}_7)$, we have:

$$
\begin{aligned}
\| \nabla F(x_1) \| &\geq \| \nabla F(x_0) \| - \| \nabla F(x_1) - \nabla F(x_0) \| \\
&\geq \varsigma_0 - M \| x_1 - x_0 \| \\
&\geq \varsigma_0 - M (1 + \frac{1}{2} e_0) \| y_0 - x_0 \| \\
&\geq \left(1 - \gamma e_0 (1 + \frac{1}{2} e_0)\right) \varsigma_0 = g_0(e_0) \varsigma_0 = \varsigma_1.
\end{aligned}
$$

By $(\mathcal{A}_6)$, we have:
$$|\nabla F(x_1) \cdot d_1| \geq \gamma \parallel \nabla F(x_1) \parallel > 0.$$

That is y_1, and x_2 are well defined.

Using (20.11) for $n = 0$, we get in turn:

$$
\begin{aligned}
|F(x_1)| &\leq \frac{M}{2} \parallel x_1 - y_0 \parallel (\parallel y_0 - x_0 \parallel + \parallel x_1 - x_0 \parallel) \\
&\leq (4 + e_0) \frac{M \gamma \varsigma_0 e_0^2}{8} \parallel y_0 - x_0 \parallel \\
&\leq \frac{1}{8}(4 + e_0) c_0 e_0^2 = c_1,
\end{aligned}
$$

so,

$$
\begin{aligned}
\parallel y_1 - x_1 \parallel &= \frac{|F(x_1)|}{|\nabla F(x_1) \cdot d_1|} \\
&\leq \frac{|F(x_1)|}{\gamma \parallel \nabla F(x_1) \parallel} \\
&\leq \frac{(4 + e_0) e_0^2}{8 g_0(e_0)} \parallel y_0 - x_0 \parallel = \frac{g_1(e_0)}{g_0(e_0)} \parallel y_0 - x_0 \parallel,
\end{aligned}
$$

and

$$\parallel y_1 - x_0 \parallel \leq \parallel y_1 - x_1 \parallel + \parallel x_1 - x_0 \parallel \leq \left(1 + \frac{1}{2} e_0 + g_0(e_0) g_2(e_0)\right) \parallel y_0 - x_0 \parallel < r^\star,$$

which shows $y_1 \in U(x_0, r^\star)$.

Moreover, we have:

$$
\begin{aligned}
\frac{M \parallel y_1 - x_1 \parallel}{|\nabla F(x_1) \cdot d_1|} &\leq \frac{M \parallel y_1 - x_1 \parallel}{\gamma \parallel \nabla F(x_1) \parallel} \\
&\leq \frac{M g_1(e_0)}{\gamma \varsigma_0 g_0^2(e_0)} \parallel y_0 - x_0 \parallel \leq \frac{g_1(e_0)}{g_0^2(e_0)} e_0 = e_1,
\end{aligned}
$$

so

$$
\begin{aligned}
\parallel x_2 - x_1 \parallel &\leq \frac{|F(y_1)|}{\gamma \parallel \nabla F(x_1) \parallel} \\
&\leq \frac{M \parallel y_1 - x_1 \parallel^2}{2 \gamma \varsigma_1} \leq \frac{1}{2} e_1 \parallel y_1 - x_1 \parallel,
\end{aligned}
$$

$$\parallel x_2 - x_1 \parallel \leq \parallel x_2 - y_1 \parallel + \parallel y_1 - x_1 \parallel \leq \left(1 + \frac{1}{2} e_1\right) \parallel y_1 - x_1 \parallel,$$

and

$$
\begin{aligned}
\parallel x_2 - x_0 \parallel &\leq \parallel x_2 - y_1 \parallel + \parallel y_1 - x_1 \parallel + \parallel x_1 - x_0 \parallel \\
&\leq \left(1 + \frac{1}{2} e_1\right) \parallel y_1 - x_1 \parallel + \left(1 + \frac{1}{2} e_0\right) \parallel y_0 - x_0 \parallel \\
&\leq \left(1 + g_0(e_0) g_2(e_0) + \frac{1}{2} e_1 \left(1 + g_0(e_0) g_2^2(e_0)\right)\right) \parallel y_0 - x_0 \parallel < r^\star,
\end{aligned}
$$

which shows $x_2 \in U(x_0, r^\star)$.

The rest follows by a simple induction argument.

254   Ioannis K. Argyros, Saïd Hilout and Mohammad A. Tabatabai

That completes the proof of Lemma 20.5.   ◇

We can show the main semilocal convergence result for (TSDNM) using recurrent sequences. In particular, we establish (see also, Lemmas 20.1–20.3) that the R–order of the method is at least three.

Theorem 20.1. *Let $F : \mathcal{D} \subseteq \mathcal{H} \longrightarrow \mathbb{R}$ be a continuously differentiable function, and let $x_0 \in \mathcal{D}$.*

Assume condition (C), and hypotheses of Lemma 20.1(e) hold.

Furthermore, assume

$$\overline{U}(x_0, r^\star) = \{x \in \mathcal{H} : \| x - x_0 \| \le r^\star\} \subseteq \mathcal{D}. \tag{20.13}$$

Then, sequence $\{x_n\}$ generated by (TSDNM) is well defined, remains in $\overline{U}(x_0, r^\star)$ for all $n \ge 0$, and converges to a zero $x^\star$ of function F.

Moreover, the following estimates hold for all $n \ge 0$:

$$\| x_n - x^\star \| \le \ell_\infty^n, \tag{20.14}$$

where,

$$\begin{aligned}
\ell_\infty^n &= \left((g_2(e_0)^{1/2})^{3^n-1} \frac{1}{1 - g_0(e_0)\, g_2(e_0)^{3^n}} + \right. \\
&\quad \left. \frac{1}{2} e_0\, g_2(e_0)^{3^n-1} \frac{1}{1 - g_0(e_0)\, g_2(e_0)^{2\times 3^n}} \right) \frac{\lambda}{\gamma a}\, g_0(e_0)^n.
\end{aligned} \tag{20.15}$$

Proof. We shall show $\{x_n\}$ is a complete sequence. We have in turn:

$$\begin{aligned}
\| x_{n+1} - x_n \| &\le (1 + \frac{1}{2} e_n) \| y_n - x_n \| \\
&\le (1 + \frac{1}{2} e_n) \frac{g_1(e_{n-1})}{g_0(e_{n-1})} \| y_{n-1} - x_{n-1} \| \\
&\le \left((1 + \frac{1}{2} e_0\, (g_2(e_0)^{1/2})^{3^n-1}) \right) \left(\prod_{i=0}^{n-1} \frac{g_1(e_i)}{g_0(e_i)} \right) \| y_0 - x_0 \| \\
&\le \frac{\lambda}{\gamma a} \left((g_2(e_0)^{1/2})^{3^n-1} + \frac{1}{2} e_0\, g_2(e_0)^{3^n-1} \right) g_0(e_0)^n,
\end{aligned}$$

so,

$$\begin{aligned}
\| x_{n+k} - x_n \| &\le \sum_{i=n}^{n+k-1} \| x_{i+1} - x_i \| \\
&\le \frac{\lambda}{\gamma a} \sum_{i=n}^{n+k-1} g_0(e_0)^i \left((g_2(e_0)^{1/2})^{3^i-1} + \frac{1}{2} e_0\, g_2(e_0)^{3^i-1} \right) \\
&\le \ell_k^n,
\end{aligned} \tag{20.16}$$

where,

$$\begin{aligned}
\ell_k^n &= \frac{\lambda}{\gamma a} \left((g_2(e_0)^{1/2})^{3^n-1} \frac{1 - g_0(e_0)^k\, ((g_2(e_0))^{1/2})^{3^n\,(3^{k-1}+1)}}{1 - g_0(e_0)\, g_2(e_0)^{3^n}} + \right. \\
&\quad \left. \frac{1}{2} e_0\, g_2(e_0)^{3^n-1} \frac{1 - g_0(e_0)^k\, g_2(e_0)^{3^n\,(3^{k-1}+1)}}{1 - g_0(e_0)\, g_2(e_0)^{2\times 3^n}} \right).
\end{aligned}$$

Chapter 20. Directional Two–Step Methods
255

In view of (20.3), $\{x_n\}$ is a Cauchy sequence in a Hilbert space $\mathcal{H}$, and as such it converges to some $x^\star \in \overline{U}(x_0, r^\star)$ (since $x^\star \in \overline{U}(x_0, r^\star)$ is a closed set).

If we let $n = 0, k \longrightarrow \infty$, we get $\| x^\star - x_0 \| \leq r^\star$, which shows $x_0 \in \overline{U}(x_0, r^\star)$. By Lemma 20.5, we have $|F(x_n)| \longrightarrow 0$, which shows (by the continuity of F): $F(x^\star) = 0$.

Finally, by letting $k \longrightarrow \infty$ in (20.16), we obtain (20.14).

That completes the proof of Theorem 20.1. $\diamond$

It is convenient for us to define some parameters, and sequences.

Definition 20.2. Let $\gamma > 0$, $a > 0$, $\lambda \geq 0$, $L_0 > 0$, $L > 0$ be given constants. Define parameters:

$$\eta = s_0 = \frac{\lambda}{a},$$

$$t_1 = s_0 + \frac{L s_0^2}{2},$$

$$\delta_1 = \frac{2L}{2L_0 + L + \sqrt{(2L_0 + L)^2 + 8L_0 L}},$$

$$\delta_2 = \begin{cases} \dfrac{L - 2L_0}{L + L_0} & if \quad L > 2L_0 \\ 0 & if \quad L \leq 2L_0, \end{cases}$$

$$\delta_3 = \frac{L s_0}{2},$$

$$\delta_4 = \frac{(s_0 + t_1) L^2 s_0^2}{4(1 - L_0 t_1)}, \quad L_0 t_1 \neq 0,$$

$$w_\infty = \frac{1 - L_0 \eta}{1 + L_0 \eta},$$

$$w_\infty^1 = \frac{2}{3 + L_0 \eta + \sqrt{(3 + L_0 \eta)^2 - 8}},$$

$$\alpha = \max \{\delta_1, \delta_2, \delta_3, \delta_4\},$$

$$\beta = \min \{w_\infty, w_\infty^1\},$$

and scalar sequences $\{t_n\}$, $\{s_n\}$ $(n \geq 0)$ by:

$$t_0 = 0, \quad s_0 = \eta,$$

$$t_{n+1} = s_n + \frac{L(s_n - t_n)^2}{2(1 - L_0 t_n)}, \tag{20.17}$$

$$s_{n+1} = t_{n+1} + \frac{L((s_n - t_n) + (t_{n+1} - t_n))}{2(1 - L_0 t_{n+1})}(t_{n+1} - s_n). \tag{20.18}$$

We shall show the following result on majorizing sequences for (TSDNM):

Lemma 20.6. *Assume:*

$$\alpha \leq \beta, \tag{20.19}$$

and for

$$\eta^\star = \frac{1}{L+L_0}, \tag{20.20}$$

$$\eta < \eta^\star. \tag{20.21}$$

Then, sequences $\{t_n\}$, $\{s_n\}$ $(n \geq 0)$ generated by (20.17) and (20.18), are well defined, non–decreasing, bounded from above by:

$$t^{\star\star} = \left(1 + \frac{2\,\delta}{2-\delta}\right)\eta, \tag{20.22}$$

and converge to their common and unique least upper bound $t^\star \in [0, t^{\star\star}]$.
Moreover, the following estimates hold for all $n \geq 0$:

$$t_n \leq s_n \leq t_{n+1} \leq s_{n+1}, \tag{20.23}$$

and

$$0 \leq s_{n+1} - t_{n+1} \leq \frac{\delta}{2}\,(s_n - t_n) \leq \left(\frac{\delta}{2}\right)^{n+1}\eta, \tag{20.24}$$

where $\delta = 2\,\alpha$.

Proof. We shall show using induction on k: (20.23), (20.24),

$$0 \leq t_{k+1} - s_k \leq \frac{\delta}{2}\,(s_k - t_k), \tag{20.25}$$

$$0 \leq s_{k+1} - t_{k+1} \leq \frac{\delta}{2}\,(s_k - t_k), \tag{20.26}$$

and

$$L_0\,t_{k+1} < 1. \tag{20.27}$$

Estimates (20.23)–(20.27) hold for $k = 0$ by the choices of δ_3, δ_4, and (20.17), (20.18) (for $n = 0$).

Assume (20.23)–(20.27) hold for all $m \leq k$.

By the induction hypotheses, we have in turn:

$$
\begin{aligned}
s_m \;&\leq\; t_m + \frac{\delta}{2}\,(s_{m-1} - t_{m-1}) \\
&\leq\; s_{m-1} + \frac{\delta}{2}\,(s_{m-1} - t_{m-1}) + \frac{\delta}{2}\,(s_{m-1} - t_{m-1}) \\
&\leq\; \eta + 2\left(\frac{\delta}{2}\eta + \left(\frac{\delta}{2}\right)^2\eta + \cdots + \left(\frac{\delta}{2}\right)^m\eta\right) \\
&=\; \eta + \frac{1 - \left(\frac{\delta}{2}\right)^m}{1 - \frac{\delta}{2}}\,\delta\eta,
\end{aligned}
\tag{20.28}
$$

and

$$t_{m+1} \leq s_m + \frac{\delta}{2}\,(s_m - t_m) = \eta + \frac{1 - \left(\frac{\delta}{2}\right)^m}{1 - \frac{\delta}{2}}\,\delta\,\eta + \left(\frac{\delta}{2}\right)^{m+1}\eta. \tag{20.29}$$

Instead of (20.25), (by (20.17)), we shall show:

$$0 \leq \frac{L\,(s_m - t_m)}{2\,(1 - L_0\,t_m)} \leq \frac{\delta}{2} \tag{20.30}$$

or

$$L\left(\frac{\delta}{2}\right)^m \eta + \delta\,L_0\left\{1 + \frac{1 - \left(\frac{\delta}{2}\right)^{m-1}}{1 - \frac{\delta}{2}}\,\delta + \left(\frac{\delta}{2}\right)^m\right\}\eta - \delta \leq 0. \tag{20.31}$$

Estimate (20.31) motivates us to define functions f_m on $[0,1)$ for $w = \frac{\delta}{2}$ by:

$$f_m(w) = L\,w^{m-1}\,\eta + 2\,L_0\left(1 + 2\,w\,(1 + w + \cdots + w^{m-2}) + w^m\right)\eta - 2. \tag{20.32}$$

We need a relationship between two consecutive f_m:

$$
\begin{aligned}
f_{m+1}(w) & \\
= \;& L\,w^m\,\eta + 2\,L_0\left(1 + 2\,w\,(1 + w + \cdots + w^{m-1}) + w^{m+1}\right)\eta - 2 \\
= \;& L\,w^m\,\eta + L\,w^{m-1}\,\eta - L\,w^{m-1}\,\eta + \\
& + 2\,L_0\left(1 + 2\,w\,(1 + w + \cdots + w^{m-2}) + w^m + w^m + w^{m+1}\right)\eta - 2 \\
= \;& f_m(w) + L\,w^n\,\eta - L\,w^{n-1}\,\eta + 2\,L_0\,(w^m + w^{m+1})\,\eta \\
= \;& f_m(w) + G_1(w)\,w^{m-1}\,\eta,
\end{aligned} \tag{20.33}
$$

where,

$$G_1(w) = 2\,L_0\,w^2 + (2\,L_0 + L)\,w - L. \tag{20.34}$$

Note that δ_1 given in Definition 20.2 is the unique positive zero of function G_1. We have by (20.20), and (20.32):

$$f_m(0) = 2\,((L + L_0)\,\eta - 1) < 0, \tag{20.35}$$

and for sufficiently large $w > 0$:

$$f_m(w) > 0 \qquad (w > 0). \tag{20.36}$$

It then follows from (20.35), (20.36), and the intermediate value theorem that there exists a positive zero w_m for each function f_m $(m \geq 2)$. The zero's w_m are unique, since $f_m'(w) > 0$ $(w > 0)$.

Define function f_∞ on $[0,1)$ by

$$f_\infty(w) = \lim_{m \to \infty} f_m(w).$$

Then, using (20.32), we get:

$$f_\infty(w) = 2 \left(\frac{1+w}{1-w} L_0 \, \eta - 1 \right).$$

Number w_∞ given in Definition 20.2 solves equation $f_\infty(w) = 0$.
Function f_∞ is increasing, since

$$f'_\infty(w) = \frac{4 L_0 \, \eta}{(1-w)^2} > 0 \qquad w \in [0,1).$$

We shall show for all $m \geq 2$:

$$f_m(\delta_1) \leq 0, \tag{20.37}$$

and

$$w^\star \leq w_{m+1} \leq w_m. \tag{20.38}$$

Using (20.33), and (20.34), we obtain for all $m \geq 2$:

$$f_m(\delta_1) = f_{m-1}(\delta_1) = \cdots = f_2(\delta_1).$$

Let i be any fixed but arbitrary natural number. Then, we get:

$$f_i(\delta_1) = \lim_{m \to \infty} f_m(\delta_1) = f_\infty(\delta_1) \leq f_\infty(w_\infty) = 0, \tag{20.39}$$

since, function f_∞ is increasing, and $\delta_1 \leq w_\infty$ by hypthesis (20.19). It then follows from the definition of the zeros w_m, and (20.39) that

$$\delta_1 \leq w_m \qquad (m \geq 2). \tag{20.40}$$

Functions f_m are increasing, which together with (20.40) imply

$$f_m(\delta_1) \leq f_m(w_m) = 0.$$

In particular, we get:

$$f_\infty(\delta_1) = \lim_{m \to \infty} f_m(\delta_1) \leq 0. \tag{20.41}$$

Hence, we showed (20.37).
Using (20.33), and (20.40), we have:

$$f_{m+1}(w_{m+1}) = f_m(w_{m+1}) + G_1(w_{m+1}) \, w_{m+1}^{m-1} \, \eta,$$

or

$$f_m(w_{m+1}) \leq 0,$$

since $f_{m+1}(w_{m+1}) = 0$, and $G_1(w_{m+1}) \, w_{m+1}^{m-1} \, \eta \geq 0$, which imply $w_{m+1} \leq w_m$.

Chapter 20. Directional Two–Step Methods

It then follows sequence $\{w_m\}$ is non–increasing, bounded below by zero, and as such it converges to its unique maximum lowest bound $w^\star$.

We shall show $w_\infty \leq w^\star$. Using (20.33), and (20.34), we get:

$$
\begin{aligned}
f_{i+1}(w_i) &= f_i(w_i) + G_1(w_i)\, w_i^{i-1}\, \eta \\
&= G_1(w_i)\, w_i^{i-1}\, \eta \geq 0,
\end{aligned}
$$

so,

$$
f_{i+2}(w_i) = f_{i+1}(w_i) + G_1(w_i)\, w_i^{i}\, \eta \geq 0.
$$

If, $f_{i+k}(w_i) \geq 0$, $k \geq 0$, then

$$
f_{i+k+1}(w_i) = f_{i+k}(w_i) + G_1(w_i)\, w_i^{i+k-1}\, \eta \geq 0.
$$

Hence, by the definition of function f_∞, we get $f_\infty(w_m) \geq 0$ $(m \geq 2)$.

But we also have $f_\infty(0) = 2\,(L_0\,\eta - 1) < 0$. That is $w_\infty \leq w_m$ for all $m \geq 2$, and consequently, $w_\infty \leq w^\star$.

Hence, we showed (20.31), and (20.38).

In order for us to show (20.26), and (20.27), we first need the estimate:

$$
\frac{L\,[(s_m - t_m) + (t_{m+1} - t_m)]\,(t_{m+1} - s_m)}{2\,(1 - L_0\,t_{m+1})}
$$

$$
= \frac{L\,[2\,(s_m - t_m) + (t_{m+1} - s_m)]\,(t_{m+1} - s_m)}{2\,(1 - L_0\,t_{m+1})}
$$

$$
= \frac{L\left(2 + \dfrac{L\,(s_m - t_m)}{2\,(1 - L_0\,t_m)}\right) \dfrac{L\,(s_m - t_m)^3}{2\,(1 - L_0\,t_m)}}{2\,(1 - L_0\,t_m)} \tag{20.42}
$$

$$
= \left(\frac{L}{2}\,\frac{s_m - t_m}{1 - L_0\,t_m}\right)^2 \frac{4 + L\,(s_m - t_m)}{2\,(1 - L_0\,t_{m+1})}\,(s_m - t_m)
$$

$$
\leq \left(\frac{\delta}{2}\right)^2 \frac{4 + L\,(s_m - t_m)}{2\,(1 - L_0\,t_{m+1})}\,(s_m - t_m).
$$

Then, in view of (20.42), and (20.18), estimates (20.26), and (20.27) hold if

$$
0 \leq \left(\frac{\delta}{2}\right)^2 \frac{4 + L\,(s_m - t_m)}{2\,(1 - L_0\,t_{m+1})} \leq \frac{\delta}{2}, \tag{20.43}
$$

or

$$
\frac{\delta}{4}\,\frac{4 + L\,(s_m - t_m)}{1 - L_0\,t_{m+1}} \leq 1, \tag{20.44}
$$

or

$$
\delta\,L\left(\frac{\delta}{2}\right)^m \eta + 4\,L_0\left\{1 + \frac{1 - \left(\dfrac{\delta}{2}\right)^m}{1 - \dfrac{\delta}{2}}\,\delta + \left(\frac{\delta}{2}\right)^{m+1}\right\}\eta + 4\,\delta \leq 4. \tag{20.45}
$$

Set $w = \dfrac{\delta}{2}$. Then, (20.45) holds if:

$$
\begin{aligned}
p_m(w) &= L\, w^{m+1}\, \eta + 2\, L_0 \left(1 + 2\, w\, (1 + w + \cdots + w^{m-1}) + \right. \\
&\quad \left. w^{m+1}\right) \eta + 4\, w - 2 \leq 0.
\end{aligned}
\tag{20.46}
$$

We have:

$$
p_{m+1}(w) = p_m(w) + G_2(w)\, w^{m+1}\, \eta,
\tag{20.47}
$$

where,

$$
G_2(w) = (L + L_0)\, w + 2\, L_0 - L.
\tag{20.48}
$$

Note that δ_2 given in Definition 20.2 solves equation $G_2(w) = 0$.
We have:

$$
p_m(0) = 2\, (L_0\, \eta - 1) < 0.
\tag{20.49}
$$

By (20.21), and for sufficiently large w

$$
p_m(w) > 0.
\tag{20.50}
$$

It follows from (20.49), (20.50), and the intermediate value theorem that there exists a positive zero w_m^1 for each function p_m. These zero's w_m^1 are unique, since:

$$
p_m'(w) > 0 \qquad (w > 0), \quad (m \geq 1).
\tag{20.51}
$$

Then, in an analogous way, we show

$$
p_m(\delta_2) \leq 0, \quad (m \geq 1),
\tag{20.52}
$$

$$
w^{\star 1} \leq w_{m+1}^1 \leq w_m^1
\tag{20.53}
$$

and

$$
p_\infty(\delta_2) \leq 0.
\tag{20.54}
$$

The induction is now completed.

It follows that sequences $\{t_n\}$, $\{s_n\}$ are non–decresing, bounded above by $t^{\star\star}$, and as such they converge to their common, and unique least upper bound $t^\star \in [0, t^{\star\star}]$.

That completes the proof of Lemma 20.6. $\qquad\qquad\qquad\qquad \diamond$

Note that $\beta \in [0, 1)$, and the verification of sufficient convergence condition (20.21) only requires computations at the original points.

We can state the main semilocal convergence result for (TSDNM) using recurrent functions.

Theorem 20.2. *Let $F : \mathcal{D} \subseteq \mathcal{H} \longrightarrow \mathbb{R}$ be a continuously differentiable function, and let $x_0 \in \mathcal{D}$.*

Assume hypotheses: $(C^\star)$, and of Lemma 20.6 hold for

$$
L_0 = \frac{M_0}{a}, \qquad L = \frac{M}{\gamma a}, \qquad \eta = \frac{\lambda}{a}.
$$

Chapter 20. Directional Two–Step Methods

Furthermore, assume

$$\overline{U}(x_0, t^\star) \subseteq \mathcal{D}. \tag{20.55}$$

Then, sequence $\{x_n\}$ generated by (TSDNM) is well defined, remains in $\overline{U}(x_0, t^\star)$ for all $n \geq 0$, and converges to a zero of function F.

Moreover, the following estimates hold for all $n \geq 0$:

$$\| x_{n+1} - y_n \| \leq t_{n+1} - s_n, \tag{20.56}$$

$$\| y_{n+1} - x_{n+1} \| \leq s_{n+1} - t_{n+1}, \tag{20.57}$$

$$\| y_{n+1} - y_n \| \leq s_{n+1} - s_n, \tag{20.58}$$

$$\| x_{n+1} - x_n \| \leq t_{n+1} - t_n, \tag{20.59}$$

and

$$\| x_{n+1} - x^\star \| \leq t^\star - t_n, \tag{20.60}$$

where, sequences $\{t_k\}$, and $\{s_n\}$, and $t^\star$ are given in Definition 20.2.

Proof. Using induction on n, we shall show (20.56), and (20.57). Estimates (20.58), (20.59) will then follow from:

$$\begin{aligned} \| y_{n+1} - y_n \| &\leq \| y_{n+1} - x_{n+1} \| + \| x_{n+1} - y_n \| \\ &\leq s_{n+1} - t_{n+1} + t_{n+1} - s_n = s_{n+1} - s_n. \end{aligned}$$

and

$$\begin{aligned} \| x_{n+1} - x_n \| &\leq \| x_{n+1} - y_n \| + \| y_n - x_n \| \\ &\leq t_{n+1} - s_n + s_n - t_n = t_{n+1} - t_n. \end{aligned}$$

Moreover, sequence $\{t_n\}$ is Cauchy. Consequently $\{x_n\}$ is Cauchy too, and the same arguments used at the end of the proof of Theorem 20.1 will also complete the proof of Theorem 20.2.

Furthemore, estimate (20.60) follows from (20.59) by using standard majorization techniques [151], [189], [568].

Estimates (20.56), and (20.57) hold for $n = 0$ by the initial conditions. We assume they hold for all $k \leq n$. Let $x_n \in U(x_0, t^\star)$. Using $(\mathcal{A}_5)$, we obtain in turn:

$$\begin{aligned} \| \nabla F(x_n) \| &\geq \| \nabla F(x_0) \| - \| \nabla F(x_n) - \nabla F(x_0) \| \\ &\geq a - M_0 \| x_n - x_0 \| \\ &\geq a - M_0 t_n = a (1 - L_0 t_n) > 0 \end{aligned} \tag{20.61}$$

(by Lemma 20.6).

Using the induction hypotheses, Lemma 20.4, we get:

$$\| F(y_n) \| \leq \frac{M}{2} \| y_n - x_n \|^2 \leq \frac{M}{2} (s_n - t_n)^2 \tag{20.62}$$

$$\begin{aligned} \| F(x_{n+1}) \| &\leq \frac{M}{2} \| x_{n+1} - y_n \| \| y_n - x_n \| + \\ &\quad \frac{M}{2} \| x_{n+1} - y_n \| \| x_{n+1} - x_n \| \\ &\leq \frac{M}{2} \left(\| y_n - x_n \| + \| x_{n+1} - x_n \| \right) \| x_{n+1} - y_n \| \\ &\leq \frac{M}{2} \left((s_n - t_n) + (t_{n+1} - t_n) \right) (t_{n+1} - s_n). \end{aligned} \tag{20.63}$$

262 Ioannis K. Argyros, Saïd Hilout and Mohammad A. Tabatabai

Using (TSDNM), (20.61)–(20.63), and the definition of sequences $\{t_n\}$, $\{s_n\}$, we obtain in turn:

$$
\begin{aligned}
\| x_{n+1} - y_n \| &= \left\| \frac{F(y_n)}{\nabla F(x_n) \cdot d_n} d_n \right\| \\
&\leq \frac{M (s_n - t_n)^2}{2\, a\, \gamma\, (1 - L_0\, t_n)} = \frac{L (s_n - t_n)^2}{2\, (1 - L_0\, t_n)} = t_{n+1} - s_n,
\end{aligned}
$$

and

$$
\begin{aligned}
\| y_{n+1} - x_{n+1} \| &= \left\| \frac{F(x_n)}{\nabla F(x_n) \cdot d_n} d_n \right\| \\
&\leq \frac{M \left[(s_n - t_n) + (t_{n+1} - t_n) \right] (t_{n+1} - s_n)}{2\, a\, \gamma\, (1 - L_0\, t_{n+1})} \\
&= \frac{L \left[(s_n - t_n) + (t_{n+1} - t_n) \right] (t_{n+1} - s_n)}{2\, (1 - L_0\, t_{n+1})} = s_{n+1} - t_{n+1},
\end{aligned}
$$

which completes the induction for estimates (20.56), and (20.57), and consequently the proof of Theorem 20.2. $\diamond$

Remark 20.1. (a) Note that $t^{\star\star}$ given in closed form by (20.22) can replace $t^\star$ in condition (20.55).

(b) The directions $d_n \in \mathcal{H}$ are given by the following

(b$_1$)

$$
d_n = \frac{\nabla F(x_n)}{\| \nabla F(x_n) \|}, \quad (n \geq 0).
$$

In this case, condition ($\mathcal{A}_6$) holds as equality for $\gamma = 1$.

(b$_2$) If $\mathcal{H} = \mathbb{R}^i$ (i a natural number), let d_n be the unit vector $e^{m(n)}$, where $m(n)$ is the index of component of $\nabla F(x_n)$ of maximal modulus

$$
|\nabla F(x_n)[m(n)]| := \max_{j=1,\cdots,i} |\nabla F(x_n)[j]|.
$$

For this choice of d_n, (TSDNM) becomes

$$
y_n = x_n - \frac{F(x_n)}{\nabla F(x_n)[m(n)]} e^{m(n)}
$$

$$
x_{n+1} = y_n - \frac{F(y_n)}{\nabla F(x_n)[m(n)]} e^{m(n)}, \quad (n \geq 0).
$$

(c) A direct comparison between the sufficient convergence conditions Lemma 20.4 (e) (Theorem 20.1), and (20.19), (20.21) of Lemma 20.6 (Theorem 20.2) does not seem possible.

Moreover, we note that if $M = M_0$ (i.e. $L_0 = L$), majorizing sequence $\{t_n\}$ given by (20.17), (20.18) is essentially reduced to the one used in Theorem 20.1.

Let us denote by $\{\bar{t}_n\}$, such a sequence. Therefore, if $M_0 < M$, then, a simple induction argument shows (under the hypotheses of Theorems 20.1 and 20.2):

$$
t_n < \bar{t}_n \quad (n > 1),
$$

Chapter 20. Directional Two–Step Methods 263

$$s_n < \bar{s}_n \qquad (n > 1),$$

$$t_{n+1} - s_n < \bar{t}_{n+1} - \bar{s}_n \qquad (n \geq 0),$$

$$t_{n+1} - s_{n+1} < \bar{t}_{n+1} - \bar{s}_{n+1} \qquad (n \geq 0),$$

$$t^\star - s_n \leq \bar{t}^\star - \bar{s}_n \qquad (n \geq 0), \quad \bar{t}^\star = \lim_{n \to \infty} \bar{t}_n$$

and

$$t^\star \leq \bar{t}^\star.$$

Finally, note that under the hypotheses of Theorem 20.1, $\{t_n\}$ is a also majorizing sequence for $\{x_n\}$, which can always replace $\{\bar{t}_n\}$.

We provide an alternative to Lemma 20.6. It is convenient for us to define numbers:

$$\delta_0 = \max\{\delta_1, \delta_2\}, \tag{20.64}$$

$$\eta_1 = \frac{2}{L\,\delta_0 + 2\,L_0\,(1+\delta_0)^2}, \tag{20.65}$$

$$\eta_2 = \frac{2\,(1-2\,\delta_0)}{L\,\delta_0^2 + 2\,L_0\,(1+\delta_0)^2}, \tag{20.66}$$

η_3, η_4 to be the minimal positive numbers, such that

$$\delta_3 \leq \delta_0 \tag{20.67}$$

and

$$\delta_4 \leq \delta_0, \tag{20.68}$$

respectively. These numbers certainly exist, for η sufficiently small.

Note also that δ_0 is a positive number independent of η, whereas δ_3, δ_4 approach zero as $\eta \to 0$.

Define

$$\eta_0^\star = \min\{\eta_i, \frac{1}{L+L_0}, i = 0, 1, 2, 3, 4\}. \tag{20.69}$$

Note that by the choice of η_1, and η_2, we have:

$$f_2(\delta_1) \leq f_2(\delta_0) \leq 0, \tag{20.70}$$

and

$$p_1(\delta_2) \leq p_1(\delta_0) \leq 0, \tag{20.71}$$

respectively.

Hence, we arrived at:

Lemma 20.7. *Assume:*

$$\eta < \eta_0^\star. \tag{20.72}$$

Then, the conclusions of Lemma 20.6 hold.

Proof. We only need to show:

$$f_m(\delta_1) \leq 0, \qquad m \geq 2 \tag{20.73}$$

and

$$p_m(\delta_2) \leq 0, \qquad m \geq 1. \tag{20.74}$$

In view of (20.33), (20.34), (20.47), and (20.48), it suffices to only show:

$$f_2(\delta_1) \leq 0, \tag{20.75}$$

and

$$p_1(\delta_2) \leq 0, \tag{20.76}$$

since,

$$f_m(\delta_1) = f_2(\delta_1) \leq 0, \qquad (m \geq 2)$$

and

$$p_m(\delta_2) = p_1(\delta_2) \leq 0, \qquad (m \geq 1).$$

But (20.75), and (20.76) are true by (20.70)–(20.72).

That completes the proof of Lemma 20.7 $\qquad \diamondsuit$

Clearly, the conclusions of Theorem 20.2 hold true if hypotheses of Lemma 20.6 are replaced by the ones of Lemma 20.7.

Chapter 21

τ–estimation for Nonlinear Regression Models

Presence of influential observations such as outliers or leverages can lead to erroneous error rates and false parameter estimates for linear as well as nonlinear regression models. In regression analysis, any observation with a high value of residual can be considered as an outlier. In practice, this situation may happen when we have a very small or a very large value for the response variable. On the other hand, when an observation on an explanatory variable deviates from its center value for that explanatory variable, then we call that observation a leverage. In the analysis of regression, one may encounter cases where outliers or leverage points may be present. There are very many types of robust regression methods such as M–estimation which was originally introduced by Huber [459]. The S–estimation by Rousseau and Yohai [641] and a high breakdown high efficiency method called MM estimation of Yohai [759] which can handle linear regression models. Conceiçao et al. [319] did a comparative study of some popular nonlinear robust regression by using simulation and concluded that the τ–robust and MM–estimators showed an overall best performance when compared to other robust regression estimators. The τ–robust for linear regression has been introduced by Yohai and Zammar [760] and the τ–robust for nonlinear regression method has been developed by Tabatabai and Argyros [676]. Pernia–Espinoza et al. [582] used τ–robust for nonlinear regression to create a TAO–robust learning algorithm in neural networks and it has been proven to be a very useful method in neural networks. A more flexible package called the AMORE package was introduced to provide the user an efficient and flexible simulation environment for research and other application of neural networks: a highly flexible environment that should allow the user to get direct access to the network parameters, providing more control over the learning details and allowing the user to customize the available functions in order to suit their needs.

Consider the general nonlinear regression model

$$y_i = g(x_i, \beta^0) + \varepsilon_i, \quad i = 1, \dots, n \tag{21.1}$$

where y_i is dependent on k–dimensional independent input vector $x_i = (x_{i1}, \dots, x_{ik})$, $\beta^0 = (\beta_1^0, \dots, \beta_p^0)$ is an unknown p–dimensional parameter vector, and ε_i is a random error term with mean $E(\varepsilon_i) = 0$ and variance $\sigma^2(\varepsilon_i) = \sigma^2$; ε_i and ε_j are uncorrelated so that the covari-

266   Ioannis K. Argyros, Saïd Hilout and Mohammad A. Tabatabai

ance $\sigma_{ij} = 0$ for all $i, j; i \neq j, i, j = 1, \ldots, n$. We also assume that the response function $g(x_i, \beta^0)$ is twice continuously differentiable on some open convex set C.

The above model can also be applied to solving a finite system of nonlinear equations, since by setting $\varepsilon_i = 0, i = 1, \ldots, n$, the model (21.1) becomes a system of nonlinear equations

$$\tilde{y}_i = g(x_i, \tilde{\beta}^0), \quad i = 1, \ldots, n. \tag{21.2}$$

Model (21.1) may be written in a form

$$y = g(\beta) + \varepsilon \tag{21.3}$$

where $y = (y_1, \ldots, y_n)', g(\beta) = (g(x_1, \beta), \ldots, g(x_n, \beta))'$ and $\varepsilon = (\varepsilon_1, \ldots, \varepsilon_n)'$.

The least squares estimator of β^0 is the vector $\hat{\beta}$ that minimizes

$$Q(\beta) = \|y - g(\beta)\|^2. \tag{21.4}$$

Let

$$G(\beta) = \begin{bmatrix} \frac{\partial}{\partial \beta_1} g(x_1, \beta) & \frac{\partial}{\partial \beta_2} g(x_1, \beta) \cdots \frac{\partial}{\partial \beta_p} g(x_1, \beta) \\ \frac{\partial}{\partial \beta_1} g(x_2, \beta) & \frac{\partial}{\partial \beta_2} g(x_2, \beta) \cdots \frac{\partial}{\partial \beta_p} g(x_2, \beta) \\ \vdots & \vdots & \vdots \\ \frac{\partial}{\partial \beta_1} g(x_n, \beta) & \frac{\partial}{\partial \beta_2} g(x_n, \beta) \cdots \frac{\partial}{\partial \beta_p} g(x_n, \beta) \end{bmatrix} = \begin{bmatrix} \frac{\partial}{\partial \beta} g(x_1, \beta) \\ \frac{\partial}{\partial \beta} g(x_2, \beta) \\ \vdots \\ \frac{\partial}{\partial \beta} g(x_n, \beta) \end{bmatrix} \tag{21.5}$$

be the Jacobian of $g(\beta)$. The more commonly used methods of computing nonlinear least squares estimates are Hartley's modified Gauss–Newton method [421], the Levenberg-Marquardt method (Levenberg [503]; Marquardt [516]) and Newton's method. The modified Gauss–Newton method uses a Taylor series approximation to $g(\beta)$ about an initial vector $\hat{\beta}_0$. The approximation function is

$$Q(\beta) = \left\| y - g(\hat{\beta}_0) - G(\hat{\beta}_0)(\beta - \hat{\beta}_0) \right\|^2 \tag{21.6}$$

The modified Gauss-Newton method is as follows:

1. Choose an initial estimate $\hat{\beta}_0$ and

    Evaluate
    $$A_0 = [G'(\hat{\beta}_0) \, G(\hat{\beta}_0)]^{-1} [G'(\hat{\beta}_0)] \, e(\hat{\beta}_0) \tag{21.7}$$
    where $e = e(\hat{\beta}_0) = (e_1(\hat{\beta}_0), \ldots, e_n(\hat{\beta}_0))'$ and $e_i = e_i(\hat{\beta}) = y_i - g(x_i, \hat{\beta}), i = 1, \ldots, n$.

2. Find $0 < \gamma_0 < 1$ such that $Q(\hat{\beta}_0 + \gamma_0 A_0) < Q(\hat{\beta}_0)$.

3. Set $\hat{\beta}_1 = \hat{\beta}_0 + \gamma_0 A_0$ and evaluate A_1 by the method of step 1 with $\hat{\beta}_0$ being replaced by $\hat{\beta}_1$.

4. Repeat the process until you reach the stopping rule. In general, the recursive Algorithm can be written in the form

    $$\hat{\beta}_{j+1} = \hat{\beta}_j + [G'(\hat{\beta}_j) \, G(\hat{\beta}_j)]^{-1} [G'(\hat{\beta}_j)] \, e(\hat{\beta}_j) \tag{21.8}$$

Chapter 21. τ–estimation for Nonlinear Regression Models 267

Marquardt's algorithm [516] occupies a middle ground between the modified Gauss–Newton's method and the method of steepest descent and can be written in the form

$$\widehat{\beta}_{j+1} = \widehat{\beta}_j + [G'(\widehat{\beta}_j)\,G(\widehat{\beta}_j) + \alpha_{j+1}I]^{-1}[G'(\widehat{\beta}_j)]\,e(\widehat{\beta}_j) \tag{21.9}$$

For all α_{j+1} sufficiently large, $\widehat{\beta}_{j+1}$ is an improvement over $\widehat{\beta}_j$ $(Q(\widehat{\beta}_{j+1}) < Q(\widehat{\beta}_j)$ under appropriate conditions).

Newton's Algorithm is an improvement over the modified Gauss–Newton. It is based on a second order Taylor series approximation to $Q(\beta)$ and can be written recursively in the form

$$\widehat{\beta}_{j+1} = \widehat{\beta}_j + \left[-\frac{\partial^2}{\partial\beta\partial\beta'}Q(\widehat{\beta}_j)\right]^{-1}\frac{\partial}{\partial\beta}Q(\widehat{\beta}_j). \tag{21.10}$$

As with the modified Gauss–Newton algorithm, one finds α_j, such that $Q[\widehat{\beta}_j + \alpha_j(\widehat{\beta}_{j+1} - \widehat{\beta}_j)] < Q(\widehat{\beta}_j)$. Then, the new estimate is $\widehat{\beta}_j + \alpha_j(\widehat{\beta}_{j+1} - \widehat{\beta}_j)$. As it is well known, the choice of initial values is very important in all of the above mentioned algorithms. A poor choice may result in slow convergence, convergence to a local minimum, or even divergence. Good initial values will result in faster convergence, and if multiple minima exist, will lead to a solution that is global rather than local. Another problem with the least squares is that it may become very inefficient in the presence of a few influential observations. For an exact definition of efficiency we refer the reader to Huber [459]. In this case, least squares is no longer optimal since the variances of the parameter estimators depend on the error variance σ^2, such solutions will be unreliable and perhaps unstable in different samples (see, [459]). Since the least squares is frequently used, there is always a possibility of keypunch errors, wrong recording of data, faulty experimentation with inadequate controls, or exceptional phenomena such as natural disasters or strikes, and sometimes, they go unnoticed because of heavy usage of computers for data processing.

To remedy the problem, several authors have proposed estimation methods which have up to 50% breakdown point, which is defined as the maximum fraction of outliers that can be added to a given sample without spoiling the estimate completely (see, Yohai and Zamar [760]).

Unfortunately, most processes for estimating the parameters of regression are under the assumption of linearity for the function $g(x_i, \beta^0)$ and the majority of them have low efficiency under the assumption of normality for the probability distribution of errors ε_i. For linear regression model Tan and Tabatabai [679] proposed modified estimates with high efficiency (see, Huber [459]). Rousseeuw and Yohai [641] defined a class of estimators called S–estimators for the regression model. They are defined by minimization of the dispersion function S of residuals $(e_1(\beta), \ldots, e_n(\beta))$, where $S = S(e_1(\beta), \ldots, e_n(\beta))$ and the dispersion function $S(e_1(\beta), \ldots, e_n(\beta))$ is defined as the solution of Huber equation [459] of the form

$$\sum_{i=1}^{n} h\left(\frac{e_i(\beta)}{S}\right) = nE_\phi(h) \tag{21.11}$$

where $E_\phi(h)$ is the mean of the real valued function h and ϕ is the standard normal density function.

The real valued function h satisfies the following properties:

1. $h(0) = 0$;

2. $h(-v) = h(v)$;

3. If $0 \leq v_1 \leq v_2$, then $h(v_1) \leq h(v_2)$;

4. h is continuous;

5. $0 < \sup h(v) < \infty$;

6. If $h(v_1) < \sup h(v_1)$ and $0 \leq v_1 < v_2$, then $h(v_1) < h(v_2)$.

The S–estimates have a high breakdown point but low efficiency under normality (see, [459]). To solve this problem, Yohai and Zamar [760] defined a broader class of scale estimates (τ–estimates). Their estimates achieve a high breakdown point and a high efficiency for the linear regression model, even under the assumption of normality for errors. They also showed that the τ–estimates of the parameters of the linear model are consistent and asymptotically normal. They defined the τ–estimate of scale τ_n by

$$\tau_n^2(v) = \frac{s^2(v)}{n} \sum_{i=1}^{n} h_2 \left(\frac{v_i}{s(v)} \right) \tag{21.12}$$

where $v = (v_1, \ldots, v_n)$ is a sample, h_1 and h_2 are real valued functions satisfying properties 1-6 and $S(v)$ is the solution to the equation (21.11). A detailed discussion regarding the existence of the solution of (21.11) can be found in [760].

Consider model (21.1). Let $e(\beta) = (e_1(\beta), \ldots, e_n(\beta))$, where,

$$e_i(\beta) = y_i - g(x_i, \beta), \quad i = 1, \ldots, n.$$

The τ–estimate for the parameter vector β^0 of nonlinear regression model (21.2) is defined by the value $\widehat{\beta}$ which minimizes the scale estimate $\tau_n(e(\beta))$, that is

$$\tau_n(e(\widehat{\beta})) = \min_{\beta \in C} \tau_n(e(\beta))$$

where

$$\tau_n(e(\beta)) = S(e(\beta)) \left[\frac{1}{n} \sum_{i=1}^{n} h_2 \left(\frac{e_i(\beta)}{S(e(\beta))} \right) \right]^{\frac{1}{2}} \tag{21.13}$$

and $S(e(\beta))$ is the root of the equation (21.11), where $h = h_1$.

Differentiating $\tau_n(e(\beta))^2$ with respect to β and setting the derivative equal to zero gives

$$\frac{\partial}{\partial \beta} \tau_n^2(e(\beta)) = \sum_{i=1}^{n} \frac{1}{n} \left\{ 2S(e(\beta)) \cdot \frac{\partial}{\partial \beta} S(e(\beta)) \cdot h_2 \left(\frac{e_i(\beta)}{S(e(\beta))} \right) + \right.$$
$$\left. \psi_2 \left(\frac{e_i(\beta)}{S(e(\beta))} \right) \left[-S(e(\beta)) \frac{\partial}{\partial \beta} g(x_i, \beta) - e_i(\beta) \cdot \frac{\partial}{\partial \beta} S(e(\beta)) \right] \right\} = 0 \tag{21.14}$$

where

$$\frac{\partial}{\partial \beta} S(e(\beta)) = - \frac{\sum_{i=1}^{n} \psi_1 \left(\frac{e_i(\beta)}{S(e(\beta))} \right) \frac{\partial}{\partial \beta} g(x_i, \beta)}{\sum_{i=1}^{n} \psi_1 \left(\frac{e_i(\beta)}{S(e(\beta))} \right) \left(\frac{e_i(\beta)}{S(e(\beta))} \right)} \tag{21.15}$$

Chapter 21. τ–estimation for Nonlinear Regression Models

by substituting (21.15) in (21.14),

$$\sum_{i=1}^{n}\left[w(\beta)\psi_1\left(\frac{e_i(\beta)}{S(e(\beta))}\right)+\psi_2\left(\frac{e_i(\beta))}{S(e(\beta))}\right)\right]\frac{\partial}{\partial\beta}g(x_i,\beta)=0 \tag{21.16}$$

where,

$$\psi_i(x)=\frac{d}{dx}h_i(x),\ i=1,2 \tag{21.17}$$

and thus

$$w(\beta)=\frac{\sum_{i=1}^{n}\left[2h_2\left(\frac{e_i(\beta)}{S(e(\beta))}\right)-\psi_2\left(\frac{e_i(\beta)}{S(e(\beta))}\right)\left(\frac{e_i(\beta)}{S(e(\beta))}\right)\right]}{\sum_{i=1}^{n}\psi_1\left(\frac{e_i(\beta)}{S(e(\beta))}\right)\left(\frac{e_i(\beta)}{S(e(\beta))}\right)} \tag{21.18}$$

Asymptotically, $\sqrt{n}(\widehat{\beta}-\beta^0)$ converges in distribution to a p–dimensional normal random vector with zero mean and covariance matrix $COV(\widehat{\beta})$ defined as

$$COV(\widehat{\beta})=\frac{E_D[\psi^2(v)]}{[E_D(\psi'(v))]^2}[G'\cdot G]^{-1} \tag{21.19}$$

where

$$\psi(v)=\frac{2E_D\left[h_2\left(\frac{v}{\sigma}\right)\right]-E_D\left[\psi_2\left(\frac{v}{\sigma}\right)\frac{v}{\sigma}\right]}{E_D\left[\psi_1\left(\frac{v}{\sigma}\right)\frac{v}{\sigma}\right]}\cdot\psi_1\left(\frac{v}{\sigma}\right)+\psi_2\left(\frac{v}{\sigma}\right)$$

and $E_D(\boldsymbol{\cdot})$ denotes the expectation with respect to the probability function D.

Let

$$\widehat{\psi}(\widehat{v})=w(\widehat{\beta})\psi_1\left(\frac{\widehat{v}}{S(e(\widehat{\beta}))}\right)+\psi_2\left(\frac{\widehat{v}}{S(e(\widehat{\beta}))}\right)$$

then an estimate $\widehat{cov}(\widehat{\beta})$ of $cov(\widehat{\beta})$ is defined as

$$\widehat{cov}(\widehat{\beta})=\frac{\sum_{i=1}^{n}\frac{1}{n}\widehat{\psi}[(e_i(\widehat{\beta}))^2}{\sum_{i=1}^{n}\left\{\frac{1}{n}[\widehat{\psi}'(e_i(\widehat{\beta}))]\right\}^2}\cdot[G'(\widehat{\beta})\cdot G(\widehat{\beta})]^{-1}, \tag{21.20}$$

where $\widehat{var}(\widehat{\beta}_i)$, $i=1,\ldots,p$, is the i–th diagonal element of the matrix $\widehat{COV}(\widehat{\beta})$.

Let f be a function from R^p into R^m, which has continuous partial derivative with respect to β. Then, one might be interested in testing the hypothesis $H_0:f(\beta^0)=0$ against the alternative $H_1:f(\beta^0)\neq0$. Define $D^*(\widehat{\beta})$ by

$$D^*(\widehat{\beta})$$

$$=\frac{f'(\widehat{\beta})\left\{\left[\frac{\partial}{\partial\beta}f(\beta)\Big|\beta=\beta^0\right][G'(\beta^0)G(\beta^0)]^{-1}\left[\frac{\partial}{\partial\beta}f(\beta)\Big|\beta=\beta^0\right]\right\}^{-1}f(\widehat{\beta})}{\sigma^2}. \tag{21.21}$$

Then, $D^*(\widehat{\beta})$ is approximately a noncentral chi-square variant with m degrees of freedom and noncentrally parameter

$$\delta^*=\frac{f'(\beta^0)\left\{\left[\frac{\partial}{\partial\beta}f(\beta)\Big|\beta=\beta^0\right][G'(\beta^0)G(\beta^0)]^{-1}\left[\frac{\partial}{\partial\beta}f(\beta)\Big|\beta=\beta^0\right]\right\}^{-1}f(\beta^0)}{2\sigma^2}.$$

Now, consider the statistic $F^*(\widehat{\beta})$ defined by

$$F^*(\widehat{\beta}) = \frac{f'(\widehat{\beta}) \left\{ \left[\frac{\partial}{\partial \beta} f(\beta) \Big| \beta = \widehat{\beta} \right] [G'(\widehat{\beta})G(\widehat{\beta})]^{-1} \left[\frac{\partial}{\partial \beta} g(\beta) \Big| \beta = \widehat{\beta} \right] \right\}^{-1} f(\widehat{\beta})}{ms^2(e(\widehat{\beta})}$$ (21.22)

where $F^*(\widehat{\beta})$ is approximately distributed as F–distribution with m numerator degrees of freedom and $n - p$ denominator degrees of freedom. The decision rule is:

If $F^*(\widehat{\beta}) > F(1 - \alpha; m, n - p)$, reject $H^0 : f(\beta^0) = 0$, where $F(1 - \alpha; m, n - p)$ is the upper $(100)\alpha\%$ critical point of F-distribution with m and $n - p$ degrees of freedom and α is the level of significance.

Yohai and Zamer [760] introduced an algorithm to compute τ–estimates for the parameters of the linear regression model. For nonlinear regression models, we proceed as follows.

Let

$$\chi(\beta) = \frac{\partial}{\partial \beta} \tau_n^2(e(\beta))$$

and

$$W_i(\beta) = \frac{\omega(\beta)\psi_1\left[e_i(\beta))/S(e(\beta))\right] + \psi_2\left[e_i(\beta))/S(e(\beta))\right]}{e_i(\beta)/S(e(\beta))}$$

Then,

$$\chi(\beta) = \sum_{i=1}^{n} \frac{-2}{n} \left[W_i(\beta) \cdot e_i(\beta) \cdot \frac{\partial'}{\partial \beta} g(x_i, \beta) \right]$$

Define the matrix

$$M(\beta) = \sum_{i=1}^{n} \frac{2}{n} \left[W_i(\beta) \frac{\partial'}{\partial \beta} g(x_i, \beta) \cdot \frac{\partial}{\partial \beta} g(x_i, \beta) \right]$$

Then

$$\beta_{j+1} = \beta_j + \triangle(\beta_j)$$ (21.23)

Where

$$\triangle(\beta_j) = -M^{-1}(\beta)\chi(\beta)$$ (21.24)

Relation (21.23) does not guarantee that $\tau_n(e(\beta_{j+1})) \leq \tau_n(e(\beta_j))$; however, the following modification given by Yohai and Zamer [760] will guarantee the validity of the above inequality.

Take $0 < \alpha < 1$. Find an integer k such that

$$\tau_n \left\{ e \left[\beta_j + \frac{\triangle(\beta_j)}{2^k} \right] \right\} \leq \tau_n[e(\beta_j)] + \frac{\alpha}{2^k} \chi'(\beta_j)\triangle(\beta_j).$$

Since $M(\beta)$ is a positive definite matrix,

$$\tau_n \left\{ e \left[\beta_j + \frac{\triangle(\beta_j)}{2^k} \right] \right\} < \tau_n[e(\beta_j)].$$ (21.25)

Chapter 21. τ–estimation for Nonlinear Regression Models

Let k_{1j} be the first of such $k's$ and let k_{2j} be the value of $k(0 \leq k \leq k_{1j})$ that minimizes

$$\tau_n \left\{ e \left[\beta_j + \frac{\triangle(\beta_j)}{2^k} \right] / 2^k \right\}.$$

Then the modified recursion is

$$\beta_{j+1} = \beta_j + \frac{\triangle(\beta_j)}{2^{k_{2j}}} \tag{21.26}$$

The modified recursion (21.26) satisfies

1. $\tau_n[e(\beta_{j+1})] \leq \tau_n[e(\beta_j)]$;

2. The sequence β_j is bounded;

3. Any accumulation point of β_j satisfies (21.11) where $h = h_1$;

4. If β^* and β^{**} are two accumulation points of β_j, then

$$\tau_n[e(\beta^*)] = \tau_n[e(\beta^{**})].$$

To find $S(e(\beta_{j+1}))$ at each iteration, start with initial estimates β_0 and S_0 and then use the recursion

$$S(e(\beta_{j+1})) = \left\{ \sum_{i=1}^{n} \left[S^2(e(\beta_j)) \cdot h_1 \left(\frac{e_i(\beta_j)}{S(e(\beta_j))} \right) / (nb) \right] \right\}^{\frac{1}{2}}$$

where $b = E_\phi(h_1)$.

The functions ψ_1 and ψ_2 can be selected to satisfy the conditions imposed by (21.17). For instance, the bisquare family of ψ functions

$$\psi(v) = \begin{cases} v \left(1 - \frac{v^2}{c^2} \right)^2 & \text{if } |v| < c \\ 0 & \text{otherwise.} \end{cases}$$

The corresponding family of h functions is

$$h(v) = \begin{cases} \frac{v^2}{2} \left(1 - \frac{v^2}{c^2} + \frac{v^4}{3c^4} \right) & \text{if } |v| \leq c \\ \frac{c^2}{6} & \text{otherwise.} \end{cases}$$

If one takes $c = c_1$, then

$$\begin{aligned} h_1(v) &= h(v) \text{ and} \\ \psi_1(v) &= \psi(v). \end{aligned}$$

Similarly, if $c = c_2$, then

$$\begin{aligned} \psi_2(v) &= \psi(v). \\ h_2(v) &= h(v) \text{ and} \end{aligned}$$

If one takes $c_1 = 1.56$ which is the solution of $\frac{E_\phi(h_1)}{h_1(c_1)} = 50\%$, then $b = .203$. Also if one chooses $c_2 = 6.08$, then the resulting τ–estimates have a breakdown point of 50% and an efficiency of 95% under the assumption of normality for ε_i's. For more detailed information regarding selecting ψ functions see Rousseeuw and Leroy [641], Huber [459].

One may consider replacing the above function $h(v)$ by a redescending differentiable cost function of the form (see also Fig. 21.1)

$$h_c(v) = 1 - c\,v\,Csch(c\,v)$$

where,

$$\psi_c(v) = \frac{d(h_c(v))}{dv} = c\,(-Csch(c\,v) + c\,v\,Coth(c\,v)\,Csch(c\,v)), \quad \text{(see also Fig. 21.2)}$$

and $Csch(x), Coth(x)$ represent the hyperbolic cosecant and the hyperbolic cotangent functions of x respectively. If $c = 0.75$, then the efficiency under the normal distribution would be 95%. If someone wants 85% efficiency, then the c–value is $c = 1.13$.

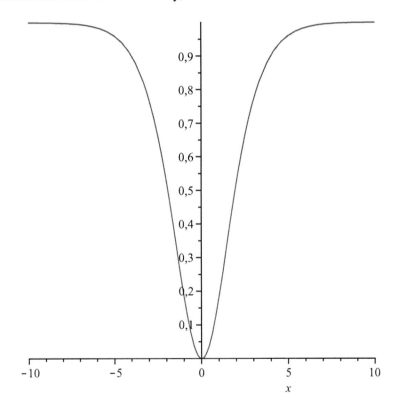

Figure 21.1. Graph of function h_c for $c = 1.13$.

We will look at a more general computing algorithm. It can easily be seen that in a Banach space setting equation (21.2) can be approached as follows: Consider mapping F defined on a subset of a Banach space B (Note that F can be chosen in particular to be $F(\beta) = \beta - g(\beta)$, and $B = R^p$). In particular, set $U(0,R) = \{z \in B \mid \|z\| \leq R\}$, for some

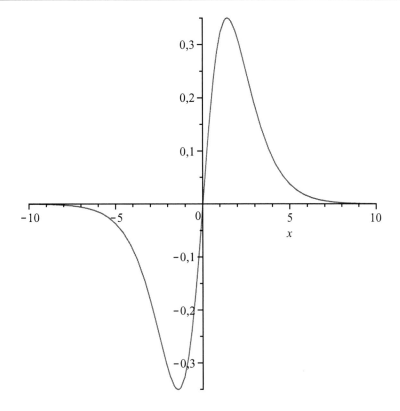

Figure 21.2. Graph of function ψ_c for $c = 1.13$.

$R > 0$, and consider the Newton–like iteration

$$z_{n+1} = z_n - L(z_n)^{-1} F(z_n) \qquad (21.27)$$

for approximating a fixed point $z^* = 0$ of equation

$$F(z) = 0. \qquad (21.28)$$

Here F is a mapping defined on $U(0,R)$ with values in B, and $L(\cdot)$ denotes a linear mapping which approximates the Fréchet–derivative $F^1(z)$ of F at $z \in U(0,R)$.

Assumption $z^* = 0$, involves no loss of generality, since any solution z^* can be transformed into 0 by introducing transformed mapping $G(z) = F(z+z^*) - z^*$.

We assume that for each fixed $r \in [0, R]$

(C) $L(0)^{-1}$ exists and for all $x, y \in U(0,r) \leq U(0,R)$

$$\left\| L(0)^{-1}(L(z) - L(0)) \right\| \leq W_0(\|z\|) + \gamma, \qquad (21.29)$$

$$\left\| L(0)^{-1}(F^1(z+t(y-z)) - A(z)) \right\| \leq W(\|z\| + t\|y-z\|) + \delta, \qquad (21.30)$$

for all $t \in [0,1]$, where W_0, W are nondecreasing nonnegative functions and the constants γ, and δ satisfy $\gamma \geq 0$, $\delta \geq 0$ and $\gamma + \delta < 1$.

Relevant results with stronger (Lipschitz–like) assumptions have been considered by Argyros [39], Dennis [335], Yamamoto and Chen [303], Zabrejko and Nguen [764].

274   Ioannis K. Argyros, Saïd Hilout and Mohammad A. Tabatabai

Define now the functions

$$p(r) = \int_0^r W(s)ds + (\gamma + \delta - W_0(r))r, \tag{21.31}$$

and

$$q(r) = 1 - \gamma - W_0(r) \text{ for all } r \in [0, R]. \tag{21.32}$$

Introduce the difference equation

$$s_{n+1} = s_n + p(s_n)q(s_n)^{-1}(n \geq 0), \; s_0 = R. \tag{21.33}$$

We can now formulate the main result:

Theorem 21.1. *Under condition* (C), *assume there exist* $z \in B$, $R > 0$ *such that* 0 *is unique zero of function* $p(r)$ *given by* (21.31) *in* $[0, R]$. *Moreover suppose* $\|z_0\| \leq a \leq R$ *and* $p(R) < 0$. *Then iterates generated by* (21.27) *are well defined for all* $n \geq 0$ *which belong to* $U(0, R)$, *and converge to* 0. *Moreover*

$$\|z_n\| \leq s_n (n \geq 0). \tag{21.34}$$

The sequence s_n *given by* (21.33) *is monotonically decreasing and converges to* 0.

Proof. We will first show that the sequence generated by (21.33) is monotonically decreasing and converges to 0. Since 0 is the unique zero of a function $p(r)$ in $[0, R]$ and $p(R) \leq 0$,

$$p(r) < 0 \text{ for all } r \in [0, R]. \tag{21.35}$$

Using (21.31),

$$0 \leq \int_0^1 W(s)ds < (1 - \gamma - \delta - W_0(r))r$$

which implies

$$g(r) > 0 \text{ for all } r \in [0, R]. \tag{21.36}$$

Using relations (21.33), (21.35), (21.36) and finite induction, it can be seen that the sequence s_n monotonically decreasing. Furthermore, (21.33) can also be written as

$$S_{k+1} = \left[\int_0^{S_k} W(r)dr + cs_k \right] p(s_k)^{-1} \geq 0 \text{ for all } k \geq 0$$

which implies $0 \leq s_{n+1} \leq s_n \; (n \geq 0)$.

Hence, there exists an $s^* \in [0, R]$ with $s_n \to s^*$ as $n \to \infty$. Note that from (21.33) and uniqueness of 0 as a zero of $p(r)$ in $[0, R]$ it follows that $s^* = 0$. By induction on n it can be shown that (21.34) holds: For $n = 0$, (21.34) becomes $\|z_0\| \leq s_0 = R$, which is true since $a \leq R$ by hypothesis. Assume (21.34) holds for n. From (21.29) and (21.36),

$$\left\| L(0)^{-1}(L(z_n) - L(0)) \right\| \leq W_0(s_n) + \gamma < 1.$$

By the Banach lemma on invertible mappings, $L(z_n)$ is invertible. By using the identity

$$L(z_n) = L(0) \left[I + L(0)^{-1}(L(z_n) - L(0)) \right]$$

Chapter 21. τ–estimation for Nonlinear Regression Models 275

we see
$$\left\| L(z_n)^{-1}L(0)) \right\| \le q(s_n)^{-1} \ (n \ge 0).$$

Using relations (21.29), (21.34), we get

$$
\begin{aligned}
\|z_{n+1}\| &= \left\| \left[L(z_n)^{-1}L(0) \right] \left\{ \int_0^1 L(0)^{-1} \left[F^1(tz_n) - L(z_n)z_n dt \right] \right\} \right\| \\
&\le \int_0^1 W(t\|z_n\| + \delta)\|z_n\| \, dt \, (q(\|z_n\|))^{-1} \tag{21.37} \\
&\le \left(\int_0^{s_n} W(t)dt + \delta s_n \right) q(s_n)^{-1} = s_{n+1}.
\end{aligned}
$$

Hence (21.34) holds for $n+1$. From relation (21.37) we conclude $z_n \in U(0,R)$. Finally, by letting $n \to \infty$ in (21.34), $z_n \to 0$ as $n \to \infty$, which completes the proof. $\diamondsuit$

In practical cases we can select $L(z_n)$ to be either $F^1(z_n)$ or $F^1(z_0)$ or $F^1(0)$ or $S(z_{n-1}, z_n)$ (secant mappings) or any other linear mappings satisfying relations (21.29) and (21.30).

It can easily be seen that the results obtained by Algorithm for commuting τ–estimates are a generalization of those obtained before. Set $B = R^P$, $L(\cdot) = M^{-1}, F = \chi$ in (21.27) to obtain (21.23). A priori estimates of the form (21.34) are also not available for (21.23) or (21.26). Moreover under the assumptions of the Theorem above, the evaluation of $k \in \mathbf{N}$ by (21.25) can be avoided using (21.27). Furthermore, these results are considered to be important when the nonlinear model (21.2) is defined in some Banach space B other than R^P.

Chapter 22

Tabaistic Regression

Tabaistic family of distributions is an important family of distributions. It was introduced in [238], the motivation behind the introduction of such family of probability distributions is to have an asymmetric family which can play a dual role as a unimodal or bimodal distribution with a host of potential real world applications. Depending on the parameter values, its probability density function can be unimodal or bimodal. The bimodal case of this family can be applied to a wide range of science and engineering problems including biosciences and bioengineering. For instance, Seeman et al. [649] realized that the distribution of dopamine receptors in brains of Schizophrenics is a bimodal. Zhang et al. [768] approximated the probability density function of tropical water vapour by a bimodal curve. Schmig et al. [647] realized that the probability density function of Minimal Luminal diameter six months after stent implantation can be approximated by a bimodal distribution. The unimodal case of this family has also numerous applications in science and engineering via the analysis of the categorical response and more specifically the analysis of binary or polytomous response when the assumption of symmetry for the link function does not hold. In the analysis of binary or polytomous response data, when symmetry assumption holds, the logit and probit links are appropriate functions to use. More information can be found in Hosmer and Lemeshow [453] and Finney [387]. The probit link function has the form

$$F_X(x) = \frac{1}{2} \left[Erf \left(\frac{x}{\sqrt{2}} \right) + 1 \right]$$

and the logistic link function is given by

$$F_X(x) = \frac{1}{1 + Exp[-x]}$$

Definition 22.1. (Tabaistic Distribution) We say a continuous random variable X follows the family of one parameter tabaistic distribution with parameters γ if the cumulative distribution function $F_X : (-\infty, \infty) \rightarrow [0, 1]$ is defined as

$$F_X(x) = \frac{1}{1 + arcsinh[Exp[-x + \gamma arcsinh(x)]]} \tag{22.1}$$

where $arcsinh(.)$ is the inverse hyperbolic sine function and the parameter γ can take values

-1, 0, or 1. The probability density function for the family (22.1) is given by

$$f_X(x) \tag{22.2}$$

$$= -\frac{Exp[-x + \gamma arcsinh(x)](-1 + \frac{\gamma}{\sqrt{1+x^2}})}{(1 + arcsinh[Exp[-x + \gamma arcsinh(x)]])^2(\sqrt{1 + Exp[-2x + 2\gamma arcsinh(x)]})}$$

If $\gamma = 1$, then the tabaistic probability density function would have a bimodal shape but for $\gamma = 0$ or 1, the tabaistic probability density function would be a unimodal function. Thus, the tabaistic distribution can cope with skewed as well as bimodal data sets.

If $\gamma = 0$, then the cumulative distribution function is given by

$$F_X(x) = \frac{1}{1 + arcsinh[Exp[-x]]} \tag{22.3}$$

In certain modeling problems one may want to consider the generalization of the cumulative distribution function (22.1) as

$$F_X(x : \alpha, \lambda) = \frac{1}{1 + arcsinh[Exp[-\frac{x-\alpha}{\lambda} + \gamma arcsinh[\frac{x-\alpha}{\lambda}]]]} \tag{22.4}$$

where the location parameter $-\infty < \alpha < \infty$ and the scale parameter $\lambda > 0$.

The probability density function for (22.4) is given by

$$f_X(x) = -\frac{(-1 + \frac{\gamma}{\sqrt{1 + \frac{(x-\alpha)^2}{\lambda^2}}})Exp[-\frac{x-\alpha}{\lambda} + \gamma arcsinh[\frac{x-\alpha}{\lambda}]]}{\lambda(1 + arcsinh[Exp[-\frac{x-\alpha}{\lambda} + \gamma arcsinh[\frac{x-\alpha}{\lambda}]]])^2(\sqrt{1 + Exp[-2\frac{x-\alpha}{\lambda} + 2\gamma arcsinh[\frac{x-\alpha}{\lambda}]]})}$$

We use in the following the tabaistic link function of the form (22.3) to develop a method for analyzing the effect of covariates on the probability of the binary response data.

In the analysis of categorical data, we frequently encounter a situation in which we desire to relate one or more variables to a binary or polytomous response. A common technique is to assume that a linear combination of the covariates determines the probability of response through a link function such as logit, probit, or complementary log–log function (McCullagh and Nelder [519]).

For each integer i, where $1 \leq i \leq n$, we define the tabaistic conditional mean of response $y_i = 1$ given p explanatory variables $x_{1i}, x_{2i}, \ldots, x_{pi}$ as a function π of the parameter vector β by

$$\pi(X_i) = P(y_i = 1 | X_i) = \frac{1}{1 + arcsinh[Exp[-\eta(X_i)]]} \tag{22.5}$$

where $1 - \pi(X_i)$ is equal to $P(y_i = 0 | X_i)$ which is the conditional probability of response $y_i = 0$ given the vector of explanatory variables X_i, and the $(p+1)$ dimensional vector $X_i = (x_{0i}, x_{1i}, \ldots, x_{pi})'$, $x_{0i} = 1$ for $1 \leq i \leq n$ and the parameter vector $\beta = (\beta_0, \beta_1, \ldots, \beta_p)'$. In general, $\eta(X_i)$ can be any linear function of the components of the parameter vector β but here we consider the case where $\eta(X_i) = \sum_{j=0}^{p} \beta_j x_{ji}$.

If $\eta(X_i) = \beta_0 + \beta_1 x_{1i}$, then the logistic regression model has its steepest slope at $x = -\dfrac{\beta_0}{\beta_1}$, where $\pi(\dfrac{\beta_0}{\beta_1}) = \dfrac{1}{2}$, but for the tabaistic regression the steepest slope is at x value such that

$$sinh\left[\frac{2(1-\pi(x))}{\pi(x)}\right] = \frac{1}{\pi(x)},$$

and the inflection point of the graph of $\pi(x)$ is at the point where

$$x = \frac{.391}{\beta_1} - \frac{\beta_0}{\beta_1} \quad \text{and} \quad \pi\left(\frac{.391}{\beta_1} - \frac{\beta_0}{\beta_1}\right) \approx 0.612282.$$

The tabit of the tabaistic regression is defined as

$$\eta(X_i) = \ln\left[csch\left(\frac{1-\pi(X_i)}{\pi(X_i)}\right)\right].$$

Now, if we consider a sample of n independent Bernoulli random variables $y_1, y_2, \ldots, y_n$, where

$$y_i = \pi(X_i) + \varepsilon_i \qquad 1 \le i \le n,$$

and $\varepsilon_1, \varepsilon_2, \ldots, \varepsilon_n$ are independent random variables with mean and variance $E(\varepsilon_i) = 0$, and $Var(\varepsilon_i) = \pi(X_i)(1 - \pi(X_i))$. Then, the random variable y_i would have a mean and variance as $E(y_i) = \pi(X_i)$, and $Var(y_i) = \pi(X_i)(1 - \pi(X_i))$.

There are various estimation methods for the estimation of the parameter vector β. The most popular method is the logistic regression [453] which is used to analyze the effects of explanatory variables on a binary response y. In the logistic regression, the function $\pi(X_i)$ is written as

$$\pi(X_i) = P(y_i = 1 | X_i) = \frac{1}{1 + Exp[-\eta(X_i)]}$$

Let $\phi(X_i) = \dfrac{\pi(X_i)}{1 - \pi(X_i)}$, then $\phi(X_i)$ can be considered as the odds of response $y_i = 1$ given vector X_i. For the tabaistic regression the $\phi(X_i)$ can be expressed as

$$\phi(X_i) = \frac{1}{arcsinh[Exp[-\eta(X_i)]]},$$

and for the ith observation, the rate of change of the tabaistic odds ratio $\phi(X_i)$ with respect to change in independent variable X_{ji} is given by

$$\frac{\partial}{\partial x_{ji}}\phi(X_i) = -\beta_j tanh\left(\frac{1}{\phi(X_i)}\right)\phi(X_i)^2$$

and the rate of change of tabaistic $\phi(X_i)$ with respect to change in the value of the parameter β_j is

$$\frac{\partial}{\partial \beta_j}\phi(X_i) = -x_{ji} tanh\left(\frac{1}{\phi(X_i)}\right)\phi(X_i)^2.$$

Let $L(\beta_0, \beta_1, \ldots, \beta_p) = \prod_{i=1}^{n} [(\pi(X_i))^{y_i} (1 - \pi(X_i))^{1-y_i}]$ be the likelihood function of n independent observations y_i. Since the maxima are not affected by monotonic mappings, we take the logarithm of the likelihood function which is equal to

$$LL(\beta_0, \beta_1, \ldots, \beta_p) = \sum_{i=1}^{n} (y_i \ln[\pi(X_i)] + (1 - y_i) \ln[1 - \pi(X_i)]). \qquad (22.6)$$

The principle of maximum likelihood is ordinarily used to estimate model parameters by maximizing the log–likelihood function $LL(\beta_0, \beta_1, \ldots, \beta_p)$. The maximum likelihood estimate of (p+1) dimensional parameter vector $\beta = (\beta_0, \beta_1, \ldots, \beta_p)'$ is

$$\widehat{\beta} = \arg \max_{\beta} (LL(\beta_0, \beta_1, \ldots, \beta_p)).$$

To find the estimate $\widehat{\beta}$ of the parameter vector β one may need to calculate

$$\frac{\partial}{\partial \beta_j} \ln[LL(\beta_0, \beta_1, \ldots, \beta_p)] = \sum_{i=1}^{n} \left[\phi(X_i)(y_i - \pi(X_i)) x_{ji} tanh \left(\frac{1}{\phi(X_i)} \right) \right].$$

The maximum likelihood estimate $\widehat{\beta}$ is the solution to the following nonlinear system of (p+1) equations of the form

$$\sum_{i=1}^{n} \left[\phi(X_i)(y_i - \pi(X_i)) x_{ji} tanh \left(\frac{1}{\phi(X_i)} \right) \right] = 0 \qquad j = 0, 1, \ldots, p. \qquad (22.7)$$

For more information regarding efficient iterative procedures for solving the nonlinear system (22.7) (see, Argyros [155]). The maximum likelihood estimate $\widehat{\beta}$ of the parameter vector β is asymptotically unbiased and efficient. The distribution of $\widehat{\beta}$ is asymptotically normal with mathematical expected value $E\left(\widehat{\beta}\right) = \beta$ and the variance–covariance matrix

$$Var\left(\widehat{\beta}\right) = \left(-E\left[\frac{\partial^2}{\partial \beta_l \partial \beta_j} \ln[LL(\beta_0, \beta_1, \ldots, \beta_p)]\right]\right) \qquad 0 \le l, j \le p.$$

An estimate of the $Var\left(\widehat{\beta}\right)$ is

$$V\widehat{a}r\left(\widehat{\beta}\right)$$

$$= \left(-\left(\begin{array}{ccc} \frac{\partial^2}{\partial \beta_0 \partial \beta_0} \ln[LL(\beta_0, \beta_1, \ldots, \beta_p)] & \cdots & \frac{\partial^2}{\partial \beta_0 \partial \beta_p} \ln[LL(\beta_0, \beta_1, \ldots, \beta_p)] \\ \vdots & \ddots & \vdots \\ \frac{\partial^2}{\partial \beta_p \partial \beta_0} \ln[LL(\beta_0, \beta_1, \ldots, \beta_p)] & \cdots & \frac{\partial^2}{\partial \beta_p \partial \beta_p} \ln[LL(\beta_0, \beta_1, \ldots, \beta_p)] \end{array}\right)\right)^{-1}.$$

To estimate $\pi(X_i)$ one may use $\widehat{\pi}(X_i)$ which is

$$\widehat{\pi}(X_i) = \frac{1}{1 + arcsinh[Exp[-\sum_{j=0}^{p} \widehat{\beta}_j x_{ji}]]}, \qquad (22.8)$$

Chapter 22. Tabaistic Regression

and an estimate of the variance of $\widehat{\pi}(X_i)$ is

$$Va\widehat{r}(\widehat{\pi}(X_i)) = \widehat{\pi}(X_i)(1 - \widehat{\pi}(X_i))$$

A $100(1-\alpha)\%$ confidence interval estimate for $\pi(X_i)$ is given by

$$\widehat{\pi}(X_i) \pm Z_{\frac{\alpha}{2}} \sqrt{\widehat{\pi}(X_i)(1 - \widehat{\pi}(X_i))}$$

To test a hypothesis about a function of a subset of parameters, one can use the well known statistical tests such as the Likelihood Ratio test, Wald test, or the Score test.

Now an attempt is made to extend the tabaistic regression to the polytomous tabaistic regression where the response variable is multinomial. Consider the vector

$$(y_{0i}, y_{1i}, \ldots, y_{ji}, \ldots, y_{k-1i})'.$$

We say that the response $y_i = j$ if every entry of the vector $(y_{0i}, y_{1i}, \ldots, y_{ji}, \ldots, y_{k-1i})'$ is equal to zero, except y_{ji} is equal to 1. For $i = 1, 2, \ldots, n$, $\sum_{j=0}^{k-1} y_{ji} = 1$.

Let $\eta_j(X_i) = \sum_{l=0}^{p} \beta_{jl}x_{li}$ for $j = 1, 2, \ldots, k-1$ and $\beta_j = (\beta_{j0}, \beta_{j1}, \ldots, \beta_{jp})'$, with β_0 a zero vector in a $p+1$ dimensional space, and $\beta = (\beta_0, \beta_1, \ldots, \beta_{k-1})'$. Then, the polytomous likelihood function is defined by

$$PL(\beta) = \prod_{i=1}^{n} \prod_{j=0}^{k-1} [(\pi_j(X_i))^{y_{ji}}]$$

and the logarithm of the polytomous likelihood function is

$$PLL(\beta) = \sum_{i=1}^{n} \left(y_{0i} \ln[\pi_0(X_i)] + y_{1i} \ln[\pi_1(X_i)] + \cdots + y_{k-1i} \ln\left[1 - \sum_{s=0}^{k-2} \pi_s(X_i)\right] \right),$$

where,

$$\pi_0(X_i) = P(y_i = 0 | X_i) = \frac{1}{1 + \sum_{s=1}^{k-1} arcsinh[Exp[-\eta_s(X_i)]]},$$

and

$$\pi_j(X_i) = P(y_i = j | X_i) = \frac{arcsinh[Exp[-\eta_j(X_i)]]}{1 + \sum_{s=1}^{k-1} arcsinh[Exp[-\eta_s(X_i)]]} \quad \text{for } j = 1, 2, \ldots, k-1.$$

The principle of maximum likelihood can used to estimate model parameters by maximizing the polytomous log–likelihood function $PLL(\beta)$. The maximum likelihood estimate of vector β is

$$\widehat{\beta} = \arg\max_{\beta}(PLL(\beta)).$$

For $j = 1, 2, \ldots, k-1$, the Tabit function $\eta_j(X_i)$ is equal to

$$\eta_j(X_i) = \ln \left[csch \left(\frac{\pi_j(X_i)}{\pi_0(X_i)} \right) \right].$$

The computation for the tabaistic analysis of binary and polytomous data model can be easily done using software packages such as SAS, SPSS, and Mathematica. The user only needs to maximize the log likelihood function of (22.6) with respect to unknown parameters of the model. We do recommend that if your distribution is skewed, tabaistic regression model seems an appropriate model of choice in the analysis of binary or polytomous data.

Chapter 23

Hyperbolastic Growth Models and Applications

In engineering, medical and biological sciences, the analysis of growth is usually characterized by a rate at which the population size changes. The definition of size depends on the investigator's objectives. The size could be considered as height, mass, volume, area, number of cells, etc. The choice of an appropriate growth model is an integral part of the analysis of the growth and will eventually aide the researcher in having a better understanding of the progression and regression of the population size and its associate velocity and acceleration. For instance in cancer research, the growth analysis of multicellular tumor spheroids may shed lights into the mutation and proliferation rates of cells. Lala [499] argued that the deceleration in tumor growth rate as a function of time may result from a decrease in proliferation fraction, increase in mitotic cell cycle or cell loss. In 2005, Tabatabai et al. [677] introduced three flexible growth dynamic models called hyperbolastic growth models H1, H2, and H3. The hyperbolastic growth models have been used as informative tools that provide assessments of growth dynamics of population size. These models give an accurate estimate of model parameters with low estimates of standard deviation. The hyperbolastic models have been applied as a tool to analyze a wide range of biomedical problems and always performed with a high degree of accuracy and precision [365], [282], [677], [5]. In this chapter, we give a brief description of hyperbolastic growth models H1, H2, and H3. We also show the fitting ability of the hyperbolastic growth models by using the data on the growth dynamics of embryonic stem cells [281], and compare the results with Deasy et al. [325], [324], Gompertz, logistic, Richards, and Weibull. Also an attempt is made to generalize the hyperbolastic growth models H1, H2, and H3 to the multivariable case in which one can investigate the effect of several explanatory variables on the population size. For instance, one can use the 2-variate hyperbolastic growth model H3 as a predictive tool to analyze the growth behavior of phytoplankton as a function of time and concentration of nutrients. Nowadays, growth models are becoming increasingly useful in explaining the spread of diseases, the dynamic behavior of diseases such as cancer and AIDS and the time course of healing of dermal or epidermal wounds. It also can be used to analyze the time course of bone healing. The multivariable hyperbolastic models can be used to analyze growth models when considering the effects of some explanatory variables may influence the growth dynamics of the disease. Such a model can help to formulate appropriate strategies to treat diseases. It also would enable researcher to evaluate the efficacy of different treatment modalities during the phases of disease progression or regression. It can be used in pharmacodynamic studies, plant growth, ecology, organ growth and modeling of tissue regeneration.

Our concern here is to present mathematical models which are highly accurate in representing the time course of data. We believe the hyperbolastic models should be very useful to researchers who are engaged in researching in this highly interesting and very diverse field of study. Even in the field of mathematical modeling of cellular growth, the range of the models has been broad. A broad class of models is based on PDE's, or systems of PDE's, often representing migration of cells and diffusion of biochemical. These systems may contain a variety of terms representing a variety of factors under consideration. For instance, in the modeling of dermal wound healing , other aspects of wound healing have also been modeled, including models of vascularization in the wound healing process and models of fibroblast formation and collagen deposition. Certain cases have also focused on the sequence of events of wound healing in relation to scar formation in the wounds. Note that the mathematical modeling has been fairly diverse, with several different types of models have been used, including differential equations, systems of differential equations, partial differential equations, often in the form of reaction diffusion equations whose solutions give rise to travelling waves.

The hyperbolastic models can be used to predict the growth kinetics of mean body weights of male and female chickens [3], plant growth, ecological problems, cellular growth including all types of stem cell, tumor growth and suppression models and the analysis of wound healing.

The papers of Ahmadi and Mottaghitalab [6] and Ahmadi and Golian [5] characterize the hyperbolastic models as a "new powerful tool" and describe their effectiveness in modeling broiler growth, in comparison with the classical Gompertz and Richard models. Furthermore, hyperbolastic growth models, especially H3, have been demonstrated to be the most accurate models in representing cellular growth, such as stem cell growth and growth of tumors in brain cancer.

The first of the three differential equations is called the hyperbolastic growth rate of type I (H1) which has the equation of the form

$$\frac{dP(t)}{dt} = \frac{P(t)}{M} \left(M - P(t)\right) \left(\delta + \frac{\theta}{\sqrt{1+t^2}}\right) \tag{23.1}$$

with the initial condition

$$P(t_0) = P_0.$$

$P(t)$ is the population size at time t, the constant M is the parameter representing carrying capacity, and δ and θ are constants.

Solving the equation (23.1) for the population size P gives

$$P(t) = \frac{M}{1 + \alpha \, exp(-\delta \, t - \theta \, arcsinh\,(t))} \tag{23.2}$$

where,

$$\alpha = \frac{M - P_0}{P_0} \, exp\left(\delta \, t_0 + \theta \, arcsinh\,(t_0)\right)$$

Chapter 23. Hyperbolastic Growth Models and Applications 285

and $arcsinh(t)$ is the inverse hyperbolic sine function of t. We call the function $P(t)$ of equation (23.2) the hyperbolastic growth model of type I or simply H1. The doubling time t for Model H1 is the solution to the following equation

$$\delta t + \theta\, arcsinh(t) + \ln \frac{M - 2P_0}{2P_0\, \alpha} = 0.$$

The second differential equation which was developed earlier is called the hyperbolastic growth rate of type II (H2) and has the form

$$\frac{dP(t)}{dt} = \frac{\alpha\, \delta\, \gamma\, P^2(t)\, t^{\gamma-1}}{M} \tanh\left(\frac{M - P(t)}{\alpha\, P(t)}\right) \tag{23.3}$$

with the initial condition $P(t_0) = P_0$ and $\gamma > 0$. Here, $tanh$ stands for hyperbolic tangent function, and M, δ, and γ are parameters.

Solving the equation (23.3) for population size P gives

$$P(t) = \frac{M}{1 + \alpha\, arcsinh(exp(-\delta\, t^{\gamma}))} \tag{23.4}$$

where,

$$\alpha = \frac{M - P_0}{P_0\, arcsinh(exp(-\delta\, t_0^{\gamma}))}$$

The doubling time t for H2 model is equal to

$$t = \left(\ln\left(csch\, \frac{M - 2P_0}{2P_0\, \alpha}\right)\right)^{1/\gamma}$$

Finally, we consider the third growth curve through the following nonlinear hyperbolastic differential equation of the form

$$\frac{dP(t)}{dt} = (M - P(t))\left(\delta\, \gamma\, t^{\gamma-1} + \frac{\theta}{\sqrt{1 + \theta^2\, t^2}}\right) \tag{23.5}$$

with the initial condition $P(t_0) = P_0$, M, δ, γ, and θ are parameters. We refer to the model (23.5) as the hyperbolastic ordinary differential equation of type III or H3.

The solution to the equation (23.5) is

$$P(t) = M - \alpha\, exp\left(-\delta\, t^{\gamma} - arcsinh(\theta\, t)\right) \tag{23.6}$$

where,

$$\alpha = (M - P_0)\, exp\left(\delta\, t_0^{\gamma} + arcsinh(\theta\, t_0)\right).$$

We call the function $P(t)$ of equation (23.6) the hyperbolastic growth model of type III or simply H3. If necessary, one can introduce shift or delay parameters in any or all hyperbolastic growth models. The doubling time t for the model H3 is the solution to the equation

$$\ln \frac{M - 2P_0}{\alpha} + \delta\, t^{\gamma} + arcsinh(\theta\, t) = 0.$$

Bursac et al. [280] performed a comparative study to investigate the fitting performance of hyperbolastic growth models H1, H2, and H3 in comparison with classical models such as Gompertz, Logistic, Weibull, and Richards. The data under study was the National Institute for Health BG01 embryonic stem cell growth data [454]. Their analysis resulted in the conclusion that the hyperbolastic growth model H3 was the best fit with a mean Absolute Relative Error of approximately 0.007 followed by the hyperbolastic growth model H2 with 0.043, Weibull with 0.063, hyperbolastic growth model H1 with 0.073, Gompertz with 0.121, logistic with 0.131, and finally Richards model with Absolute Relative Error of about 0.134. Notice that the Richards model has an Absolute Relative Error which is over nineteen times the Absolute Relative Error of the best model fit, hyperbolastic H3. The Sherley JL et al. [655] and B.M. Deasy et al. [325], [324]. Sherley defined a growth model which takes into account the generation of both dividing and nondividing cells. The Sherley model has the form

$$P(t) = P_0 \left(.5 + \frac{1 - (2\alpha)^{\frac{t}{Dt}+1}}{2(1-2\alpha)} \right) \tag{23.7}$$

where $P(t)$ is the population size at time t, P_0 is the initial number of cells, α is the mitotic fraction, and DT is the division time. Deasy et.al. [324] applied the Sherley model to describe the mechanisms of muscle stem cell expansion with Cytokines. Jankowski et.al. [466] used this model to investigate the role of CD34 expression and cellular fusion in the regression capacity of cells. Deasy et al. [325] expanded the Sherley growth model by incorporating terms into equation (23.7) to account for cell loss and cell differentiation. Their growth model has a form

$$P(t) = P_0 \left(.5 + \frac{1 - (2\alpha)^{\frac{t}{Dt}+1}}{2(1-2\alpha)} \right) - M \tag{23.8}$$

where M is added to take into account the cell loss and one may consider $P(t)$ as the sum of two terms, one corresponds to proliferating cells and the other is associated with differentiated cells.

In this chapter, we refer to formula (23.8) as the Deasy growth model. Figure 23.1 depicts the scatter plot of the observed and estimated number of stem cells using the hyperbolastic H3 model. Graphically, there is virtually no visible difference between the observed and estimated values. The R–Squared for the hyperbolastic H3 model is almost 1.00 while for the expanded Deasy is 0.971.

Evidence suggests that many complex biomedical growth dynamics of population size P may depend not only on time as a variable but also on a set of explanatory variables which may affect the growth in a variety of ways such as acceleration or deceleration of the growth dynamics. For instance, the commercial broiler breeders may be interested in investigating the effects of dietary interventions on the growth dynamics of both male and female broilers. Aggrey [4] discussed the use of spline linear regression in the analysis of chicken growth using different protein diets. Jennifer Perrone et al. [583] administered antidepressant drugs Fluoxetine and Olanzapine in rats in order to analyze the growth dynamics of their body weight. The body weight is not only a function of time but it is also a function of type of administered drug dosages or drug types.

Therefore there may be a need to measure the population growth while allowing for the effects of explanatory variables. We generalize the hyperbolastic growth models to

Chapter 23. Hyperbolastic Growth Models and Applications

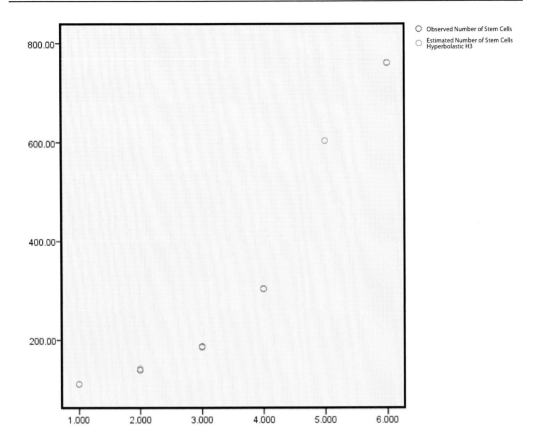

Figure 23.1. Scatter Plot of the Observed and the Estimated Number of Stem Cells using Hyperbolastic H3 Model.

accommodate the case when one has a multiple of predictors and wants to analyze the effects of these predictors in predicting the population size.

Let X be a k–dimensional vector of explanatory variables and λ be a p–dimensional vector of parameters. We define the size function P of the vector X as the generalized hyperbolastic model of type H1 if P takes the functional form

$$P(X;M,\delta,\theta,\lambda) = \frac{M}{1 + \alpha \exp\left(-\delta g(X;\lambda) - \theta \, arcsinh\left(g(X;\lambda)\right)\right)} \qquad (23.9)$$

where, M and θ are parameters, and

$$\alpha = \frac{M - P_0}{P_0} \exp\left(\delta g(X_0;\lambda) + \theta \, arcsinh\left(g(X_0;\lambda)\right)\right)$$

with $P_0 = P(X_0;M,\delta,\theta,\lambda)$. The function $g(X;\lambda)$ is a link function whose functional form depends on the nature of the problem under study. For instance, the explanatory variables may accelerate the growth or decay through a link function $g(X;\lambda)$ of the form

$$g(X;\lambda) = \exp\left(\sum_{i=1}^{k} \lambda_i X_i\right).$$

The rate of growth of $P(X;M,\delta,\theta,\lambda)$ with respect to change in explanatory variable X_j is given by

$$\frac{\partial P(X;M,\delta,\theta,\lambda)}{\partial X_j}$$
$$= \frac{P(X;M,\delta,\theta,\lambda)}{M}(M - P(X;M,\delta,\theta,\lambda))\left(\delta + \frac{\theta}{\sqrt{1+g(X;\lambda)^2}}\right)\frac{\partial g(X;\lambda)}{\partial X_j} \qquad (23.10)$$

where $j = 1, 2, \cdots, k$.

If $g(t;\lambda) = t$, and $k = 1$, then the model (23.10) reduces to the hyperbolastic growth model H1.

The second model is called the generalized hyperbolastic growth rate function H2 with a vector of explanatory variables X and a carrying capacity parameter M. the growth rate function has a form

$$\frac{\partial P(X;M,\lambda)}{\partial X_j} = \frac{\alpha}{M}P(X;M,\lambda)^2\, tanh\left(\frac{M-P(X;M,\lambda)}{\alpha\,P(X;M,\lambda)}\right)\frac{\partial g(X;\lambda)}{\partial X_j} \qquad (23.11)$$

where λ is a vector of parameters. As stated before, depending on the nature of wound, the link function may take a variety of different forms. One possible choice for the link function is

$$g(X;\lambda) = \lambda_1\, exp\left(\lambda_2 \sum_{i=3}^{k+2} \lambda_i\, X_{i-2}\right).$$

where $P_0 = P(X_0;M,\lambda)$.

Solving equation (23.11) for population size P gives

$$P(X;M,\lambda) = \frac{M}{1 + \alpha\,arcsinh\,(-g(X;\lambda))} \qquad (23.12)$$

where,

$$\alpha = \frac{M - P_0}{P_0\,arcsinh\,(exp(-g(X_0;\lambda)))}.$$

Finally, we consider the third nonlinear generalized hyperbolastic equation of type H3 which has the form

$$\frac{\partial P(X;M,\theta_0,\theta_1,\lambda)}{\partial X_j} = (M - P(X;M,\theta_0,\theta_1,\lambda))\left(\frac{\partial g_1(X;\theta_0,\lambda)}{\partial X_j} + \right.$$
$$\left. \frac{1}{\sqrt{1+g_2^2(X;\theta_1,\lambda)}}\frac{\partial g_2(X;\theta_1,\lambda)}{\partial X_j}\right) \qquad (23.13)$$

with parameters θ_0, θ_1, initial condition $P_0 = P(X_0;M,\theta_0,\theta_1,\lambda)$, and the parameter vector λ. The vector X is a vector of explanatory variables and t is the explanatory variable time. The solution to the equation (23.13) is

$$P(X;M,\theta_0,\theta_1,\lambda) = M - \alpha\,exp(-g_1(X,\theta_0,\lambda) - arcsinh\,(g_2(X;\theta_1,\lambda))) \qquad (23.14)$$

where

$$\alpha = (M - P_0)\,(g_1(X_0,\theta_0,\lambda) + arcsinh\,(g_2(X_0;\theta_1,\lambda))).$$

We call the function $P(X; M, \theta_0, \theta_1, \lambda)$ of equation (23.13) the generalized hyperbolastic model of type H3. The choice of $g_1(X; \theta_1, \lambda)$ and $g_2(X; \theta_1, \lambda)$ depends on the nature of the problem under study. One possible choice of link functions may have a form

$$g_1(X; \theta_0, \lambda) = \theta_0 \, exp \left(\lambda_1 \left(\lambda_2 t + \sum_{i=3}^{k+2} \lambda_i X_{i-2} \right) \right)$$

and

$$g_2(X; \theta_1, \lambda) = \theta_1 \, exp(t) + \sum_{i=3}^{k+2} \lambda_i X_{i-2}.$$

In Summery, the hyperbolastic growth models and their generalizations seem to be very accurate in predicting the growth dynamics of population size. They can be used to understand the growth behavior of cell populations such as cell proliferation and quiescence rates both in vivo and in vitro. It can also be used to analyze stem cell proliferations. These stem cells have a potential to differentiate into a variety of cell types in human body and can eventually be used to treat serious diseases which have no definite cure or perhaps can be used in gene therapy treatments. The hyperbolastic models can tell us how and how well an organism is growing. They can help the scientists to monitor the growth velocity and acceleration and be able to devise an efficient way of producing stem cells.

References

[1] Adly, S., Perturbed algorithms and sensitivity analysis for a general class of variational inclusions, *J. Math. Anal. Appl.*, **201**, (2), (1996), 609–630.

[2] Adly, S., A perturbed iterative method for a general class of variational inequalities, *Serdica Math. J.*, **22**, (2), (1996), 69–82.

[3] Aggrey, S.E., Comparison of Three Nonlinear and Spline Regression Models for Describing Chicken Growth Curves, *Poultry Science,* **81**, (2002), 1782–1788.

[4] Aggrey, S.E., Modelling the Effect of Nutritional Status on Pre–Asymptotic and relative Growth Rates in a Random-Bred Chicken population, *Journal of Animal, Breeding and Genetics,* **121**(4), (2004), 260–268.

[5] Ahmadi, H., Golian, A., Non–linear hyperbolast growth models for describing growth curve in classical strain of broiler chicken, *Res. J. Bio. Sci.,* **3**, (2008), 1300–1304.

[6] Ahmadi, H., Mottaghitalab, M., Hyperbolastic Models as a New Powerful: Tool to Describe Broiler Growth Kinetics, *Poult Sci*, 2007, 2461–2465.

[7] Ahues, M., A note on perturbed fixed slope iterations, *Appl. Math. Lett.,* **18**, 4, (2005), 375–380.

[8] Ahues, M., Newton Methods with Hölder derivative, *Numer. Funct. Anal. Optim.,* **(5–6)**, (2004), 379–395.

[9] Alefeld, G., Herzberger, J., Introduction to interval computations. Translated from the German by Jon Rokne, *Computer Science and Applied Mathematics,* Academic Press, New York, 1983.

[10] Alizadeh, F., Haeberly, J.–P.A., Overton, M.L., Complementarity and nondegeneracy in semidefinite programming, *Math. Progr.,* **77**, (1997), 29-43.

[11] Alizadeh, F., Haeberly, J.–P.A., Overton, M.L., Primal-dual interior point algorithms for semidefinite programming: stability, convergence and numerical results, *SIAM J. Optim.,* **8**, (1998), 743-768.

[12] Allgower, E.L., Böhmer, K., Application of the mesh independence principle to mesh refinement strategies, *SIAM J. Numer. Anal.,* **24**, (1987), 1335–1351.

References

[13] Allgower, E.L., Böhmer, K., Potra, F.A., Rheinboldt, W.C., A mesh independence principle for operator equations and their discretizations, *SIAM J. Numer. Anal.*, **23**, 1, (1986), 160–169.

[14] Alvarez, F., Bolte, J., Munier, J., *A unifying local convergence result for Newton's method in Riemannian manifolds*, Institut National de Recherche en informatique et en automatique, Theme Num-Numeriques, Project, Sydoco, Rapport de recherche No. 5381, November 2004, France.

[15] Amat, S., Blanda, J., Busquier, S., A Steffensen's type method with modified functions, *Riv. Mat. Univ. Parma*, (**7**), 7, (2007), 125–133.

[16] Amat, S., Busquier, S., Convergence and numerical analysis of a family of two–step Steffensen's method, *Comput. Math. Appl.*, **49**, (2005), 13–22.

[17] Amat, S., Busquier, S., On a Steffensen's type method and its behavior for semismooth equations, *Appl. Math. Comput.*, **177**, (2), (2006), 819–823.

[18] Amat, S., Busquier, S., Third–order iterative methods under Kantorovich conditions, *J. Math. Anal. Appl.*, **336**, (1), (2007), 243–261.

[19] Amat, S., Busquier, S., Candela, V.F., A class of quasi–Newton generalized Steffensen's methods on Banach spaces, *J. Comput. Appl. Math.*, **149**, 2, (2002), 397–406.

[20] Amat, S., Busquier, S., Gutiérrez, J.M., On the local convergence of secant–type methods, *Intern. J. Comput. Math.*, **81**, 9, (2004), 1153–1161.

[21] Amat, S., Busquier, S., Plaza, S., Dynamics of the King and Jarratt iterations, *Aequationes Math.*, **69**, (3), (2005), 212–223.

[22] Amat, S., Busquier, S., Salanova, M.A., A fast Chebyshev's method for quadratic equations, *Appl. Math. Comput.*, **148**, (2004), 461–474.

[23] Amer, S.M., On solution of nonlinear singular integral equations with shift in generalized Hölder space, *Chaos Solitions Fractals*, **12**, (7), (2001), 1323–1334.

[24] Amer, S.M., Dardery, S., About an approximation method for the solution of nonlinear singular integral equations with shift, *Bull. Fac. Sci. Assiut Univ. C*, **29**, (2), (2000), 1–15.

[25] Amer, S.M., Dardery, S., On the application of Newton–Kantorovich method to nonlinear singular integral equations, *Proc. Pakistan Acad. Sci.*, **37**, (1), (2000), 79–86.

[26] Amer, S.M., Nagdy, A.S., On the modified Newton's approximation method for the solution of non–linear singular integral equations, *Hokkaido Math. J.*, **29**, (1), (2000), 59–72.

[27] An, H.B., Bai, Z–Z., Directional secant method for nonlinear equations, *J. Comput. Appl. Math.*, **175**, (2005), 291–304.

References

[28] Anitescu, M., Coroian, D.I., Nashed, M.Z., Potra, F.A., Outer inverses and multi-body system simulation, *Numer. Funct. Anal. Optim.*, **17** (1996), 661–678.

[29] Appell, J., The Newton–Kantorovich method for nonlinear integral operators, *World Congress of Nonlinear Analysts*, Vol. I–IV (Tampa, FL, 1992), de Gruyter, Berlin, 1996, 2419–2422.

[30] Appell, J., De Pascale, E., Evkuta, N.A., Zabrejko, P.P., On the two–step Newton method for the solution of nonlinear operator equations, *Math. Nachr.*, **172**, (1995), 5–14.

[31] Appell, J., De Pascale, E., Kalitvin, J.V., Zabrejko, P.P., On the application of the Newton–Kantorovich method to nonlinear partial integral equations, *Z. Anal. Anwendungen*, **15**, (2), (1996), 397–418.

[32] Appell, J., De Pascale, E., Lysenko, J.V., Zabrejko, P.P., New results on Newton–Kantorovich approximations with applications to nonlinear integral equations, *Numer. Funct. Anal. Optim.*, **18**, (1997), 1–17.

[33] Appell, J., De Pascale, E., Zabrejko, P.P., On the application of the Newton–Kantorovich method to nonlinear integral equations of Uryson type, *Numer. Funct. Anal. Optim.*, **12**, (3–4), (1991), 271–283.

[34] Appell, J., De Pascale, E., Zabrejko, P.P., On the application of the method of successive approximations and the Newton–Kantorovich method to nonlinear functional-integral equations, *Adv. Math. Sci. Appl.*, **2**, (1), (1993), 25–38.

[35] Argyros, I.K., Quadratic equations and applications to Chandrasekhar's and related equations, *Bull. Austral. Math. Soc.*, **32**, (1985), 275–292.

[36] Argyros, I.K., On the cardinality of solutions of multilinear differential equations and applications, *Int. J. Math. Math. Sci.*, **9**, 4, (1986), 757–766

[37] Argyros, I.K., On the approximation of some nonlinear equations, *Aequationes Mathematicae*, **32**, (1987), 87–95.

[38] Argyros, I.K., On polynomial equations in Banach space, perturbation techniques and applications, *Int. J. Math. Math. Sci.*, **10**, 1, (1987), 69–78.

[39] Argyros, I.K., Newton–like methods under mild differentiability conditions with error analysis, *Bull. Austral. Math. Soc.*, **37**, (1987), 131–147.

[40] Argyros, I.K., On Newton's method and nondiscrete mathematical induction, *Bull. Austral. Math. Soc.*, **38**, (1988), 131–140.

[41] Argyros, I.K., On a class of nonlinear integral equations arising in neutron transport, *Aequationes Math.*, **36**, (1988), 99–111.

[42] Argyros, I.K., The secant method and fixed points of nonlinear operators, *Monatshefte für Math.*, **106**, (1988), 85–94.

[43] Argyros, I.K., On the number of solutions of some integral equations arising in radiative transfer, *Int. J. Math. Math. Sci.,* **12**, 2, (1989), 297–304

[44] Argyros, I.K., Improved error bounds for a certain class of Newton–like methods, *J. Approx. Th. and Its Appl.,* **61**, (1990), 80–98.

[45] Argyros, I.K., Error founds for the modified secant method, *BIT,* **20**, (1990), 92–200.

[46] Argyros, I.K., On the solution of equations with nondifferentiable operators and the Pták error estimates, *BIT,* **30**, (1990), 752–754.

[47] Argyros, I.K., On some projection methods for the approximation of implicit functions, *Appl. Math. Letters,* **32**, (1990), 5–7.

[48] Argyros, I.K., The Newton–Kantorovich method under mild differentiability conditions and the Pták error estimates, *Monatsh. Math.,* **101**, (1990), 175–193.

[49] Argyros, I.K., The secant method in generalized Banach spaces, *Appl. Math. Comp.,* **39**, (1990), 111–121.

[50] Argyros, I.K., A mesh independence principle for operator equations and their discretizations under mild differentiability conditions, *Computing,* **45**, (1990), 265–268.

[51] Argyros, I.K., On the convergence of some projection methods with perturbation, *J. Comp. Appl. Math.,* **36**, (1991), 255–258.

[52] Argyros, I.K., On an application of the Zincenko method to the approximation of implicit functions, *Public. Math. Debrecen,* **39**, (3–4), (1991), 1–7.

[53] Argyros, I.K., On an iterative algorithm for solving nonlinear equations, *Beitrage zür Numerischen Math.,* **10**, 1 (1991), 83–92.

[54] Argyros, I.K., On a class of quadratic equations with perturbation, *Funct. et Approx. Comm. Math.,* **XX**, (1992), 51–63.

[55] Argyros, I.K., Improved error bounds for the modified secant method, *Int. J. Computer Math.,* **43**, (1–2), (1992), 99–109.

[56] Argyros, I.K., Some generalized projection methods for solving operator equations, *J. Comp. Appl. Math.,* **39**, 1, (1992), 1–6.

[57] Argyros, I.K., On the convergence of generalized Newton–methods and implicit functions, *J. Comp. Appl. Math.,* **43**, (1992), 335–342.

[58] Argyros, I.K., On the convergence of inexact Newton–like methods, *Publ. Math. Debrecen,* **42**, (1&2), (1992), 1–7.

[59] Argyros, I.K., On a mesh–independance principle for operator equations and the Secant method, *Acta Math. Hung.,* **60**, (1992), 7–19.

[60] Argyros, I.K., Newton–like methods in generalized Banach spaces, *Funct. Approx. Comment. Math.,* **22** (1993), 13–20.

References

[61] Argyros, I.K., On the convergence of a Chebysheff–Halley–type method under Newton–Kantorovich hypothesis, *Appl. Math. Letters*, **5**, 5, (1993), 71–74.

[62] Argyros, I.K., Newton–like methods in partially ordered linear spaces, *J. Approx. Th. and its Appl.*, **9**, 1, (1993), 1–10.

[63] Argyros, I.K., On the solution of undetermined systems of nonlinear equations in Euclidean spaces, *Pure Math. Appl.*, **4**, 3, (1993), 199–209.

[64] Argyros, I.K., A convergence theorem for Newton–like methods under generalized Chen–Yamamato–type assumptions, *Appl. Math. Comp.*, **61**, 1, (1994), 25–37.

[65] Argyros, I.K., On the discretization of Newton–like methods, *Internat. J. Computer. Math.*, **52**, (1994), 161–170.

[66] Argyros, I.K., On the midpoint method for solving nonlinear operator equations and applications to the solution of integral equations, *Rev. Anal. Numer. Theor. Approx.*, *Tome* 23, fasc. 2, (1994), 139–152.

[67] Argyros, I.K., A multipoint Jarratt–Newton–type approximation algorithm for solving nonlinear operator equations in Banach spaces, *Functiones et approximatio, Commentarii Matematiki, XXIII*, (1994), 97–108.

[68] Argyros, I.K., On Stirling's method, *Tamkang J. Math.*, **27**, 1, (1995), 37–52.

[69] Argyros, I.K., Stirling's method and fixed points of nonlinear operator equations in Banach space, *Publ. Inst. Math. Acad. Sin.*, **23**, 1, (1995), 13–20.

[70] Argyros, I.K., A unified approach for constructing fast two–step Newton–like methods, *Monatsh. Math.*, **119**, (1995), 1–22.

[71] Argyros, I.K., Results on controlling the residuals of perturbed Newton–like methods on Banach spaces with aconvergence structure, *Southwest J. Pure Appl. Math.*, **1**, (1995), 32–38.

[72] Argyros, I.K., On the method of tangent hyperbolas, *J. Appr. Th. Appl.*, **12**, 1 (1996), 78–96.

[73] Argyros, I.K., Generalized conditions for the convergence of inexact Newton methods on Banach spaces with a convergence structure and applications, *Pure Math. Appl.*, **7** (1996), 197–214.

[74] Argyros, I.K., On an extension of the mesh–independence principle for operator equations in Banach space, *Appl. Math. Lett.*, **9**, 3, (1996), 1–7.

[75] Argyros, I.K., A generalization of Edelstein's theorem on fixed points and applications, *Southwest J. Pure Appl. Math.*, **2**, (1996), 60–64.

[76] Argyros, I.K., Chebysheff–Halley–like methods in Banach spaces, *Korean J. Comp. Appl. Math.*, **4**, 1, (1997), 83–107.

[77] Argyros, I.K., Concerning the convergence of inexact Newton methods, *J. Comp. Appl. Math.*, **79**, (1997), 235–247.

[78] Argyros, I.K., General ways of constructing accelerating Newton–like iterations on partially ordered topological spaces, *Southwest J. Pure Appl. Math.*, **2**, (1997), 1–12.

[79] Argyros, I.K., On a new Newton–Mysovskii–type theorem with applications to inexact Newton–like methods and their discretizations, *IMA J. Num. Anal.*, **18**, (1997), 37–56.

[80] Argyros, I.K., On the convergence of two–step methods generated by point–to–point operators, *Appl. Math. Comput.*, **82**, 1, (1997), 85–96.

[81] Argyros, I.K., Improved error bounds for Newton–like iterations under Chen–Yamamoto assumptions, *Appl. Math. Letters*, **10**, 4, (1997), 97–100.

[82] Argyros, I.K., Inexact Newton methods and nondifferentiable operator equations on Banach spaces with a convergence structure, *Approx. Th. Appl.*, **13**, 3, (1997), 91–104.

[83] Argyros, I.K., A mesh independence principle for inexact Newton–like methods and their discretizations under generalized Lipschitz conditions, *Appl. Math. Comp.*, **87**, (1997), 15–48.

[84] Argyros, I.K., Concerning the convergence of inexact Newton methods, *J. Comp. Appl. Math.*, **79**, (1997), 235–247.

[85] Argyros, I.K., Smoothness and perturbed Newton–like methods, *Pure Math. Appl.*, **8**,1, (1997), 13–28.

[86] Argyros, I.K., The asymptotic mesh independence principle for inexact Newton–Galerkin–like methods, *Pure Math. Appl.*, **8**, (2–3), (1997), 169–194.

[87] Argyros, I.K., Improving the rate of convergence of Newton methods on Banach spaces with a convergence structure and applications, *Appl. Math. Lett.*, **10** (1997), 21–28.

[88] Argyros, I.K., A new convergence theorem for steffenson's method on Banach spaces and applications, *Southwest J. Pure Appl. Math.*, **01**, (1997), 23–29.

[89] Argyros, I.K., *Polynomial operator equations in abstract spaces and applications,* CRC Press LLC, Boca raton Florida USA, 1998.

[90] Argyros, I.K., On the convergence of a certain class of iterative procedures under relaxed conditions with applications, *J. Comp. Appl. Math.*, **94**, (1998), 13–21.

[91] Argyros, I.K., *The theory and application of abstract polynomial equations*, St.Lucie/CRC/Lewis Publ. Mathematics series, 1998, Boca Raton, Florida, U.S.A.

[92] Argyros, I.K., Sufficient conditions for constructing methods faster than Newton's, *Appl. Math. Comp.*, **93**, (1998), 169–181.

References 297

[93] Argyros, I.K., A new convergence theorem for the Jarratt method in Banach spaces, *Computers and Mathematics with Applications,* **36**, 8, (1998), 13–18.

[94] Argyros, I.K., Improving the order and rates of convergence for the Super–Halley method in Banach spaces, *Comp. Appl. Math.,* **5**, 2, (1998), 465–474.

[95] Argyros, I.K., Improved error bounds for a Chebysheff–Halley–type method, *Acta Math. Hungarica,* **84**, 3, (1999), 211–221.

[96] Argyros, I.K., Convergence domains for some iterative processes in Banach spaces using outer and generalized inverses, *Comput. Anal. Appl.,* **1**, 1, (1999), 87–104.

[97] Argyros, I.K., Concerning the convergence of a modified Newton–like method, Journal for Analysis and its Applications (ZAA), **18**, 3, (1999), 1–8.

[98] Argyros, I.K., Concerning the radius of convergence of Newton's method and applications, *Korean J. Comp. Appl. Math.,* **6**, 3, (1999), 451–462.

[99] Argyros, I.K., Convergence rates for inexact Newton–like methods of singular points and applications, *Appl. Math. Comp.,* **102**, (1999), 185–201.

[100] Argyros, I.K., Relation between forcing sequences and inexact Newton iterates in Banach spaces, *Computing,* **63**, 2, (1999), 134–144.

[101] Argyros, I.K., An error analysis for the midpoint method, *Tamkang J. Math.,* **30**, (1999), 71–83.

[102] Argyros, I.K., A new convergence theorem for the inexact Newton method based on assumptions involving the second Fréchet–derivative, *Comput. Appl. Math.,* **37**, 7, (1999), 109–115.

[103] Argyros, I.K., Forcing sequences and inexact Newton iterates in Banach space, *Appl. Math. Lett.,* **13**, 1, (2000), 77–80.

[104] Argyros, I.K., Local convergence of inexact Newton–like iterative methods and applications, *Comput. Math. Appl.,* **39**, (2000), 69–75.

[105] Argyros, I.K., *Advances in the Efficiency of Computational Methods and Applications,* World Scientific Publ. Co., River Edge, NJ, 2000.

[106] Argyros, I.K., A mesh independence principle for perturbed Newton–like methods and their discretizations, *Korean J. Comp. Appl. Math.,* **7**, 1, (2000), 139–159.

[107] Argyros, I.K., Newton methods on Banach spaces with a convergence structure and applications, *Computers Math. with Appl., Intern. J., Pergamon Press,* **40**, 1, (2000), 37–48.

[108] Argyros, I.K., Semilocal convergence theorems for a certain class of iterative procedures using outer or generalized inverses, *Korean J. Comp. Appl. Math.,* **7**, 1, (2000), 29–40.

References

[109] Argyros, I.K., The effect of rounding errors on a certain class of iterative methods, *Applicationes Mathematicae*, **27**, 3, (2000), 369–375.

[110] Argyros, I.K., Local convergence of Newton's method for nonlinear equations using outer or generalized inverses, *Chechoslovak Math. J.*, **50**, (125), (2000), 603–614.

[111] Argyros, I.K., On a class of nonlinear implicit quasivariational inequalities, *Pan American Math. J.*, **10**, 4, (2000), 101–109.

[112] Argyros, I.K., On the radius of convergence of Newton's method, *Intern. J. Comput. Math.*, **77**, (2001), 389–400.

[113] Argyros, I.K., A Newton–Kantorovich theorem for equations involving m-Fréchet-differentiable operators and applications in radiative transfer, *J. Comp. Appl. Math.*, **131**, 1–2, (2001), 149–159.

[114] Argyros, I.K., A mesh independence principle for inexact Newton–type methods and their discretizations, *Annales Univ. Sci. Budapest, Sect. Comp.*, **20**, (2001), 31–53.

[115] Argyros, I.K., Error bounds for the midpoint method in Banach spaces, *Comm. Appl. Nonlinear Anal.*, **08**, (2001), 103–117.

[116] Argyros, I.K., A new semilocal convergence theorem for Newton's method in Banach space using hypotheses on the second Fréchet–derivative, *J. Comput. Appl. Math.*, **139**, (2001), 369–373.

[117] Argyros, I.K., On the radius of convergence of Newton's method, *Int. J. Comput. Math.*, **77**, 3, (2001), 389–400.

[118] Argyros, I.K., Semilocal convergence theorems for Newton's method using outer inverses and hypotheses on the second Fréchet–derivative, *Monatshefte fur Mathematik*, **132**, (2001), 183–195.

[119] Argyros, I.K., On general auxiliary problem principle and nonlinear mixed variational inequalities, *Nonlinear Functional Analysis and Applications*, **6**, 2, (2001), 247–256.

[120] Argyros, I.K., On an iterative procedure for approximating solutions of quasi variational inequalities, *Advances in Nonlinear Variational Inequalities*, **4**, 2, (2001), 39–42.

[121] Argyros, I.K., On generalized variational inequalities, *Advances in Nonlinear Variational Inequalities*, **4**, 2, (2001), 75–78.

[122] Argyros, I.K., On a semilocal convergence theorem for a class of quasi variational inequalities, *Advances in Nonlinear Variational Inequalities*, **4**, 2, (2001), 43–46.

[123] Argyros, I.K., On the convergence of a Newton–like method based on m-Fréchet-differentiable operators and applications in radiative transfer, *J. Comput. Anal. Appl.*, **4**, 2, (2002), 141–154.

References   299

[124] Argyros, I.K., A unifying semilocal convergence theorem for Newton–like methods based on center Lipschitz conditions, *Comput. Appl. Math.*, **21**, 3, (2002), 789–796.

[125] Argyros, I.K., A semilocal convergence analysis for the method of tangent hyperbolas, *Journal of Concrete and Applicable Analysis*, **1**, 2, (2002), 135–144.

[126] Argyros, I.K., The asymptotic mesh independence principle for Newton–Galerkin methods using twice Fréchet differentiable operators without Lipschitz conditions, *Commu. Appl. Nonlinear. Anal.*, **9**, 4, (2002), 67–75.

[127] Argyros, I.K., On the solution of generalized equations using m ($m \geq 2$) Fréchet differential operators, *Comm. Appl. Nonlinear Anal.*, **09**, (2002), 85–89.

[128] Argyros, I.K., Results on the solution of generalized equations, *Comm. Appl. Nonlinear Anal.*, **09**, (2002), 103–107.

[129] Argyros, I.K., On the convergence and application of Newton's method under weak Hölder continuity assumptions, *Intern. J. Comput. Math.*, **5**, (2003), 767–780.

[130] Argyros, I.K., An improved error analysis for Newton–like methods under generalized conditions, *J. Comput. Appl. Math.*, **157**, 1, (2003), 169–185.

[131] Argyros, I.K., An improved convergence analysis and applications for Newton-like methods in Banach space, *Numer. Funct. Anal. Optim.* **24**, 7 and 8 (2003), 653–672.

[132] Argyros, I.K., On a Multistep Newton Method in Banach Spaces and the Pták Error Estimates, *Advances in Nonlinear Variational Inequalities*, **6**, 2, (2003), 121–135.

[133] Argyros, I.K., New and generalized convergence conditions for the Newton–Kantorovich method, *J. Appl. Anal.*, **9**, 2, (2003), 287–299.

[134] Argyros, I.K., On a theorem of L.V. Kantorovich concerning Newton's method, *J. Comp. Appl. Math.*, **155**, (2003), 223–230.

[135] Argyros, I.K., A convergence analysis and applications for the Newton-Kantorovich method in K–normed spaces. *Rend. Circ. Mat. Palermo* **(2)**, 53 (2004), 251–271.

[136] Argyros, I.K., On the Newton–Kantorovich hypothesis for solving equations, *J. Comput. Appl. Math.*, **169**, (2004), 315–332.

[137] Argyros, I.K., A unifying local–semilocal convergence analysis and applications for two–point Newton–like methods in Banach space, *J. Math. Anal. Appl.*, **298**, (2004), 374–397.

[138] Argyros, I.K., *Newton Methods*, Nova Science Publ. Corp., New York, 2005.

[139] Argyros, I. K., Concerning the "terra incognita" between convergence regions of two Newton methods, *Nonlinear Analysis,* **62**, (2005), 179–194.

[140] Argyros, I.K., An improved approach of obtaining good starting points for solving equations by Newton's method, *Adv. Nonlinear Var. Ineq.* **8**, (2005), 133–142.

[141] Argyros, I.K., On a two–point Newton–like method of convergent order two, *Int. J. Comput. Math.*, **88**, 2, (2005), 219–234.

[142] Argyros, I.K., *Approximate solution of operator equations with applications,* World Scientific Publ. Co., Hackensack, New Jersey, 2005, USA.

[143] I.K. Argyros, New sufficient convergence conditions for the Secant method, *Chechoslovak Math. J.*, **55**, (2005), 175–187.

[144] Argyros, I.K., On the Newton–Kantorovich method in Riemannian manifolds, Adv. Nonlinear Var. Ineq., 8, 2, (2005), 81–85.

[145] Argyros, I.K., On the semilocal convergence of the Gauss–Newton method, Adv. Nonlinear Var. Inequal., 8 (2005), 93–99.

[146] I.K. Argyros, A convergence analysis of Newton–like methods for singular equations using outer or generalized inverses, Appl. Math. (Warsaw), 32 (2005), 37–49.

[147] Argyros, I.K., A refined Newton's method mesh independence principle for a class of optimal shape design problems, *Central Eur. J. Math.,* **4** ,(2006), 562–572.

[148] Argyros, I.K., Relaxing the convergence conditions for Newton–like methods, *J. Appl. Math. Comput.,* **21**, (2006), 119–126.

[149] Argyros, I.K., An improved convergence analysis of a superquadratic method for solving generalized equations, *Rev. Colombiana Math.*, **40**, (2006), 65–73.

[150] Argyros, I.K., A Kantorovich–type analysis for a fast iterative method for solving equations, *J. Math. Anal. Appl.,* **332**, (2007), 97–108.

[151] Argyros, I.K., Computational theory of iterative methods. *Series: Studies in Computational Mathematics,* **15**, Editors: C.K. Chui, and L. Wuytack, 2007, Elsevier Publ. Co. New York, U.S.A.

[152] Argyros, I.K., On the gap between the semilocal convergence domain of two Newton methods, *Applicationes Mathematicae,* **34**(2), (2007), 193–204.

[153] Argyros, I.K., On the convergence of the Secant method under the gamma condition, *Cent. Eur. J. Math.*, **5**, (2007), 205–214.

[154] Argyros, I.K., Approximating solutions of equations using Newton's method with a modified Newton's method iterate as a starting point, *Revue d'Analyse Numérique et de la Théorie de l'approximation,* **36** (2007), 123–137.

[155] Argyros, I.K., *Convergence and applications of Newton–type iterations,* Springer–Verlag, 2008, New York.

[156] Argyros, I.K., Approximating solutions of equations by combining Newton–like methods, *J. Korea Soc. Educ. Ser. B: Pure and Appl. Math.*, **15** (2008), 35–45.

References
301

[157] Argyros, I.K., On a class of Newton–like methods for solving nonlinear equations, *J. Comput. Appl. Math.*, **228** (2009), 115–122.

[158] Argyros, I.K., On the semilocal convergence of inexact Newton methods in Banach spaces, *J. Comput. Appl. Math.*, **228**, 1, (2009), 434–443.

[159] Argyros, I.K., On Newton's method for solving equations containing Fréchet–differentiable operators of order at least two, *Appl. Math. Comput*, **215**, 4, (2009), 1553–1560.

[160] Argyros, I.K., On Ulm's method using divided differences of order one, *Numer. Algorithms*, **52**, 3, (2009), 295–320.

[161] Argyros, I.K., Newton's method on Lie groups, *J. Appl. Math. Comput.*, **31**, 1–2, (2009), 217–228.

[162] Argyros, I.K., On Ulm's method for Fréchet differentiable operators, *J. Appl. Math. Comput.*, **31**, 1–2, (2009), 97–111.

[163] Argyros, I.K., On the convergence of Stirling's method in Banach spaces under gamma–type condition, *Adv. Nonlinear Var. Inequal.*, **12**, 2, (2009), 17–23.

[164] Argyros, I.K., On the convergence of Newton's method and locally Hölderian inverses of operators, *J. Korea Soc. Math. Educ. Ser. B Pure Appl. Math.*, **16**, 1, (2009), 13–18.

[165] Argyros, I.K., On the local convergence of a midpoint method in Banach spaces under a gamma–type condition, *Proyecciones*, **28**, 2, (2009), 155–167.

[166] Argyros, I.K., On the Newton–Kantorovich theorem and nonlinear finite element methods, *Appl. Math. (Warsaw)*, **36**, 1, (2009), 75–81

[167] Argyros, I.K., On the radius of convergence of Newton's method under average mild differentiability conditions, *J. Appl. Math. Comput.*, **29**, 1–2, (2009), 429–435.

[168] Argyros, I.K., Concerning the convergence of Newton's method and quadratic majorants, *J. Appl. Math. Comput.*, **29**, 1–2, (2009), 391–400.

[169] Argyros, I.K., On the comparison of a Kantorovich–type and Moore theorems, *J. Appl. Math. Comput.*, **29**, 1–2, (2009), 117–123.

[170] Argyros, I.K., A semilocal convergence analysis for directional Newton methods, Mathematics of Computation, *A.M.S*, **80**, (2011), 327–343.

[171] Argyros, I.K., On the convergence region of Newton's method under Hölder continuity conditions, *Inter. J. Computers Math.*, to appear.

[172] Argyros, I.K., Chen, D., The midpoint method for solving equations in Banach spaces, *Appl. Math. Letters*, **5**, 4, (1992), 7–9.

References

[173] Argyros, I.K., Chen, D., The midpoint method in Banach spaces and the Pták error estimates, *Appl. Math. Comp.*, **62**, (1994), 1–15.

[174] Argyros, I.K., Chen, D., On the midpoint iterative method for solving nonlinear operator equations and applications to the solution of integral equations, *Revue D'analyse Numérique et de Théorie de l'Approximation, Tome* **23**, fasc. 2, (1994), 139–152.

[175] Argyros, I.K., Chen, D., An inverse–free Jarratt type approximation in a Banach space, *Approx. Theory Appl.*, **12**, (1996), 19–30.

[176] Argyros, I.K., Chen, J., On local convergence of a Newton–type method in Banach space, *Int. J. Comput. Math.*, **86**, 8, (2009), 1366–1374.

[177] Argyros, I.K., Chen, D., Qian, Q., The Jarratt method in a Banach space setting, *J. Comput. Appl. Math.*, **51**, (1994), 103–106.

[178] Argyros, I.K., Cho, Y–J., Qin, X., On the implicit iterative process for strictly pseudo–contractive mappings in Banach spaces, *J. Appl. Math. Comput.*, **233**, 2, (2009), 208–216.

[179] Argyros, I.K., Hilout, S., Newton's methods for variational inclusions under conditioned Fréchet derivative, *Applicationes Mathematicae*, **34**, (2007), 349–357.

[180] Argyros, I.K., Hilout, S., An improved local convergence analysis for Secant–like method, *East Asian Math. J.*, **23**, (2), (2007), 261–270.

[181] Argyros, I.K., Hilout, S., Local convergence of Newton–like methods for generalized equations, *Appl. Math. Comput.*, **197**, (2), (2008), 507–514.

[182] Argyros, I.K., Hilout, S., Multipoint method for generalized equations under mild differentiability conditions, *Functiones et Approximatio*, **XXXVIII**, (2008), 7–19.

[183] Argyros, I.K., Hilout, S., A Fréchet derivative–free cubically convergent method for set–valued maps, *Numerical Algorithms*, **48**, (4), (2008), 361–371.

[184] Argyros, I.K., Hilout, S., Steffensen method for solving generalized equations, *Serdica Math. J.*, **34**, (2), (2008), 455–466.

[185] Argyros, I.K., Hilout, S., On the midpoint method for solving generalized equations, *Punjab University J. Math.*, **40**, (2008), 63–70.

[186] Argyros, I.K., Hilout, S., On the local convergence of Newton–type method in Banach spaces under a gamma–type condition, Proyecciones *J. Math.*, **27**, (1), (2008), 1–14.

[187] Argyros, I.K., Hilout, S., On a Secant–like method for solving generalized equations, *Math. Bohemica*, **133**, 3 (2008), 313–320.

[188] Argyros, I.K., Hilout, S., A cubically convergent method without second order derivative for solving generalized equations, *Inter. J. Modern Math.*, **3**, 2, (2008), 187–195.

References   303

[189] Argyros, I.K., Hilout, S., *Efficient methods for solving equations and variational inequalities,* Polimetrica Publisher, Milano, Italy, 2009.

[190] Argyros, I.K., Hilout, S., *Aspects of the computational theory for certain iterative methods,* Polimetrica Publisher, Milano, Italy, 2009.

[191] Argyros, I.K., Hilout, S., On the weakening of the convergence of Newtons method using recurrent functions, *J. Complexity,* **25**, (2009), 530–543.

[192] Argyros, I.K., Hilout, S., Enclosing roots of polynomial equations and their applications to iterative processes, *Surveys Math. Appl.,* **4**, (2009), 119–132.

[193] Argyros, I.K., Hilout, S., On the convergence of a Jarratt–type method using recurrent functions, *J. Pure Appl. Math.: Adv. and Appl.,* **2**(2), (2009), 121–144.

[194] Argyros, I.K., Hilout, S., On the convergence of Steffensen's method on Banach spaces under the gamma condition, *Comm. Appl. Nonlinear Anal.,* **16**(4), (2009), 73–84.

[195] Argyros, I.K., Hilout, S., On the local convergence of the Gauss–Newton method, *Punjab Univ. Math. J.,* **41**, (2009), 23–33.

[196] Argyros, I.K., Hilout, S., On the convergence of Newton–type methods under mild differentiability conditions, *Numerical Algorithms,* **52**(4), (2009), 701–726.

[197] Argyros, I.K., Hilout, S., On the convergence of two–step Newton–type methods of high efficiency index, *Appl. Math.,* **36**(4), (2009), 465–499.

[198] Argyros, I.K., Hilout, S., An improved convergence analysis of Newton's methods for systems of equations with constant rank derivatives, *Mathematica,* **51**(74), (2009), 99–110.

[199] Argyros, I.K., Hilout, S., On the convergence of some iterative procedures under regular smoothness, *PanAmericain Math. J.,* **19**(2), (2009), 17–34.

[200] Argyros, I.K., Hilout, S., On multipoint iterative processes of efficiency index higher than Newton's method, *J. Nonlinear Sci. Appl.,* **2**, (2009), 195–203.

[201] Argyros, I.K., Hilout, S., Newton's method for approximating zeros of vector fields on Riemannian manifolds, *J. Appl. Math. Comput.,* **29**(1–2), (2009), 417–427.

[202] Argyros, I.K., Hilout, S., An improved local convergence analysis for a two–step Steffensen–type method, *J. Appl. Math. Comput.,* **30**(1–2), (2009), 237–245.

[203] Argyros, I.K., Hilout, S., On the convergence of inexact Newton–type methods using recurrent functions, *PanAmericain Math. J.,* **19**(1), (2009), 79–96.

[204] Argyros, I.K., Hilout, S., Local convergence analysis of inexact Newton–like methods, *J. Nonlinear Sci. Appl.,* **2**(1), (2009), 11–18.

[205] Argyros, I.K., Hilout, S., On the semilocal convergence of a Newton–type method of order three, *J. Korea Soc. Math. Educ. Ser. B: Pure Appl. Math.*, **27**(1), (2010), 1–27.

[206] Argyros, I.K., Hilout, S., An improved local convergence analysis for Newton–Steffensen–type method, *J. Appl. Math. Comput.*, **32**(1), (2010), 111–118.

[207] Argyros, I.K., Hilout, S., Convergence conditions for the Secant method, *CUBO A Math. J.*, **12**(1), (2010), 163–176.

[208] Argyros, I.K., Hilout, S., On the convergence of Stirling–type methods using recurrent functions, *PanAmericain Math. J.*, **20**(1), (2010), 93–10.

[209] Argyros, I.K., Hilout, S., A Newton–like method for nonsmooth variational inequalities, *Nonlinear Anal.: T.M.A.*, **72**, (2010), 3857–3864.

[210] Argyros, I.K., Hilout, S., A convergence analysis of Newton–like method for singular equations using recurrent functions, *Numer. Funct. Anal. and Optimization*, **31**(2), (2010), 112–130.

[211] Argyros, I.K., Hilout, S., Inexact Newton methods and recurrent functions, *Appl. Math.*, **37**, (2010), 113–126.

[212] Argyros, I.K., Hilout, S., Newton–like method for nonsmooth subanalytic variational inclusions, *Mathematica*, **52**(75), (2010), 5–13.

[213] Argyros, I.K., Hilout, S., A Kantorovich–type analysis of Broyden's method using recurrent functions, *J. Appl. Math. Comput.*, **32**(2), (2010), 353–368.

[214] Argyros, I.K., Hilout, S., On Newton's method defined on not necessarily bounded domains, *J. Pure Appl. Math.: Adv. and Appl.*, **3**(1), (2010), 1–16.

[215] Argyros, I.K., Hilout, S., Convergence conditions for Secant–type methods, *Czechoslovak Math. J.*, **60**(1), (2010), 253–272.

[216] Argyros, I.K., Hilout, S., Improved generalized differentiability conditions for Newton–like methods, *J. Complexity*, **26**(3), (2010), 316–333.

[217] Argyros, I.K., Hilout, S., Hummel–Seebeck method for generalized equations under conditioned second Fréchet derivative, *J. Nonlinear Funct. Anal. Appl.*, to appear.

[218] Argyros, I.K., Hilout, S., Newton–Steffensen–type method for perturbed nonsmooth subanalytic variational inequalities, *J. Nonlinear Funct. Anal. Appl.*, to appear.

[219] Argyros, I.K., Hilout, S., *On the convergence of Newton–like methods for solving equations using slantly differentiable operators,* submitted.

[220] Argyros, I.K., Hilout, S., Traub–Potra–type method for set–valued maps, *The Australian J. Math. Anal. Appl.*, to appear.

[221] Argyros, I.K., Hilout, S., A convergence analysis for directional two–step Newton methods, *Numerical Algorithms*, **55**(4), (2010), 503–528.

References 305

[222] Argyros, I.K., Hilout, S., A Kantorovich–type convergence analysis of the Newton–Josephy method for solving variational inequalities, *Numerical Algorithms*, **55**(4), (2010), 447–466.

[223] Argyros, I.K., Hilout, S., On the Gauss–Newton method, *J. Appl. Math. Comput.*, in press.

[224] Argyros, I.K., Hilout, S., On Newton–like methods of "bounded deterioration" using recurrent functions, *Aequationes Math.*, **9**(1–2), (2010), 61–82.

[225] Argyros, I.K., Hilout, S., On the semilocal convergence of Newton–like methods using recurrent polynomials, *Rev. Anal. Numér. Théor. Approx.*, to appear.

[226] Argyros, I.K., Hilout, S., Secant–like method for solving generalized equations, *Methods and Appl. of Analysis,* **16**(4), (2009), 469–478.

[227] Argyros, I.K., Hilout, S., On Newton's method using recurrent functions and hypotheses on the first and second Fréchet derivatives, *Atti Semin. Mat. Fis. Univ. Modena,* to appear.

[228] Argyros, I.K., Hilout, S., On Newton's method for solving equations and function splitting, *Numer. Math.: T.M.A.,* **4**(1), (2011), 53–67.

[229] Argyros, I.K., Hilout, S., On the convergence of Newton–type methods using recurrent functions, *International J. Comput. Math.*, in press.

[230] Argyros, I.K., Hilout, S., Superquadratic method for generalized equations under relaxed conditions on the second derivative, *J. Nonlinear Funct. Anal. Appl.*, to appear.

[231] Argyros, I.K., Ren, H., On an improved local convergence analysis for the secant method, *Numer. Algorithms*, **52**, 2, (2009), 257–371.

[232] Argyros, I.K., Ren, H., On the convergence of modified Newton methods for solving equations containing a non–differentiable term, *J. Comput. Appl. Math.*, **231**, 2, (2009), 897–906.

[233] Argyros, I.K., Ren, H., On convergence of the modified Newton's method under Hölder continuous Frchet derivative, *Appl. Math. Comput.*, **213**, 2, (2009), 440–448.

[234] Argyros, I.K., Szidarovszky, F., Convergence of general iteration schemes, *J. Math. Anal. Appl.*, **168**, 1, (1992), 42–52.

[235] Argyros, I.K., Szidarovszky, F., On the monotone of general Newton–like methods, *Bull. Austral. Math. Soc.*, **45**, (1992), 489–502.

[236] Argyros, I.K., Szidarovszky, F., *The Theory and Applications of Iteration Methods,* C.R.C. Press, Boca Raton, Florida, 1993.

[237] Argyros, I.K., Szidarovszky, F., On the convergence of modified contractions, *J. Comput. Appl. Math.*, **55**, 2, (1994), 97–108.

References

[238] Argyros, I.K., Tabatabai, M.A., Tabaistic regression and its application to the space shuttle Challenger O–ring data, *J. Appl. Math. Comput.*, to appear, doi: 10.1007/S12190–009–0300-3.

[239] Argyros, I.K., Tabatabai, M.A., An algorithm for solving nonlinear programming problems using Karmarkar's technique, *Math. Sci. Res. Hot–Line*, **5**, 4, (2001), 59–67.

[240] Argyros, I.K., Tabatabai, M.A., Error bounds for the Halley method in Banach spaces, *Adv. Nonlinear Var. Inequal.*, **3**, 2, (2000), 1–13.

[241] Argyros, I.K., Tabatabai, M.A., Robust estimation and testing for general nonlinear regression models, *Appl. Math. Comput.*, **58**, 1, (1993), 85–101.

[242] Argyros, I.K., Tabatabai, M.A., Chen, D., The Halley–Werner method in Banach spaces, *Rev. Anal. Numér. Théor. Approx.*, **23**, 1, (1994), 1–14.

[243] Argyros, I.K., Uko, L.U., Generalized equations, variational inequalities and a weak Kantorovich theorem, *Numer. Algorithms*, **52**, 3, (2009), 321–333.

[244] Argyros, I.K., Uko, L.U., A generalized Kantorovich theorem on the solvability of nonlinear equations, *Aequationes Math.*, **77**, 1–2, (2009), 99–105.

[245] Atkinson, K.E., The numerical evaluation of fixed points for completely continuous operators, *SIAM J. Num. Anal.*, **10**, (1973), 799–807.

[246] Atkinson, K.E., A Survey of Numerical Methods for the Solution of Fredholm Integral Equations of the Second Kind, *SIAM*, Philadelphia, 1976.

[247] Aubin, J.P., Lipschitz behavior of solutions to convex minimization problems, *Mathematics of Operations Research*, **9**, (1984), 87–111.

[248] Aubin, J.P., Frankowska, H., *Set–valued analysis*, Birkhäuser, Boston, 1990.

[249] Avila, J.H., *Continuation method for nonlinear equations, Technical Report TR-142*, Computer Science Center, University of Maryland, January, 1971.

[250] Avila, J.H., The feasibility of continuation methods for nonlinear equations, *SIAM J. Numer. Anal.*, **11**, 1, (1974), 102–122.

[251] Bai, Z.Z., A class of two–stage iterative methods for systems of weakly nonlinear equations, *Numer. Algorithms*, **14**, 4, (1997), 295–319.

[252] Bai, Z.Z., Wang, D.R., On the convergence of the factorization update algorithm, *J. Comput. Math.*, **11**, (1993), 236–249.

[253] Begehr, H., Efendiev, M.A., On the asymptotics of meromorphic solutions for nonlinear Riemann–Hilbert problems, *Math. Proc. Cambridge Philos. Soc.*, **127**, 91, (1999), 159–172.

[254] Belluce, L.P., Kirk, W.A., Fixed point theorems for a certain class of nonexpansive mappings, *Proc. Amer. Math. Soc.,* **50**, (1969), 144–146.

[255] Ben–Israel, A., A Newton–Raphson method for the solution of systems of operators, *J. Math. Anal. Appl.,* **15**, (1966), 243–252.

[256] Ben–Israel, A., Greville, T.N.E., *Generalized Inverses: Theory and Applications,* John Wiley and Sons, 1974.

[257] Ben–Israel, A., Levin, Y., *Maple programs for directional Newton methods,* are available at ftp://rutcor.rutgers.edu/pub/bisrael/Newton–Dir.mws

[258] Berinde, V., Remarks on the convergence of the Newton–Raphson method, *Revue d'analyse Numér. Th. Approx.,* **24**, (1 and 2), (1995), 15–21.

[259] Bianco, A.M., Yohai V.J., *Robust regression in the logistic regression model,* In H. Rieder, Ed. Robust Statistics, Data Analysis and Computer Intensive Methods, Springer–Verlag Publishing, 1996, 17–3, New York.

[260] Blum, L., Cucker, F., Shub, M., Smale, S., *Complexity and real computation.* With a foreword by Richard M. Karp., Springer–Verlag, New York, 1998.

[261] Bolte, J., Daniilidis, A., Lewis, A., The Lojasiewicz inequality for nonsmooth subanalytic functions with applications to subgradient dynamical systems, *SIAM J. Optimization,* **17**, (2006), 1205–1223.

[262] Bolte, J., Daniilidis, A., Lewis, A., Tame functions are semismooth, *Math. Program., Ser. B,* **117**, (1–2), (2009), 5–19.

[263] Bohl, E., *Monotonie: Lösbarkeit und Numerik bei Operatorgleichungen,* (German) Springer Tracts in Natural Philosophy, Vol. 25. Springer–Verlag, Berlin–New York, 1974.

[264] Bonnans, J.F, Local study of Newton type algorithms for constrained problems, Optimization (Varetz, 1988), *Lecture Notes in Math.,* 1405, Springer, Berlin, (1989), 13–24.

[265] Bonnans, J.F, An introduction to Newton type algorithms for nonlinearly constrained optimization problems, New methods in optimization and their industrial uses (Pau/Paris, 1987), 1–17, *Internat. Schriftenreihe Numer. Math.,* **87**, Birkhuser, Basel, 1989.

[266] Bonnans, J.F, Local analysis of Newton–type methods for variational inequalities and nonlinear programming, *Applied Mathematics and Optimization,* **29**, (1994), 161–186.

[267] Bonnans, J.F, Shapiro, A., Perturbation analysis of optimization problems, *Springer Series in Operations Research,* Springer–Verlag, New York, 2000.

[268] Bosarge, W.E., Falb, P.L., A multipoint method of third order, *J. Optim. Theory Appl.,* **4**, (1969), 156–166.

[269] Bosarge, W.E., Falb, P.L., Infinite dimensional multipoint methods and the solution of two point boundary value problems, *Numer. Math.,* **14**, (1970), 264–286.

[270] Brent, R.P., *Algorithms for Minimization Without Derivatives,* Prentice Hall, Englewood Cliffs, New Jersey, 1973.

[271] Brézis, H., Opérateurs maximaux monotones et semi–groupes de contractions dans les espaces de Hilbert, North–Holland Mathematics Studies, No. 5, *Notas de Matemática* (**50**), North–Holland Publishing Co., Amsterdam–London; American Elsevier Publishing Co., Inc., New York, 1973.

[272] Browder, F.E., Petryshyn, W.V., The solution by iteration of linear functional equations in Banach spaces, *Bull. Amer. Math. Soc.,* **72**, (1996), 566–570.

[273] Browder, F.E., Nonexpansive nonlinear operators in Banach spaces, *Proc. Nat. Acad. Sci. USA,* **54**, (1965), 1011–1014.

[274] Brown, P.N., A local convergence theory for combined inexact–Newton/finite–difference projection methods, *SIAM J. Numer. Anal.,* **24**, 2, (1987), 407–434.

[275] Brown, P.N., Hindmarsh, A.C., Walker, H.F., Experiments with quasi–Newton methods in solving stiff ODE systems, *SIAM J. Sci. Statist. Comput.,* **6**, (2), (1985), 297–313.

[276] Brown, P.N., Saad, Y., Convergence theory of nonlinear Newton–Krylov algorithms, *SIAM J. Optim.,* **4**, (2), (1994), 297–330.

[277] Brown, P.N., Vassilevski, P.S., Woodward, C.S., On mesh–independent convergence of an inexact Newton–multigrid algorithm, *SIAM J. Sci. Comput.,* **25**, (2), (2003), 570–590.

[278] Broyden, C.G., Dennis, J.E., Moré, J., On the local and superlinear convergence of quasi–Newton methods, *J. Inst. Math. Appl.,* **12**, (1973), 223–246.

[279] Burmeister, W., Inversion freie verfahren zur lösung nichtlinearen operatorgleichungen, *Zeitschrift für Angewandte Mathematik und Mechanik,* **52**, (1972), 101–110.

[280] Bursac, Z., Tabatabai, M.A., Williams, D.K., Non–linear Hyperbolastic Growth Models and Applications in Cranofacial and StemCell Growth, In 2005 *Proceedings of the American Statistical Association Biometrics Section [CD-ROM],* Alexandria, VA: American Statistical Association, 2006, 190–197.

[281] Tabatabai, Eby, W., Singh, K.P., Mathematical modelling of stem cell proliferation, *Med. Biol. Eng. Comp.,* DOI: 10.1007/s11517-010-0686-y, in press.

[282] Tabatabai, M.A., Bursac, Z., Eby, W., Singh, K.P., Hyperbolastic modeling of wound healing, *Math. Comput. Modelling,* DOI: 10.1016/j.mcm.2010.10.013, in press.

[283] Byelostotskij, A.Ja.,, Some methods for the solution of functional equations (Russian), *Uspekhi Matem. Nauk,* **17**, (5), (1962), 192–193.

References 309

[284] Cabuzel, C., A multipoint iteration formula for solving a variational inclusions in the Hölder case, *An. Stiint. Univ. Al. I. Cuza Iasi. Mat. (N.S.)*, **54**(1), (2008), 147–160.

[285] Cabuzel, C., Piétrus, A., Solving variational inclusions by a multipoint iteration method under center–Hölder continuity conditions, *Applicationes Mathematicae*, **34**, (2007), 493–503.

[286] Cabuzel, C., Piétrus, A., Local convergence of Newton's method for subanalytic variational inclusions, *Positivity*, **12**, 3, (2008), 525–533.

[287] Cabuzel, C., Piétrus, A., Solving variational inclusions by a method obtained using a multipoint iteration formula, *Revista Matemática Complutense*, **22**, (1), (2009), 63–74.

[288] Candela, V., Marquina, A., Recurrence relations for rational cubic methods I: The Halley method, *Computing*, **44**, (1990), 169–184.

[289] Caponetti, D., De Pascale, E., Zabrejko, P.P., On the Newton–Kantorovich method in K–normed spaces, *Rend. Circ. Mat. Palermo*, (**2**) 49, no. 3, (2000), 545–560.

[290] Cătinaş, E., On some iterative methods for solving nonlinear equations, *Revue d'analyse numérique et de théorie de l'approximation*, **23**, 1, (1994), 47–53.

[291] Cătinaş, E., Inexact perturbed Newton methods, and applications to a class of Krylov solvers, *J. Optim. Theory Appl.*, **108**, 3, (2001), 543–570.

[292] Cătinaş, E., On the superlinear convergence of the successive approximations method, *J. Optim. Theory Appl.*, **113**, 3, (2002), 473–485.

[293] Chandrasekhar, S., *Radiative transfer*, Dover Publ., New York, 1960.

[294] Chen, B., Harker, P.T., A continuation method for monotone variational inequalities, *Math. Programming*, **69**, (1995), 237–253.

[295] Chen, D., Kantorovich–Ostrowski convergence theorems and optimal error bounds for Jarratt's iterative method, *Intern. J. Computer. Math.*, **31**, (3 and 4), (1990), 221–235.

[296] Chen, D., On the convergence of a class of generalized Steffensen's iterative procedures and error analysis, *Int. J. Comput. Math.*, **31**, (1989), 195–203.

[297] Chen, J., Li, W., Convergence behaviour of inexact Newton methods under weak Lipschitz condition, *J. Comput. Appl. Math.*, **191** (2006), 143–164.

[298] Chen, X., Superlinear convergence of smoothing quasi–Newton methods for nonsmooth equations, *J. Comput. Appl. Math.*, **80** (1997), 105–126.

[299] Chen, X., Nashed, M.Z., Convergence of Newton–like methods for singular operator equations using outer inverses, *Numer. Math.*, **66**, (1993), 235–257.

[300] Chen, X., Nashed, Z., Qi, L., Convergence of Newton's method for singular smooth and nonsmooth equations using adaptive outer inverses, *SIAM J. Optim.*, **7**, 2, (1997), 445–462.

[301] Chen, X., Nashed, M.Z., Qi, L., Smoothing methods and semismooth methods for nondifferentiable operator equations, *SIAM J. Numer. Anal.*, **38** (2000), 1200–1216.

[302] Chen, X., Qi, L., Sun, D., Global and superlinear convergence of the smoothing Newton method and its application to general box constrained variational inequalities, *Math. Comp.*, **67** (1998), 519–540.

[303] Chen, X., Yamamoto, T., Convergence domains of certain iterative methods for solving nonlinear equations, *Numer. Funct. Anal. Optim.*, **10**, (1 and 2), (1989), 37–48.

[304] Chui, C.K., Quak, F., Wavelets on a bounded interval. In: Numerical Methods of Approximation Theory, Vol. 9, (eds: Braess D. and Larry L. Schumaker, *Intern. Ser. Num. Math.*, Vol. 105), Basel: Birkhäuser Verlag, 53–75.

[305] Chow, S.N., Hale, J.K., *Methods of Bifurcation Theory*, springer–Verlag, New–York, 1962.

[306] Cianciaruso, F., Convergence theorems for Newton–like methods for operators with generalized Hölder derivative, *Fixed Point Theory*, **5**, (1), (2004), 21–35.

[307] Cianciaruso, F., Convergence of Newton–Kantorovich approximations to an approximate zero, *Numer. Funct. Anal. Optim.*, **28**, (5), (2007), 631–645.

[308] Cianciaruso, F., A further journey in the "terra incognita" of the Newton–Kantorovich method, *Nonlinear Funct. Anal. Appl.*, to appear.

[309] Cianciaruso, F., De Pascale, E., The Newton–Kantorovich approximations for the nonlinear integro–differential equations of mixed type, *Ricerche Mat.*, **51**, (2002), no. 2, 249–260 (2003).

[310] Cianciaruso, F., De Pascale, E., The Newton–Kantorovich approximations for nonlinear singular integral equations with shift, *J. Integral Equations Appl.*, **14**, (3), (2002), 223–237.

[311] Cianciaruso, F., De Pascale, E., Newton–Kantorovich aproximations when the derivative is Hölderian: Old and new results, *Numer. Funct. Anal. Optim.*, **24**, (2003), 713–723.

[312] Cianciaruso, F., De Pascale, E., Estimates of majorizing sequences in the Newton–Kantorovich method, *Numer. Funct. Anal. Optim.*, **27**, (5–6), (2006), 529–538.

[313] Cianciaruso, F., De Pascale, E., Estimates of majorizing sequences in the Newton–Kantorovich method: A further improvement, *J. Math. Anal. Appl.*, **322**, (2006), 329–335.

References   311

[314] Cianciaruso, F., De Pascale, E., Zabrejko, P.P., Some remarks on the Newton-Kantorovič approximations, *Atti Sem. Mat. Fis. Univ. Modena,* **48**, (1), (2000), 207–215.

[315] Cohen, G., Auxiliary problem principle and decomposition of optimization problems, *J. Optim. Theory Appl.,* **32**, 3, (1980), 277–305.

[316] Cohen, G., Auxiliary problem principle extended to variational inequalities, *J. Optim. Theory Appl.,* **59**, 2, (1988), 325–333.

[317] Collatz, L., *Functional Analysis and Numerisch Mathematik,* Springer–Verlag, New York, 1964.

[318] Collett, D., *Modeling Binary Data*, Chapman & Hall Publishing, 1991, New York.

[319] Conceiçao, E.L.T., Portugal, A.A.T.G., A performance comparison of some high breakdown robust estimators for nonlinear parameter estimation, *16th European Symposium on Computer Aided Process Engineering and 9th International Symposium on Process Systems Engineering,* Marquardt and Pantelides (Editors), (2006), 279–284.

[320] Danes, J., Fixed point theorems, Nemyckii and Uryson operators, and continuity of nonlinear mappings, *Comment. Math. Univ. Carolinae,* **11**, (1970), 481–500.

[321] Darbo, G., Punti uniti in trasformationa codominio non compatto, *Rend. Sem. Mat. Univ. Padova,* **24**, (1955), 84–92.

[322] Daubechies, I., *Ten Lectures in Wavelets*, (Conf. Board Math. Sci. (CBMS) Vol. 61), Society for Industrial and Applied Mathematics (SIAM), Philadelphia, PA, 1992.

[323] Davis, H.T., *Introduction to nonlinear differential and integral equations,* Dover Publications, Inc., New York, 1962.

[324] Deasy, B.M., Jankowski, R.J., Payne, T.R, Cao, B., Goff, J.P., Greenberger, J.S., Huard, J., Modeling Stem Cell Population Growth: Incorporating Terms for proliferative, Heterogeneity, *Stem Cells,* **21**, (2003), 536–545.

[325] Deasy, B.M., Qu–Peterson, Z., Grrenberger, J.S, Huard. J., Mechanisms of Muscle: Stem Cell Expansion with Cytokines, *Stem Cells,* **20**, (2002), 20, 50–60.

[326] Decker, D.W., Keller, H.B., Kelley, C.T., Convergence rates of Newton's method at singular points, *SIAM J. Numer. Anal.,* **20**, 2, (1983), 296–314.

[327] Dedieu, J.P., Penality functions in subanalytic optimization, Optimization, **26**, (1992), 27–32.

[328] Dedieu, J.P., Estimations for the separation number of a polynomial system, *J. Symbolic Comput.,* **24**, 6, (1997), 683–693.

[329] Dedieu, J.P., Kim, M–H., Newton's method for analytic systems of equations with constant rank derivatives, *J. Complexity,* **18** (2002), 187–209.

[330] Dedieu, J.P., Shub, M., Newton's method for overdetermined systems of equations, *Math. Comp.,* **69** (2000), 1099–1115.

[331] Dembo, R.S., Eisenstat, S.C., Steihaug, T., Inexact Newton methods, *SIAM J. Numer. Anal.*, **19**, 2, (1982), 400–408.

[332] Demidovich, N.T., Zabrejko, P.P., Lysenko, Ju.V., Some remarks on the Newton–Kantorovich mehtod for nonlinear equations with Hölder continuous linearizations, *Izv. Akad. Nauk Belorus*, **3**, (1993), 22–26, (in Russian).

[333] Dennis, J.E., On the Kantorovich hypothesis for Newton's method, *SIAM J. Numer. Anal.,* **6**, 3, (1969), 493–507.

[334] Dennis, J.E., On the convergence of Newton–like methods, Numerical methods for nonlinear algebraic equations, (Proc. Conf., Univ. Essex, Colchester, 1969), Gordon and Breach, London, (1970), 163–181.

[335] Dennis, J.E., Toward a unified convergence theory for Newton–like methods, in Nonlinear Functional Analysis and Applications (L.B. Rall, ed.), Academic Press, New York, (1971), 425–472.

[336] Dennis, J.E., *A brief survey of convergence results for quasi–Newton methods,* Nonlinear programming (Proc. Sympos., New York, 1975), SIAM–AMS Proc., Vol. IX, Amer. Math. Soc., Providence, R. I., (1976), 185–199.

[337] Dennis, J.E., A brief introduction to quasi–Newton methods, Numerical analysis (Proc. Sympos. Appl. Math., Atlanta, Ga., 1978), *Proc. Sympos. Appl. Math.,* XXII, Amer. Math. Soc., Providence, R.I., (1978), 19–52.

[338] Dennis, J.E., Moré, J.J., A characterization of superlinear convergence and its application to quasi–Newton methods, *Math. Comp.,* **28**, (1974), 549–560.

[339] Dennis, J.E., Moré, J.J., Quasi–Newton methods, motivation and theory, *SIAM Rev.,* **19**, (1), (1977), 46–89.

[340] Dennis, J.E., Moré, J.J., Quasi–Newton methods, motivation and theory, (Chinese) Translated from the English by Wen Yu Sun., *Appl. Math. Math. Comput.,* **1**, 1983, 1–30.

[341] Dennis, J.E., Schnabel, R.B., Least change secant updates for quasi–Newton methods, *SIAM Rev.,* **21**, (4), (1979), 443–459.

[342] Dennis, J.E., Sheng, S.B., Vu, P.A., A memoryless augmented Gauss–Newton method for nonlinear least–squares problems, *J. Comput. Math.,* **6**, (4), (1988), 355–374.

[343] Dennis, J.E., Walker, H.F., Inaccuracy in quasi–Newton methods: local improvement theorems, Mathematical programming at Oberwolfach, II (Oberwolfach, 1983), *Math. Programming Stud.,* **22**, (1984), 70–85.

[344] De Pascale, E., Zabrejko, P.P, New convergence criteria for Newton–Kantorovich method and some applications to nonlinear integral equations, *Rend. Sem. Mat. Univ. Padova*, **100**, (1998), 211–230.

[345] De Pascale, E., Zabrejko, P.P., Convergence of the Newton–Kantorovich method under Vertgeim conditions: a new improvement, *Z. Anal. Anwendvugen*, **17**, (1998), 271–280.

[346] De Pascale, E., Zabrejko, P.P., Fixed point theorems for operators in spaces of continuous functions, *Fixed Point Theory*, **5**, (1), (2004), 117–129.

[347] Deuflhard, P., Newton Methods for Nonlinear Problems. Affine invariance and adaptive algorithms, *Springer Series in Computational Mathematics,* **35**, Springer–Verlag, New–York, 2004.

[348] Deuflhard, P., Heindl, G., Affine invariant convergence theorems for Newton's method and extensions to related methods, *SIAM J. Numer. Anal.*, **16**, (1979), 1–10.

[349] Deuflhard, P., Potra, F.A., Asymptotic mesh independence of Newton–Galerkin methods via a refined Mysovskii theorem, *SIAM J. Numer. Anal.*, **29**, 5, (1992), 1395–1412.

[350] Deuflhard, P., Schiela, A., Weiser, M., Asymptotic mesh independence of Newton's method, revisited, *SIAM J. Numer. Anal.*, **42**, 5, (2005), 1830–1845.

[351] Diaconu, A., On the approximation of solutions of equations in Banach spaces using approximant sequences, Conference on Analysis, *Functional Equations, Approximation and Convexity,* in Honor of E. Popoviciv, Cluj–Napora, October 15–16, (1999), 62–72.

[352] Do Carmo, M.P., *Riemannian Geometry*, Birkhauser, 1992, Boston, USA.

[353] Dontchev, A.L., Local analysis of a Newton–type method based on partial linearization. Renegar, James (ed.) et al., The mathematics of numerical analysis. 1995 AMS-SIAM summer seminar in applied mathematics, Providence, *RI: AMS. Lect. Appl. Math.*, **32**, (1996), 295–306.

[354] Dontchev, A.L., Local convergence of the Newton method for generalized equation, *C.R.A.S.*, **322**, Serie I, (1996), 327–331.

[355] Dontchev, A.L., Uniform convergence of the Newton method for Aubin continuous maps, *Serdica Math. J.,* **22**, (1996), 385–398.

[356] Dontchev, A.L., Hager, W.W., An inverse function theorem for set–valued maps, *Proc. Amer. Math. Soc.*, **121**, (1994), 481–489.

[357] Dontchev, A.L., Quincampoix, M., Zlateva, N., Aubin criterion for metric regularity, *J. Convex Anal.*, **13**, (2006), 281–297.

314   References

[358] Dontchev, A.L., Rockafellar, R.T., Characterizations of strong regularity for variational inequalities over polyhedral convex sets, *SIAM J. Optim.* **6**, 4, (1996), 1087–1105.

[359] Dontchev, A.L., Rockafellar, R.T., Ample parameterization of variational inclusions, *SIAM J. Optim.* **12**, 1, (2001), 170–187.

[360] Dontchev, A.L., Rockafellar, R.T., Regularity and conditioning of solution mappings in variational analysis, *Set–Valued Anal.*, **12**, (1–2), (2004), 79–109.

[361] Döring, B., Iterative lösung gewisser randwertprobleme und integralgleichungen, *Appl. Mat.*, **24**, (1976), 1–31.

[362] Dunford, N., Schwartz, J.T., *Linear operators*. Part I, Int. Publ. Leyden, (1963).

[363] Duvuat, G., Lions, J.L., *Inequalities in Physics and Mechanics*, Springer–Verlag, Berlin, 1976.

[364] Eaves, B.C., *A locally quadratically convergent algorithm for computing stationary point*, Dpt of Operations Rearch, Stanford

[365] Eby, W., Tabatabai, M.A., Bursac, Z., Hyperbolastic modeling of tumor growth with a combined treatment of iodoacetate and dimethylsulfoxide, *BMC Cancer*, **10**, (2010): 509.

[366] Edelstein, M., On fixed and periodic points under contractive mappings., *J. London Math. Soc.*, **37**, (1962), 74–79.

[367] Edelstein, M., A remark on a theorem of M.A. Krasnoselskii, *Amer. Math. Monthly*, **73**, (1966), 509–510.

[368] Eisenstat, S.C., Walker, H.F., Globally convergent inexact Newton methods, *SIAM J. Optim.*, **4**, (2), (1994), 393–422.

[369] Eisenstat, S.C., Walker, H.F., Choosing the forcing terms in an inexact Newton method, Special issue on iterative methods in numerical linear algebra, (Breckenridge, CO, 1994), *SIAM J. Sci. Comput.*, **17**, (1), (1996), 16–32.

[370] Ezquerro, J.A., Gutiérrez, J.M., Hernández, M.A., Salanova, M. A., Solving nonlinear integral equations arising in radiative transfer, *Numer. Funct. Anal. Optim.*, **20**, (7 and 8), (1999), 661–673.

[371] Ezquerro, J.A., Hernández, M.A., Multipoint super–Halley type approximation algorithms in Banach spaces, *Numer. Funct. Anal. Optim.* **21**, (7 and 8), (2000), 845–858.

[372] Ezquerro, J.A., Hernández, M.A., Recurrence relations for Chebyshev–type methods, *Appl. Math. Optim.*, **41** (2000), 227–236.

[373] Ezquerro, J.A., Hernández, M.A., A special type of Hammerstein integral equations, *Int. Math. J.*, **1**, (6), (2002), 557–566.

References

[374] Ezquerro, J.A., Hernández, M.A., On an application of Newton's method to nonlinear operators with ω–conditioned second derivative, *BIT,* **42**, no. 3, (2002), 519–530.

[375] Ezquerro, J.A., Hernández, M.A., Generalized differentiability conditions for Newton's method, *IMA J. Numer. Anal.,* **22** (2002), 187–205.

[376] Ezquerro, J.A., Hernánadez, M.A., The Ulm method under mild differentiability conditions, *Numer. Math.,* **109**, (2), (2008), 193–207.

[377] Ezquerro, J.A., Hernández, M.A., An optimization of Chebyshev's method, *J. Complexity,* **25** (2009), 343–361.

[378] Ezquerro, J.A., Hernández, M.A., New iterations of R–order four with reduced computational cost, *BIT Numer. Math.,* **49** (2009), 325–342.

[379] Ezquerro, J.A., Hernández, M.A., Salanova, M.A., Remark on the convergence of the midpoint method under mild differentiability conditions, *J. Comp. Appl. Math.,* **98**, (1998), 305–309.

[380] Ezquerro, J.A., Hernández, M.A., Salanova, M. A., A discretization scheme for some conservative problems, Proceedings of the 8th Inter. Congress on Comput. and Appl. Math., ICCAM–98 (Leuven), *J. Comput. Appl. Math.,* **115** (2000), 181–192.

[381] Fahim, F.A., Esmat, A.Y., Mady, E.A., Ibrahim, E.K., Antitumor Activities of Iodoacetate and Dimethylsulphoxide Against Solid Ehrlich Carcinoma Growth in Mice, *Biol res,* **36**, (2003), 253–262.

[382] Feinstauer, M., Zernicek, A., Finite element solution on nonlinear elliptic problems, *Numer. Math.* **50**, (1987), 471–475.

[383] Ferreira, O.P., Svaiter, B.F., Kantorovich's theorem on Newton's method in Riemannian manifolds, *J. Complexity,* **18**, (2002), 304–329.

[384] Ferreira, O.P., Svaiter, B.F., Kantorovich's majorants principle for Newton's method, *Comput. Optim. Appl.,* **42**, 2, (2009), 213–229.

[385] Ferreira, O.P., Svaiter, B.F., *Kantorovich's theorem on Newton's method,* preprint.

[386] Ferris, M.C., Pang, J.S., Engineering and economic applications of complementarity problems, *SIAM Rev.,* **39**, 4, (1997), 669–713.

[387] Finney, D.J., *Probit Analysis,* Cambridge University Press, 1971, Cambridge.

[388] Foerster, H., Frommer, A., Mayer, G., Inexact Newton methods on a vector supercomputer, *J. Comp. Appl. Math,* **58**, (1995), 237–253.

[389] Fujimoto, T., Global asymptotic stability of nonlinear difference equations I, *Econ. Letters,* **22**, (1987), 247–250.

[390] Fujimoto, T., Global asymptotic stability of nonlinear difference equations II, *Econ. Letters,* **23**, (1987), 275–277.

[391] Galperin, A., Kantorovich majorization and functional equations, *Numer. Funct. Anal. Optim.*, **24**, (2003), 783–811.

[392] Galperin, A., On convergence domains of Newton's and modified Newton methods, *Numer. Funct. Anal. Optim.*, **26**, 3, (2005), 385–405.

[393] Galperin, A., Secant method with regularity continuous divided differences, *J. Comput. Appl. Math.*, **193**, (2006), 574–575.

[394] Galperin, A., Waksman, Z., Regular smoothness and Newton's method, *Numer. Funct. Anal. Optim.*, **15**, (7 and 8), (1994), 813–858.

[395] Galperin, A., Waksman, Z., Newton–type methods under regular smoothness, *Numer. Funct. Anal. Optim.*, **17**, (1996), 259–291.

[396] Gander, W., On Halley's iteration method, *Amer. Math. Monthly,* **92**, (1985), 131–134.

[397] Geoffroy, M.H., *A secant type method for variational inclusions,* Preprint.

[398] Geoffroy, M.H., Hilout, S., Piétrus, A., Acceleration of convergence in Dontchev's iterative method for solving variational inclusions, *Serdica Math. J.,* **29**, (2003), 45–54.

[399] Geoffroy, M.H., Hilout, S., Piétrus, A., Stability of a cubically convergent method for generalized equations, *Set–Valued Analysis,* **14**, (2006), 41–54.

[400] Geoffroy, M.H., Piétrus, A., Superquadratic method for solving generalized equations in the Hölder case, *Ricerche di Math. LII, fasc.* **2**, (2003), 231–240.

[401] Geoffroy, M.H., Piétrus, A., Local convergence of some iterative methods for solving generalized equations, *J. Math. Anal. Appl.*, **290**, (2004), 497–505.

[402] Gill, P.E., Murray, W., Wright, M.A., *Practical Optimization,* Academic Press, 1991, London.

[403] Glowinski, R., Lions, J.L., Trémolières, R., *Numerical analysis of variational inequalities,* North–Holland, Amsterdam, 1982.

[404] Gragg, W.B., Tapia, R.A., Optimal error bounds for the Newton–Kantorovich theorem, *SIAM J. Numer. Anal.,* **11**, (1974), 10–13.

[405] Graves, L.M., Riemann integration and Taylor's theorem in general analysis, *Trans. Amer. Math. Soc.,* **29**, 1, (1927), 163–177.

[406] Grigat, E., Sachs, G., Predictor–corrector continuation method for optimal control problems, in: Variational calculus, optimal control and applications (Trassenheide, 1996), Volume 124, *Int. Ser. Numer. Math.*, Birkhäuser, Basel, (1998), 223–232.

[407] Guo, X., On semilocal convergence of inexact Newton methods, *J. Comput. Math.*, **25**, 2, (2007), 231–242.

[408] Gutiérrez, J.M., A new semilocal convergence theorem for Newton's method, *J. Comput. Appl. Math.*, **79**, (1997), 131–145.

[409] Gutiérrez, J.M., Hernández, M.A., Newton's method under weak Kantorovich conditions, *IMA J. Numer. Anal.,* **20** (2000), 521–532.

[410] Gutiérrez, J.M., Hernández, M.A., An acceleration of Newton's method: Super–Halley method, *Appl. Math. Comp.*, **117**, (2001), 223–239.

[411] Gutiérrez, J.M., Hernández, M.A., Salanova, M.A., Accessibility of solutions by Newton's method, *Internat. J. Comput. Math.*, **57**, (1995), 239–247.

[412] Gutiérrez, T.M., Hernández, M.A., Salanova, M.A., A family of the Chebyshev–Halley type methods in Banach spaces, *Bull. Austral. Math. Soc.,* **55**, (1997), 113–130.

[413] Gwinner, J., Generalized Stirling–Newton methods, In W. Oettli, K. Ritter (eds), Optimization and Operations Research, Oberwolfach, 1975, *Lecture Notes Economics and Mathematical Systems,* **11**, (1976), 99–135.

[414] Hadeller, K.P., Shadowing orbits and Kantorovich's theorem, *Numer. Math.*, **73**, (1996), 65–73.

[415] Haeberly, J.–P.A., *Remarks on nondegeneracy in mixed semidefinite-quadratic programming.* Unpublished memorandum, available at URL:http://corky.fordham.edu/ haeberly/papers/sqldegen.ps.gz

[416] Hald, O.H., On a Newton–Moser type method, *Numer. Math.*, **23**, (1975), 411–425.

[417] Han, D., Wang, X., The error estimates of Halley's method, *Numer. Math. J. Chinese Univ. (English Ser.),* **6**, 2, (1997), 231–240.

[418] Han, S., Pang, J., Rangaraj, N., Globally convergent Newton methods for nonsmooth equations, *Math. Oper. Res.,* **17**, 3, (1992), 586–607.

[419] Han, D.F., Wang, X., Convergence of a deformed Newton method, *Appl. Math. Comput.,* **94**, (1998), 65–72.

[420] Harker, P.T., Pang, J.S., Finite dimensional variational inequality and nonlinear complementarity problems: a survey of theory, algorithms and applications, *Mathematical Programming,* **48**, (1990), 161–220.

[421] Hartley, H.O., The modified Gauss–Newton method for the fitting of nonlinear regression functions by least squares, *Technometrics,* **3**, (1961), 269–280.

[422] Hartman, P., *Ordinary differential equations,* John Wiley & Sons, Inc., New–York, London, Sydney, 1964.

[423] Häußler, W.M., A Kantorovich–type convergence analysis for the Gauss–Newton method, *Numer. Math.*, **48**, (1986), 119–125.

318   References

[424] Heinkenschloss, M., Kelley, C.T., Tran, H.T., Fast algorithms for nonsmooth compact fixed–point problems, *SIAM J. Numer. Anal.,* **29** (1992), 1769–1792.

[425] Helgason, S., *Differential Geometry*, Lie Groups and Symmetric Spaces, Pergamon Press, Oxford, 1982.

[426] Hellinger, E., Toeplitz, O., Integralgleichungen und Gleichungen mit unendlichvielen Unbekannten, (German), Chelsea Publishing Company, New–York, (1953), 1335–1616.

[427] Hernández, M.A., The Newton method for operators with Hölder continuous first derivatives, *J. Optim. Th. Appl.*, **109**, 3, (2001), 631–648.

[428] Hernández, M.A., A modification of the classical Kantorovich conditions for Newton's method, *J. Comp. Appl. Math.,* **137**, (2001), 201–205.

[429] Hernández, M.A., Rubio, M.J., Recurrence relations for the super–Halley method, *Comput. Math. Appl.,* **36**, (1998), 1–8.

[430] Hernández, M.A., Rubio, M.J., The Secant method and divided differences Hölder continuous, *Appl. Math. Comp.,* **124**, (2001), 139–149.

[431] Hernández, M.A., Rubio, M.J., Semilocal convergence of the Secant method under mild convergence conditions of differentiability, *Comput. Math. Appl.*, **44**, (2002), 277–285.

[432] Hernández, M.A., Rubio, M.J., The Secant method for nondifferentiable operators, *Appl. Math. Letters,* **15**, (2002), 395–399.

[433] Hernández, M.A., Rubio, M.J., An uniparametric family of iterative processes for solving nondifferentiable equations, *J. Math. Anal. Appl.*, **275**, (2002), 821–834.

[434] Hernández, M.A., Rubio, M.J., ω–conditioned divided differences to solve nonlinear equations, *Monografías del Semin. Matem. García de Galdeano,* **27**, (2003), 409–417.

[435] Hernández, M.A., Rubio, M.J., A modification of Newton's method for nondifferentiable equations, *J. Comp. Appl. Math.*, **164/165**, (2004), 323–330.

[436] Hernández, M.A., Rubio, M.J., Ezquerro, J.A., Secant–like methods for slving nonlinear integral equations of the Hammerstein type, *J. Comput. Appl. Math.*, **115**, (2000), 245–254.

[437] Hernández, M.A., Rubio, M.J., Ezquerro, J.A., Solving a special case of conservative problems by Secant–like method, *Appl. Math. Cmput.,* **169** (2005), 926–942.

[438] Hernández, M.A., Salanova, M.A., Modification of the Kantorovich assumptions for semilocal convergence of the Chebyshev methods, *J. Comput. Appl. Math.*, **126**, (2000), 131–143.

References 319

[439] Hernández, M.A., Salanova, M.A., A new third–order iterative process for solving nonlinear equations, *Monatshefte für Mathematick,* **133**, (2001), 131–142.

[440] Higle, J.L., Sen, S., On the convergence of algorithms with applications to stochastic and nondifferentiable optimization, *SIE Working* Paper #89-027, University of Arizona, 1989.

[441] Hille, E., Philips, R.S., *Functional Analysis and Semigroups,* Amer. Math. Soc. Coll. Publ., New–York, 1957.

[442] Hilout, S., Steffensen–type methods on Banach spaces for solving generalized equations, *Adv. Nonlinear Var. Inequa.*, **10**, 2, (2007), 105–113.

[443] Hilout, S., An uniparametric Newton–Steffensen–type methods for perturbed generalized equations, *Adv. Nonlinear Var. Inequa.*, **10**, 2, (2007), 115–124.

[444] Hilout, S., Superlinear convergence of a family of two–step Steffensen–type methods for generalized equations, *Int. J. Pure Appl. Math.*, **40**, 1, (2007), 1–10.

[445] Hilout, S., A two–step Steffensen–type method for nonsmooth variational inclusions, *Comm. Appl. Nonlinear Anal.,* **14**, Number 3, (2007), 27–34.

[446] Hilout, S., Convergence analysis of a family of Steffensen–type methods for generalized equations, *J. Math. Anal. Appl.*, **339**, (2), (2008), 753–761.

[447] Hilout, S., An uniparametric Secant–type method for nonsmooth generalized equations, *Positivity,* **12**, (2008), 281–287.

[448] Hilout, S., Piétrus, A., A semilocal convergence of a secant–type method for solving generalized equations, *Positivity*, **10**, (2006), 673–700.

[449] Hiriart–Urruty, J.B., Lemaréchal, C., Convex analysis and minimization algorithms (two volumes): I. Fundamentals II. *Advanced theory and bundle methods, Grundlehren der Mathematischen Wissenschaften* , Vol. 305 and 306, Springer–Verlag, Berlin, 1993.

[450] Homeir, H., A modified method for root finding with cubic convergence, *J. Comput. Appl. Math.*, **157**, (2003), 227–230.

[451] Homeier, H., A modified Newton method with cubic convergence: the multivariable case, *J. Comput. Appl. Math.*, **169**, (2004), 161–169.

[452] Hoppe, R.H.W., *Numerical methods for large–scale nonlinear systems,* Handouts published on–line on the web, (2005), 1–108.

[453] Hosmer, D.W., Lemeshow, S., *Applied Logistic Regression,* Wiley Publishing, 2000, New York.

[454] http://Stemcells.nih.gov/research/NIHresearch/scunit/growthCurves.asp

[455] Hu, N., Shen, W., Li, C., Kantorovich's type theorems for systems of equations with constant rank derivatives, *J. Comput. Appl. Math.,* **219** (2008), 110–122.

[456] Hu, Z., A new semilocal convergence theorem for Newton's method involving twice Fréchet–differentiability at only one point, *J. Comput. Appl. Math.,* **181**, 2, (2005), 321–325.

[457] Huang, Z., A note of Kantorovich theorem for Newton iteration, *J. Comput. Appl. Math.,* **47**, (1993), 211–217.

[458] Huang, Z.D., On the convergence of inexact Newton method, *J. Zheijiang University, Natur. Sci. Edition,* **30**, 4, (2003), 393–396.

[459] Huber, P.J., Robust Statistics, New York: John Wiley, 1981.

[460] Hummel, P.M., Seebeck, C.L.Jr., A generalized of Taylor's expansion, *Amer. Math. Monthly,* **56**, (1949), 243–247.

[461] Ioffe, A.D., Tikhomirov, V.M., *Theory of extremal problems*, North Holland, Amsterdam, 1979.

[462] Ip, Chi–M., Kyparisis, J., Local convergence for quasi–Newton methods for B–differentiable operators, *Math. Prog.,* **56**, (1992), 71–89.

[463] Isac, G., Leray–Schauder type alternative, *Complementarity problems and Variational inequalities*, Volume 87 (Nonconvex optimization and its applications), Springer, 2006.

[464] Ishikawa, S., Fixed points by a new iteration method, *Proc. Amer. Math. Soc.,* **44**, (1974), 147–150.

[465] Izmailov, A.F, Solodov, M.V., Inexact Josephy–Newton framework for generalized equations and its applications to local analysis of Newtonian methods for constrained optimization, *Comput. Optim. Appl.,* in press, DOI. 10.1007/s10589–009–9265–2.

[466] Jankowski, R.J., Deasy, B.M., Cao, B., Gates Charley, Huard Johny: The Role of CD34 Expression and Cellular fusion in the Regeneration Capacity of Myogenic Progenitor Cells, *Journal of Cell Science,* **115**, (2002), 4361–4372.

[467] Jarratt, P., Some efficient fourth order multipoint methods for solving equations, *BIT,* **9**, (1969), 119–124.

[468] Jean–Alexis, C., Piétrus, A., A variant of Newton's method for generalized equations, *Rev. Colombiana Mat.,* **39**, (2005), 97–112.

[469] Jean–Alexis, C., A cubic method without second order derivative for solving generalized equations, *Compt. Rend. Acad. Bulg. Sci.,* **59**, (2006), 1213–1218.

[470] Josephy, N.H., Newton's method for generalized equations, *Technical Report No. 1965,* Mathematics Research Center, University of Wisconsin, Madison, WI, 1979.

References 321

[471] Josephy, N.H., *Quasi–Newton method for generalized equations,* Technical Summary Report No. 1966, Mathematics Research Center, University of Wisconsin–Madison, June 1979, available from National Technical Information Service, Springfield, VA 22161, under Accession No. A077 097.

[472] Josephy, N.H., *A Newton method for the PIES energy model,* Technical Summary Report No. 1971, Mathematics Research Center, University of Wisconsin–Madison, June 1979, available from National Technical Information Service, Springfield, VA 22161, under Accession No. A077 102.

[473] Josephy, N.H., *Hogan's PIES example and Lemke's algorithm,* Technical Summary Report No. 1972, Mathematics Research Center, University of Wisconsin–Madison, June 1979, available from National Technical Information Service, Springfield, VA 22161, under Accession No. A077 103.

[474] Kanno, S., Convergence theorems for the method of tangent hyperbolas, *Math. Japon.,* **37**, 4, (1992), 711–722.

[475] Kantorovich, L.V., On Newton's method for functional equations (Russian), *Dokl. Akad. Nauk. SSSR,* **59**, (1948), 1237–1240.

[476] Kantorowitsch, L.W., Akilow, G.P., *Funktionalanalysis in normierten Räumen.* (German) Übersetzt aus dem Russischen von Heinz Langer und Rolf Kühne. In deutscher Sprache herausgegeben von P. Heinz Müller, Verlag Harri Deutsch, Thun–Frankfurt am Main, 1978.

[477] Kantorovich, L.V., Akilov, G.P., *Functional Analysis,* Pergamon Press, Oxford, 1982.

[478] Kelley, C.T., *Identification of the support of nonsmoothness,* in Large scale optimization (Gainesville, FL, 1993), Kluwer Acad. Publ., Dordrecht, 1994, 192–205.

[479] Kelley, C.T., Sachs, E.W., Multilevel algorithms for constrained compact fixed point problems, Iterative methods in numerical linear algebra (Copper Mountain Resort, CO, 1992), *SIAM J. Sci. Comput.,* **15** (1994), 645–667.

[480] Kikuchi, F., Finite element analysis of a nondifferentiable nonlinear problem related to MHD equilibria, *J. Fac. Sci. Univ. Tokyo Sect. IA Math.,* **35** (1988), 77–101.

[481] Kinderlehrer, D., Stampacchia, G., *An introduction to variational inequalities and their applications,* Academic Press, Editors: S. Eilenberg and H. Bass, 1980.

[482] King, A.J., Rockafellar, R.T., Sensitivity analysis for nonsmooth generalized equations, *Math. Programming,* **55**, (1992), 193–212.

[483] King, R.F., Tangent methods for nonlinear equations, *Numer. Math.,* **18**, (1972), 298–304.

[484] Klatte, D., Kummer, B., Nonsmooth equations in optimization. Regularity, calculus, methods and applications. *Nonconvex Optimization and its Applications,* **60**, Kluwer Academic Publishers, Dordrecht, 2002.

322   References

[485] Klatte, D., Kummer, B., Newton methods for stationary points: an elementary view of regularity conditions and solution schemes, *Optimization,* **56**, (4), (2007), 441–462.

[486] Klatte, D., Kummer, B., Optimization methods and stability of inclusions in Banach spaces, *Math. Programming, Ser. B*, **117**, (1–2), (2009), 305–330.

[487] Kojima, M., Shindo, S., Extensions of Newton and quasi–Newton methods to systems of PC^1 equations, *J. Oper. Res. Soc. Japan,* **29**, (1986), 352–374.

[488] Kornstaedt, H.J., Ein allgemeiner Konvergenzstaz für verschärfte Newton–Verfahrem, *ISNM* **28**, Birkhaüser Verlag, Basel and Stuttgart, 1975, 53–69.

[489] Krasnosel'skii, M.A., *Positive solutions of operator equations,* Goz. Isdat. Fiz. Mat. Moscow 1962; Transl. by R. Flaherty and L. Boron, P. Noordhoff, Groningen, 1964.

[490] Krasnosel'skii, M.A., *Topological Methods in the Theory of Nonlinear Integral Equations*, Pergamon Press, London, 1966.

[491] Krasnosel'skii, M.A., Rutitskii, Ya.B., On the theory of equations with concave operators, *SMZh*, **10**, 3, (1969).

[492] Krasnosel'skii, M.A., Vainikko, G.M., Zabrejko, P.P., Rutiskii, Ya.B.,Stetsenko, V.Ya., *Approximate Solution of Operator Equations*, Wolters–Noordhoff Publishing, Groningen, 1972.

[493] Krasnosel'skii, M.A., Zabrejko, P.P., *Geometrical Methods of Nonlinear Analysis*, Springer–Verlag, New–York, 1984.

[494] Kung, H.T., *The complexity of obtaining starting points for solving operator equations by Newton's metho*d, Technical Report NR 044–422, Department of Computer Science, Carnegie–Mellon University, Pittsburgh, PA, 15213, October, 1975.

[495] Kuratowski, C., Sur les espaces complets, *Fund. Math.,* **15**, (1930), 301–309.

[496] Kurchatov, V.A., Optimization with respect to the number of arithmetic operations and entropy of difference methods of linearization (Russian), *Izv. Vyssh. Uchebn. Zaved.*, **4**, (1990), 33–37.

[497] Kwon, U.K., Redheffer, R.M., Remarks on linear equations in Banach space, *Arch. Rational Mech. Anal.,* **32**, (1969), 247–254.

[498] Laasonen, P., Ein überquadratisch konvergenter iterativer algorithmus, *Ann. Acad. Sci. Fenn. Ser I*, **450**, (1969), 1–10.

[499] Lala, P.K., Age –Specific Changes in the proliferation of Ehrlich Ascites Tumor Cells Grown as Solid Tumors, *Cancer research*, **32**, (1972), 628–636.

[500] Lancaster, P., Error analysis for the Newton–Raphson method, *Numer. Math.,* **9**, (1968), 55–68.

[501] Laumen, M., Newton's mesh independence principle for a class of optimal design problems, *SIAM J. Control Optim.*, **37**, (1999), 1070–1088.

[502] Laumen, M., A Kantorovich theorem for the structured PSB update in a Hilbert space, *J. Optim. Theory Appl.*, **105**, 2, (2000), 391–415.

[503] Levenberg, K., A method for the solution of certain problems in least squares, *Quarterly Journal of Applied Mathematics*, **2**, (1944), 164–168.

[504] Levin, Y., Ben–Israel, A., Directional Newton methods in n variables, Mathematics of Computation, *A.M.S.*, **71**, 237, (2002), 251–262.

[505] Levitin, E.S., *Perturbation Theory in Mathematical Programming and Its Applications*, Wiley Interscience Series in Discrete Mathematics and Optimization, Wiley and Sons, New York, 1982.

[506] Lewis, S.R., *Challenger the Final Voyage*, New York, Columbia University Press, 1988.

[507] Lewy, H., Stampacchia, G., On the regularity of the solution of a variational inequality, *Comm. Pure Appl. Math.*, **22**, (1969), 153–188.

[508] Li, C., Zhang, W–H., Jin, X–Q., Convergence and uniqueness properties of Gauss-Newton's method, *Comput. Math. Appl.*, **47**, (2004), 1057–1067.

[509] Lions, J.–L., Stampacchia, G., Variational inequalities, *Comm. Pure Appl. Math.*, **20**, (1967), 493–519.

[510] Liskovetz, O.A., *Variational methods for solving nonstable problems*, Nauka Technika, Minsk, (1981), (in Russian).

[511] Liusternik, L.A., Sobolev, V.J., *Elements of functional analysis*, Ungar Publ., 1961.

[512] Lukács, G., The generalized inverse matrix and the surface–surface intersection problem. *Theory and practice of geometric modeling* (Blaubeuren, 1988), 167–185, Springer, Berlin, 1989.

[513] Lysenko, J.V., Conditions for the convergence of the Newton–Kantorovich method for nonlinear equations with Hölder linearizations, *Dokl. Akad. Nauk BSSR*, **38**, (1994), 20–24 (in Russian).

[514] Mann, W.R., Mean value methods in iteration, *Proc. Amer. Math. Soc.*, **4**, (1953), 506–510.

[515] Marcotte, P., Wu, J.H., On the convergence of projection methods, *J. Optim. Theory Appl.*, **85**, 2, (1995), 347–362.

[516] Marquardt, D.W., An algorithm for least–squares estimation of nonlinear parameters, *J. Soc. for Industrial and Appl. Math.*, **11**, (1963), 431–441.

[517] Matveev, A.F., Yunganns, P., On the construction of an approximate solution of a nonlinear integral equation of permeable profile (Russian), Differ. Uravn., 33 (9), (1997) 1242–1252, 1295. *Translation in Differential Equations,* **33**, (1997), no. 9, 1249–1259, (1998).

[518] McCormick, S.F., A revised mesh refinement strategy for Newton's method applied to nonlinear two–point boundary value problems, *Numer. Treat. Diff. Equ. Appl. Proc.*, **679**, Lecture Notes Math., Springer, (1978), 15–23.

[519] McCullagh, P., Nelder, J.A., *Generalized linear models,* Second Edition, Chapman & Hall, 1989, London.

[520] Meyer, P.W., *Die Anwendung Verallgemeinerter Normen zer Fehlerabschätzung Bei Iteration Sverfahren*, Dissertation, Düsseldort, 1980.

[521] Meyer, P.W., Das modifizierte Newton–Verfahren in verallgemeinerten Banach-Räumen, *Numer. Math.,* **43**, 1, (1984), 91–104.

[522] Meyer, P.W., Newton's method in generalized Banach spaces, *Numer. Funct. Anal. Optim.,* **9**, (3 and 4), (1987), 244–259.

[523] Meyer, P.W., A unifying theorem on Newton's method, *Numer. Funct. Anal. Optim.,* **13**, (5–6), (1992), 463–473.

[524] Miel, G.J., Unified error analysis for Newton–type methods, *Numer. Math.,* **33** (1979), 391–396.

[525] Miel, G.J., Majorizing sequences and error bounds for iterative methods, *Math. Comp.,* **34**, 149, (1980), 185–202.

[526] Mifflin, R., Semismooth and semiconvex functions in constrained optimization, *SIAM J. Control Optimization,* **15** (1977), 959–972.

[527] Migovich, F.M., On the convergence of projection–iterative methods for solving non-linear operator equations, *Dopov. Akad. Nauk. Ukr. RSR, Ser. A,* **1**, (1970), 20–23.

[528] Minty, G.J., Monotone (nonlinear) operators in Hilbert space, *Duke Mathematics Journal,* **29**, (1973), 341–346.

[529] Minty, G.J., On the monotonicity of the gradient of a convex function, *Pacific Journal of Mathematics,* **14**, (1964), 243–247.

[530] Miranda, C., Un osservatione su un teorema d, Brouwer, *Ball. Unione Mat. Ital., Serr.* **11**, 3, (1940), 5–7.

[531] Mirsky, L., *An Introduction to linear algebra,* Clarendon Press, Oxford, England, 1955.

[532] Moore, R.H., Approximations to nonlinear operator equations and Newton's method, *Numer. Math.,* **12**, (1968), 23–34.

References 325

[533] Moore, R.E., A test for existence of solutions to nonlinear systems, *SIAM J. Numer. Anal.*, **14**, 4, (1977), 611–615.

[534] Moore, R.E., *Methods and applications of interval analysis,* SIAM Publications, Philadelphia, Pa., 1979.

[535] Mordukhovich, B.S., Complete characterization of openness metric regularity and Lipschitzian properties of multifunctions, *Trans. Amer. Math. Soc.*, **340**, (1993), 1–36.

[536] Mordukhovich, B.S., Stability theory for parametric generalized equations and variational inequalities via nonsmooth analysis, *Trans. Amer. Math. Soc.*, **343**, (1994), 609–657.

[537] Mordukhovich, B.S., Coderivatives of set–valued mappings: calculs and applications, *Nonlinear Anal. Theo. Meth. and Appl.*, **30**, (1997), 3059–3070.

[538] Mordukhovich, B.S., *Variational analysis and generalized differentiation*, I. Basic theory, II. Applications (two volumes), Springer, Grundlehren Series, Vol. 330 and 331, 2006.

[539] Moret, I., A note on Newton–type iterative methods, *Computing* **33**, (1984), 65–73.

[540] Moret, I., On the behaviour of approximate Newton methods, *Computing*, **37**, 3, (1986), 185–193.

[541] Moret, I., On a general iterative scheme for Newton–type methods, *Numer. Funct. Anal. Optim.*, **9**, (10–12), (1987/1988), 1115–1137.

[542] Moret, I., A Kantorovich–type theorem for inexact Newton methods, *Numer. Funct. Anal. Optim.*, **10**, (3 and 4), (1989), 351–365.

[543] Morini, B., Convergence behaviour of inexact Newton methods, *Math. Comp.*, **68** (1999), 1605–1613.

[544] Moser, J., Stable and random motions in dynamical systems with special emphasis on celestial mechanics, Herman Weil Lectures, *Annals of Mathematics Studies* **77**, Princeton Univ. Press, Princeton, NJ, 1973.

[545] Moudafi, A., Proximal point algorithm extended to equilibrium problems, *J. Nat. Geom.*, **15**, (1–2), (1999), 91–100.

[546] Moudafi, A., Second–order differential proximal methods for equilibrium problems, *JIPAM. J. Inequal. Pure Appl. Math.*, **4**, (1), (2003), Article 18, 7 pp., (electronic).

[547] Moudafi, A., A perturbed inertial proximal method for maximal monotone operators, *Comm. Appl. Nonlinear Anal.*, **11**, (3), (2004), 101–107.

[548] Moudafi, A., A hybrid inertial projection–proximal method for variational inequalities, *JIPAM. J. Inequal. Pure Appl. Math.*, **5**, (3), (2004), Article 63, 5 pp., (electronic).

[549] Moudafi, A., On finite and strong convergence of a proximal method for equilibrium problems, *Numer. Funct. Anal. Optim.,* **28**, (11–12), (2007), 1347–1354.

[550] Mukaidani, H., Shimomura, T., Mizukami, K., Asymptotic expansions and a new numerical algorithm of the algebraic Riccati equation for multiparameter singularly perturbed systems, *J. Math. Anal. Appl.,* 267, (1), (2002), 209–234.

[551] Muroya, Y., Practical monotonous iterations for nonlinear equations, *Mem. Fac. Sci. Kyushu Univ., Ser. A,* **22**, (1968), 56–73.

[552] Muroya, Y., Left subinverses of matrices and monotonous iterations for nonlinear equations, *Memoirs of the Faculty of Science and Engineering,* Waseda University, **34**, (1970), 157–171.

[553] Mysovskii, I., On the convergence of Newton's method, *Trudy Mat. Inst. Steklov,* **28**, (1949), 145–147 (in Russian).

[554] Nagatou, K., Yamamoto, N., Nakao, M.T., An approach to the numerical verification of solutions for nonlinear elliptic problems with local uniqueness, *Numer. Funct. Anal. Optim.,* **20**, (5–6), (1999), 543–565.

[555] Nayakkankuppam, M.V., Overton, M.L., Conditioning for semidefinite programs, *Math. Program.,* **85**, (3, Ser. A), (1999), 525–540.

[556] Necepurenko, M.T., On Chebysheff's method for functional equations (Russian), *Usephi, Mat. Nauk,* **9**, (1954), 163–170.

[557] Nerekenov, T.K., *Necessary and sufficient conditions for uryson and nemytskii operators to satisfy a Lipschitz condition* (Russian), VINITI 1459, 81, Alma–Ata, (1981).

[558] Y. Nesterov, Y., Nemirovskii, A., Interior–point polynomial algorithms in convex programming, *SIAM Studies in Appl. Math.,* **13**, Philadelphia, PA, (1994).

[559] Nesterov, Y., Nemirovsky, A., Interior Point Polynomial Methods in Convex Programming, *SIAM,* Philadelphia, 1994.

[560] Neter, J., Wasserman, W., Kunter, M.H., *Applied Linear Statistical Models,* Homewood, Illinois: Irwin, 1985.

[561] Neumaier, A., Shen, Z., The Krawczyk operator and Kantorovich's theorem, *J. Math. Anal. Appl.,* **149**, 2, (1990), 437–443.

[562] Noble, B., *The numerical solution of nonlinear integral equations and related topics,* University Press, Madison, WI, (1964).

[563] Noor, K.I., Noor, M.A., *Iterative methods for a class of variational inequalities, in Numerical Analysis of singular perturbation problems,* Hemker and Miller, eds., Academic Press, New–York, (1985), 441–448.

[564] Noor, M.A., An iterative scheme for a class of quasivariational inequalities, *J. Math. Anal. Appl.,* **110**, 2, (1985), 463–468.

[565] Noor, M.A., Generalized variational inequalities, *Appl. Math. Letters*, **1**, (1988), 119–122.

[566] Ojnarov, R., Otel'baev, M., A criterion for a Uryson operator to be a contraction (Russian), *Dokl. Akad. Nauk. SSSR,* **255**, (1980), 1316–1318.

[567] Okuguchi, K., *Expectations and stability in oligopoly models,* Springer–Verlag, New York, (1976).

[568] Ortega, L.M., Rheinboldt, W.C., Iterative Solution of *Nonlinear Equations in Several Variables,* Academic press, New York, 1970.

[569] Ostrowski, A.M., *Solution of Equations in Euclidean and Banach Spaces*, Academic press, New York, 1973.

[570] Owren, B., Welfert, B., The Newton Iteration on Lie Groups, *BIT*, **40**, 2, (2000), 121–145.

[571] Özban, A.Y., Some new variants of Newton's method, *Appl. Math. Letters*, **17**, (2004), 677–682.

[572] Paardekooper, M.H.C., An upper and a lower bound for the distance of a manifeld to a nearby point, *J. Math. Anal. Appl.*, **150**, (1990), 237–245.

[573] Pandian, M.C., A convergence test and componentwise error estimates for Newton–type methods, *SIAM J. Num. anal.,* 22, (1985), 779–791.

[574] Pang, Jong–Shi, Newton's method for B–differentiable equations, *Math. Oper. Res.*, **15**, 2, (1990), 311–341.

[575] Păvăloiu, I., Sur la méthode de Steffensen pour la résolution des équations opérationnelles non linéaires, Rev. *Roumaine Math. Pures Appl.,* **13**, 6, (1968), 857–861.

[576] Păvăloiu, I., *Introduction in the theory of approximation of equations solutions,* Dacia Ed., Cluj–Napoca, 1976.

[577] Păvăloiu, I., *Rezolvarea equaţiilor prin interpolare.* Dacia Publ. cluj–Napoca, Romania, (1981).

[578] Păvăloiu, I., *On the convergency of a Steffensen–type method,* Seminar on Mathematical Analysis, 121–126, Preprint, 91–7, "Babeş–Bolyai" Univ., Cluj–Napoca, 1991.

[579] Păvăloiu, I., A convergence theorem concerning the method of chord, Revue d'Analyse *Numérique et de Théorie de l'Approximation,* **21**, 1, (1992), 59–65.

[580] Peng, Ji–M., Kanzow, C., Fukushima, M., A hybrid Josephy–Newton method for solving box constrained variational inequality problems via the D–gap function, *Optim. Methods Softw,* **10** (1999), 687–710.

References

[581] Pereyra, V., Iterative methods for solving nonlinear least square problems, *SIAM J. Numer. Anal.*, **4**, (1967), 27–36.

[582] Pernia–Espinoza, A.V., Joaquin, B., Martinez–de–Pison, O–M.F.J., Gonzalez–Marcos, A., TAO–Robust Backpropagation Learning Algorithm, *NeuralNetworks,* **18**, (2005), 191–204.

[583] Perrone, J.A., Chabla, J.M., Hallas, B.H., Horowitz, J.M., Torres, G., Weight Loss Dynamics During Combined Fluoxetine and Olanzapine treatment, *BMC Pharmocology,* **4**: 27, (2004).

[584] Petryshin, W.V., Williamson, T.E., Strong and weak convergence of the sequence of successive approximation for quasi–nonexpansive mappings, *J. Math. Anal. Appl.,* **43**, (1973), 459–497.

[585] Petzeltova, H., Remark on a Newton–Moser type method, *Commentationes Mathematicae Universitatis Carolinae,* **21**, (1980), 719–725.

[586] Piétrus, A., Generalized equations under mild differentiability conditions, *Rev. Real. Acad. Ciencias de Madrid,* **94 (1)**, (2000), 15–18.

[587] Piétrus, A., Does Newton's method converges uniformly in mild differentiability context?, *Rev. Colombiana Math.,* **32**, (2), (2000), 49–56.

[588] Polyak, B.T., *Introduction to optimization.* Translated from the Russian. With a foreword by D.P. Bertsekas. Translations Series in Mathematics and Engineering. Optimization Software, Inc., Publications Division, New York, 1987.

[589] Potra, F.A., On a modified secant method, *Revue d'Analyse Numérique et la Théorie de l'Approximation,* **8**, 2, (1979), 203–214.

[590] Potra, F.A., An application of the induction method of V. Pták to the study of Regula Falsi, *Aplikace Matematiky,* **26**, (1981), 111–120.

[591] Potra, F.A., On the convergence of a class of Newton–like methods, Iterative Solution of Nonlinear Systems of Equations, (Oberwolfach, 1982), *Lecture Notes in Math.,* **953**, Springer, Berlin–New York, 1982, 125–137.

[592] Potra, F.A., On an iterative algorithm of order 1.839... for solving nonlinear equations, *Numer. Funct. Anal. Optim.,* **7**, 1, (1984–85), 75–106.

[593] Potra, F.A., Sharp error bounds for a class of Newton–like methods, *Libertas Mathematica,* **5,** (1985), 71–84.

[594] Potra, F.A., The Kantorovich theorem and interior point methods, *Math. Progr.,* **102**, 1, (2005), 47–70.

[595] Potra, F.A., Pták, V., Nondiscrete induction and double step secant method, *Math. Scand.,* **46**, (1980), 236–250.

References

[596] Potra, F.A., Pták, V., On a class of modified Newton processes, *Numer. Func. Anal. Optim.*, **2**, 1, (1980), 107–120.

[597] Potra, F.A., Pták, V., Sharp error bounds for Newton's process, *Numer. Math.*, **34**, (1980), 67–72.

[598] Potra, F.A., Pták, V., Nondiscrete induction and an inversion–free modification of Newton's method, *Casopis pro pestováni matematiky,* **108**, (1983), 333–341.

[599] Potra, F.A., Pták, V., Nondiscrete induction and iterative processes, *Research Notes in Mathematics,* **103**, Pitman Advanced Publishing program, Boston, 1984.

[600] Potra, F.A., Wright, S.J., Interior–point methods, *J. Comp. Appl. Math.*, **124**, (2000), 281–302.

[601] Proinov, P.D., General local convergence theory for a class of iterative processes and its applications to Newton's method, *J. Complexity,* **25**, (2009), 38–62.

[602] Proinov, P.D., New general convergence theory for iterative processes and its applications to Newton–Kantorovich type theorems, *J. Complexity,* **26**, (2010), 3–42.

[603] Pták, V., The rate of convergence of Newton's process., *Numer. Math.*, **25**, 3, (1976), 279–285.

[604] Pousin, J., Rappaz, J., Consistency, stability, a priori and a posteriori errors for Petrov-Galerkin's method applied to nonlinear problems, *Numer. Math.*, **69**, (1994), 213–231.

[605] Qi, L., Sun, J., A nonsmooth version of Newton's method, *Math. Progr.*, **58**, (1993), 353–367.

[606] Rall, L.B., *Computational solution of nonlinear operator equations,* Wiley, New–York, (1968).

[607] Rall, L.B., *Nonlinear functional analysis and applications,* Academic Press, New York, (1971).

[608] Rall, L.B., A comparison of the existence theorems of Kantorovich and Moore, *SIAM J. Numer. Anal.,* **17**, 1, (1980), 148–161.

[609] Ralph, D., *On branching numbers of normal manifolds, Technical Report 92–1283,* Department of Computer Science, Cornell University, Ithaca, NY, May 1992.

[610] Ralston, A., Rabinowitz, P., *A first course in numerical analysis,* 2nd Edition, Mc Graw–Hill, 1978.

[611] Rappaz, J., Approximation of a nondifferentiable nonlinear problem related to MHD equilibria., *Numer. Math.,* **45**, (1984), 117–133.

[612] Redheffer, R., Walter, W., A comparison theorem for difference inequalities, *J. Diff. Eq.,* **44**, (1982), 111–117.

[613] Renegar, J., A polynomial–time algorithm, based on Newton's method, for linear programming, *Math. Progr.*, **40**, 1, Ser. A, (1988), 59–93.

[614] Renegar, J., Shub, M., Unified complexity analysis for Newton LP methods, *Math. Progr.*, **53**, 1, Ser. A, (1992), 1–16.

[615] Rheinboldt, W.C., A unified convergence theory for a class of iterative processes, *SIAM J. Numer. Anal.*, **5**, (1968), 42–63.

[616] Rheinboldt, W.C., An adaptive continuation process for solving systems of nonlinear equations, *Polish Academy of Science, Banach Ctr. Publ.*, **3**, (1977), 129–142.

[617] Rheinboldt, W.C., Solution fields of nonlinear equations and continuation methods, *SIAM J. Numer. Anal.*, **17**, (2), (1980), 221–237.

[618] Rheinboldt, W.C., Error estimation for nonlinear parametrized problems, *Proceedings of the international conference on innovative methods for nonlinear problems*, (New Orleans, La., 1984), Pineridge, Swansea, (1984), 295–311.

[619] Rheinboldt, W.C., On a theorem of S. Smale about Newton's method for analytic mappings, *Appl. Math. Lett.*, **1**, (1), (1988), 69–72.

[620] Rheinboldt, W.C., On the sensitivity of solutions of parameterized equations, *SIAM J. Numer. Anal.*, **30**, (2), (1993), 305–320.

[621] Rheinboldt, W.C., On the theory and error estimation of the reduced basis method for multi–parameter problems, *Nonlinear Anal.*, **21**, (11), (1993), 849–858.

[622] Rheinboldt, W.C., Methods for solving systems of nonlinear equations, Second edition, CBMS–NSF, *Regional Conference Series in Applied Mathematics,* **70**, Society for Industrial and Applied Mathematics (SIAM), Philadelphia, PA, 1998.

[623] Rheinboldt, W.C., Numerical continuation methods: a perspective, Numerical analysis, 2000, Vol. IV, Optimization and nonlinear equations, *J. Comput. Appl. Math.*, **124**, (1–2), (2000), 229–244.

[624] Robinson, S.M., Generalized equations and their solutions, part I: basic theory, *Math. Programming Study*, **10**, (1979), 128–141.

[625] Robinson, S.M., Strongly regular generalized equations, *Mathematics of Operations Research,* **5,** (1980), 43–62.

[626] Robinson, S.M., Generalized equations and their solutions, part II: applications to nonlinear programming, *Math. Programming Study*, **19**, (1982), 200–221.

[627] Robinson, S.M., Generalized equations, in: A. Bachem, M. Grötschel and B. Korte, eds., *Mathematical programming: the state of the art* (Springer, Berlin), (1982), 346–367.

[628] Robinson, S.M., Generalized equations. *Mathematical programming: the state of the art* (Bonn, 1982), 346–367, Springer, Berlin, 1983.

References 331

[629] Robinson, S.M., *Newton's method for a class of nonsmooth functions*, *Set–Valued Anal.*, **2**, (1994), 291–305.

[630] Robinson, S.M., Nonsmooth continuation for generalized equations, Recent advances in optimization (Trier, 1996), 282–291, *Lecture Notes in Econom. and Math. Systems*, **452**, Springer, Berlin, 1997.

[631] Robinson, S.M., A reduction method for variational inequalities, *Math. Programming*, **80**, (1998), 161–169.

[632] Robinson, S.M., Structural methods in the solution of variational inequalities, Nonlinear optimization and related topics (Erice, 1998), 369–380, *Appl. Optim.*, **36**, Kluwer Acad. Publ., Dordrecht, 2000.

[633] Robinson, S.M., Variational conditions with smooth constraints: structure and analysis, *Math. Programming, Ser. B*, **97**, (2003), 245–265.

[634] Robinson, S.M., Solution continuity in monotone affine variational inequalities, *SIAM J. Optim.*, **18**, (3), (2007), 1046–1060.

[635] Robinson, S.M., Lu, S., Solution continuity in variational conditions, *J. Glob. Optim.*, **40**, (2008), 405–415.

[636] Rockafellar, R.T., *Convex analysis*, Princeton University Press, Princeton, 1967.

[637] Rockafellar, R.T., Lipschitzian properties of multifunctions, *Nonlinear Analysis*, **9**, (1984), 867–885.

[638] Rockafellar, R.T., Wets, R. J–B., Variational analysis, *A Series of Comprehensives Studies in Mathematics*, Springer, 317, 1998.

[639] Rokne, J., Newton's method under mild differentiability conditions with error analysis, *Numer. Math.*, **18**, (1972), 401–412.

[640] Roos, C., Vial, J.–Ph., Terlaky, T., *Theory and algorithms for linear optimization: an interior point approach*, Wiley Interscience Series in Discrete Mathematics and Optimization, John Wiley and Sons, 1997.

[641] Rousseeuw, P.J., and Leroy, A.M., *Robust Regression and Outlier Detection*, New York: John Wiley, 1987.

[642] Schomber, H., Monotonically convergent iterative methods for nonlinear systems of equations, *Numer. Math.*, **32**, (1979), 97–104.

[643] Schmidt, J.W., Eine übertragung der Regula Falsi and gleichungen in Banachräumen, I, *Z. Angew. Math. Mech.*, **43**, (1963), 97–110.

[644] Schmidt, J.W., Untere Fehlerschranken fur Regula–Falsi Verhafren, *Period. Hungar.*, **9**, (1978), 241–247.

References

[645] Schmidt, J.W., Leonhardt, H., Eingrenzung von Losungen mit der Regala Falsi, *Computing*, **6**, (1970), 318–329.

[646] Schmidt, J.W., Schwetlick, H., Ableitungsfreie Verfahren mit hohere Konvergenzgeschwindigkeit, *Computing*, **3**, (1968), 215–226.

[647] Schömig, A., Kastrati, A., Elisi, S., Schhlen, H., Dirschinger, J., Dannegger, F., Whilhelm, M., Ulm, K., Bimodal Distribution of Angiographic Measures of Restenosis Six Months After Coronary Stent Placement, *Circulation*, **96**, (1997), 3880–3887.

[648] Schröder, J., Nichtlineare majoranten beim verfahren der schrittweisen näherung, *Arch. der Math.*, **7,** (1956), 471–484.

[649] Seeman, P., Ulpian, C., Bergeron, C., Riederer, P., Jellinger, K., Gabriel, E., Reynolds, G.P., Tourtellotte, W.W., Bimodal Distribution of Dopamine Receptor in Brains of Schizophrenics, *Science*, **225**, (4663), (1984), I 728–731.

[650] Shapiro, A., On concepts of directional differentiability, *J. Optim. Theory Appl.*, **66**, (1990), 477–487.

[651] Shapiro, A., Sensitivity analysis of parametrized programs via generalized equations, *SIAM J. Control Optim.*, **32**, (2), (1994), 553–571.

[652] Shapiro, A., A variational principle and its applications, Parametric optimization and related topics, IV (Enschede, 1995), *Approx. Optim.*, **9**, Lang, Frankfurt am Main, (1997), 341–357.

[653] Shapiro, A., Sensitivity analysis of generalized equations, Optimization and related topics, 1, *J. Math. Sci. (N. Y.)*, **115**, (4), (2003), 2554–2565.

[654] Shapiro, A., Sensitivity analysis of parameterized variational inequalities, *Math. Oper. Res.*, **30**, (1), (2005), 109–126.

[655] Sherley, J.L., Stadler, P.B., Stadler, J.S., A Quantitative Method for the Analysis of Mammalian Cell Proliferation in Cultuure in Terms of Dividing and Dividing Cells, *Cell Prolif,* **28**, (1995), 137–144.

[656] Shub, M., Smale, S., Complexity of Bezout's theorem 1: geometric aspects, *J. Amer. Math. Soc.*, **6,** (1993), 459–501.

[657] Slugin, S.N., Approximate solution of operator equations on the basis of Caplygin method, (Russian), *Dokl. Nauk SSSR,* **103**, (1955), 565–568.

[658] Slugin, S.N., Monotonic processes of bilateral approximation in a partially ordered convergence group, *Soviet. Math.*, **3**, (1962), 1547–1551.

[659] Smale, S., Differentiable dynamical systems, *Bull. Amer. Math. Soc.,* **73**, (1967), 747–817.

References 333

[660] Smale, S., Newton method estimates from data at one point. *The merging of disciplines: new directions in pure, applied and computational mathematics* (R. Ewing, K. Gross, C. Martin, eds), Springer–Verlag, New York, (1986), 185–196.

[661] Smale, S., Newton's contribution and the computer revolution, *Math. Medley,* **17**, (2), (1989), 51–57.

[662] Solodov, M.V., Svaiter, B.F., A globally convergent inexact Newton method for systems of monotone equations, Reformulation: nonsmooth, piecewise smooth, semismooth and smoothing methods, (Lausanne, 1997), *Appl. Optim.,* **22**, Kluwer Acad. Publ., Dordrecht, (1999), 355–369.

[663] Solodov, M.V., Svaiter, B.F., A truly globally convergent Newton–type method for the monotone nonlinear complementarity problem, *SIAM Journal on Optimization,* **10**, (2000), 605–625.

[664] Solodov, M.V., Svaiter, B.F., A new proximal–based globalization strategy for the Josephy–Newton method for variational inequalities, *Optim. Methods Softw.,* **17**, (5), (2002), 965–983.

[665] Stakgold, I., *Green's Functions and Boundary Value Problems,* Wiley, New York, 1998.

[666] Stampacchia, G., Formes bilinéaires coercitives sur les ensembles convexes, *Comptes Rendus de L'Académie des Science de Paris,* **258**, (1964), 4413–4416.

[667] Stampacchia, G., Variational inequalities, Theory and Applications of Monotone Operators, (*Proc. NATO Advanced Study Inst.,* Venice, 1968), Edizioni "Oderisi", Gubbio, (1969), 101–192.

[668] Stampacchia, G., Regularity of solutions of some variational inequalities, Nonlinear Functional Analysis, (*Proc. Sympos. Pure Math.,* Vol. XVIII, Part 1, Chicago, Ill., 1968), Amer. Math. Soc., Providence, R.I., (1970), 271–281.

[669] Stampacchia, G., On the regularity of solutions of variational inequalities, *Proc. Internat. Conf. on Functional Analysis and Related Topics,* (Tokyo, 1969), Univ. of Tokyo Press, Tokyo, (1970), 285–289.

[670] Stirling, J., *Methodus differentialis: sive tractatus de summatione et interpolations serierum infiniterum,* W. Boyer, London, 1730.

[671] Stoer, J., Bulirsch, K., Introduction to numerical analysis, Springer–Verlag, 1976.

[672] Stoffer, D., Palmer, K.J., Rigorous verification of chaotic behaviour of maps using validated shadowing, *Nonlinearity,* **12**, (6), (1999), 1683–1698.

[673] D. Stoffer, U. Kirchgraber, Verification of chaotic behaviour in the planar restricted three body problem, *Appl. Numer. Math.* **39**, (2001) nos. 3–4, 415–433.

[674] Szidarovszky, F., Yakowitz, S., *Principles and procedures in numerical analysis,* Plenum, New–York, 1978.

References

[675] Szidarovszky, F., Bahill, T., *Linear systems theory*, CRC Press, Boca Raton, FL, 1992.

[676] Tabatabai, M.A., Argyros, I.K., Robust estimation and testing for general nonlinear regression models, *Appl. Math. Comput.*, **58**, (1993), 85–101.

[677] Tabatabai, M., Williams, D.K., Bursac, Z., Hyperbolastic Growth Models: Theory and Application, *Theoretical Biology and Medical Modeling*, **2**, 2005, 1–13.

[678] Tan, K.K., Xu, H.K., Approximating fixed points of nonexpansive mappings by the Ishikawa iteration process, *J. Math. Anal. Appl.*, **178**, (1993), 301–308.

[679] Tan, W.Y. and Tabatabai, M.A., A modified winsorized regression procedure for linear models, *J. Statistical Comput. and Simul.*, **30**, (1988), 299–313.

[680] Tapia, R.A., The Kantorovich theorem for Newton's method, *Amer. Math. Monthly*, **11**, (1971), 10–13.

[681] Tapia, R.A., Newton's method for optimization problems with equality constraints, *SIAM J. Numer. Anal.*, **11**, (1974), 874–886.

[682] Tapia, R.A., A general approach to Newton's method for Banach space problems with equality constraints, *Bull. Amer. Math. Soc.*, **80**, (1974), 355–360.

[683] Tapia, R.A., A stable approach to Newton's method for general mathematical programming problems in $\mathbb{R}^n$, Collection of articles dedicated to Magnus R. Hestenes., *J. Optimization Theory Appl.*, **14**, (1974), 453–476.

[684] Taylor, A.E., *Introduction to functional analysis*, Wiley, New York, (1957).

[685] Tishyadhigama, S., Polak, E., Klessig, R., A comparative study of several convergence conditions for algorithms modeled by point–to–set maps, *Math. Programming Stud.*, **10**, (1979), 172–190.

[686] Törnig, W., Monoton konvergente Iterationsverfahren zür Lösung michtlinearer differenzen–randwertprobleme, *Beiträge zür Numer. Math.*, **4**, (1975), 245–257.

[687] Traub, J.F., *Iterative Methods for the Solution of Equations,* Englewood Cliffs, New Jersey: Prentice Hall, 1964.

[688] Traub, J.F. (editor), *Analytic Computational Complexity*, Academic Press, 1975.

[689] Traub, J.F., Woźniakowsi, H., Convergence and complexity of Newton iteration for operator equations, *J. Assoc. Comput. Mach.*, **26**, (2), (1979), 250–258.

[690] Traub, J.F., Woźniakowsi, H., A general theory of optimal algorithms, *ACM Monograph Series,* Academic Press, Inc., Harcourt Brace Jovanovich, Publishers, New–York, London, 1980.

[691] Traub, J.F., Woźniakowsi, H., Optimal radius of convergence of interpolatory iterations for operator equations, *Aequationes Math.*, **21**, (2–3), (1980), 159–172.

References   335

[692] Traub, J.F., Woźniakowsi, H., Convergence and complexity of interpolatory–Newton iteration in a Banach space, *Comput. Math. Appl.*, **6**, (4), (1980), 385–400.

[693] Tsuchiya, T., An application of the Kantorovich theorem to nonlinear finite element analysis, *Numer. Math.*, **84**, (1999), 121–141.

[694] Tsuchiya, T., Babuska, I., A priori error estimates of finite element solutions of parametrized strongly nonlinear boundary value problems, *J. Comp. Appl. Math.*, **79**, (1997), 41–66.

[695] Uko, L.U., On a class of general strongly nonlinear quasivariational inequalities, *Revista di Matematica Pura ed Applicata*, **11**, (1992), 47–55.

[696] Uko, L.U., Generalized equations and the generalized Newton method, *Mathematical Programming*, **73**, (1996), 251–268.

[697] Uko, L.U., Adeyeye, J.O., Generalized Newton–iterative methods for nonlinear operator equations, *Nonlinear Studies*, **8**, 4, (2001), 465–477.

[698] Uko, L.U., Orozco, J.C., Some p–norm convergence results for Jacobi and Gauss–Seidel iterations, *Revista Colombiana de Matemáticas*, **38**, (2004), 65–71.

[699] Uko, L.U., Velásquez Ossa, R.E., Convergence Analysis of a one–step intermediate Newton iterative scheme, *Revista Colombiana de Matematicas*, **35**, (2001), 21–27.

[700] Ulm, S. Ju., A generalization of Steffensen's method for solving non–linear operator equations, (Russian), *Ž. Vyčisl. Mat. i Mat. Fiz.*, **4** (1964), 1093–1097.

[701] Ulm, S. Ju., On iterative methods with successive approximation of the inverse operator (in Russian), *Tzv. Akad. Nauk Est. SSR*, **16**, (1967), 403–411.

[702] Urabe, M., Convergence of numerical iteration in solution of equations, *J. Sci. Hiroshima Univ., Ser. A*, **19**, (1976), 479–489.

[703] Uzawa, H., The stability of dynamic processes, *Econometrica*, **29**, (1961), 617–631.

[704] Vainberg, M.M., Variational method and method of monotone operators, *Nauka*, M. (1972), (in Russian).

[705] Van Den Dries, L., Miller, C., Geometric categories and O–minimal structures, *Duke Math.*, **84**, (1996), 497–540.

[706] Vandergraft, J.S., Newton's method for convex operators in partially ordered spaces, *SIAM J. Numer. Anal.*, **4**, (1967), 406–432.

[707] Varga, R.S., *Matrix iterative analysis*, Prentice–Hall, Englewood Cliffs, NJ, 1962.

[708] Verma, R.U., A class of projection–contraction methods applied to monotone variational inequalities, *Applied Mathematics Letters*, **13**, (2000), 55–62.

[709] Verma, R.U., General class of relaxed pseudococoercive nonlinear variational inequalities and relaxed projection methods, *Adv. Nonlinear Var. Inequal.,* **8**, (2), (2005), 131–140.

[710] Verma, R.U., Nonlinear pseudococoercive variational problems and projection methods, *Nonlinear Funct. Anal. Appl.,* **10**, (4), (2005), 641–649.

[711] Verma, R.U., General convergence analysis for two-step projection methods and applications to variational problems, *Appl. Math. Lett.,* **18**, (11), (2005), 1286–1292.

[712] Verma, R.U., General projection systems and relaxed cocoercive nonlinear variational inequalities, *ANZIAM J.,* **49**, (2), (2007), 205–212.

[713] Verma, R.U., Existence theorems for a class of generalized vector variational inequality problems, *Fixed point theory and applications,* Vol. 6, Nova Sci. Publ., New York, (2007), 165–170.

[714] Verma, R.U., Two–step models for projection methods and their applications, *Math. Sci. Res. J.,* **11**, (6), (2007), 444–453.

[715] Vertgeim, B.A., On conditions for the applicability of Newton's method, *(Russian) Dokl. Aknd. N., SSSR,* **110**, (1956), 719–722.

[716] Vertgeim, B.A., On some methods for the approximate solution of nonlinear functional equations in Banach spaces, Uspekhi Mat. Nauk, 12 (1957), 166–169 (in Russian); English transl.: *Amer. Math. Soc. Transl.,* **16**, (1960), 378–382.

[717] Vertgeim, B.A., Optimal alternation of the fundamental and the modified Newton–Kantorovich process, (Russian), Optimal. Planirovanie, 17, (1970), 10–31.

[718] Vertgeim, B.A., Certain generalizations of the Newton–Kantorovich method of approximate solution of nonlinear operator equations, (Russian), *Optimizacija Vyp.,* **16**, (33), 180, (1975), 84–90.

[719] Walker, H.F., Watson, L.T., Least–change Secant update methods for underdetermined systems, *SIAM J. Numer. Anal.,* **27**, (1990), 1227–1262.

[720] Wang, J.H., Li, C., Kantorovich's theorem for Newton's method on Lie groups, *J. Zhejiang University Science A,* **8**, 6, (2007), 978–986.

[721] Wang, X., Convergence of Newton's method and inverse function theorem, *Math. Comp.,* **68**, (1999), 169–186.

[722] Wang, X.H., Convergence on the iteration of Halley family in weak condition, *Chinese Science Bulletin,* **42**, 7, (1997), 552–555.

[723] Wang, X.H., Li, C., Convergence of Newton's method and uniqueness of the solution of equations in Banach spaces, II, *Acta Math. Sin. (Engl. Ser.),* **19**, (2003), 405–412.

	References	337

[724] Wang, Z., Semilocal convergence of Newton's method for finite–dimensional variational inequalities and nonlinear complementarity problems, Doctoral thesis, Fakultät für Mathematik, Universität Karlsruhe, Germany, June 2005.

[725] Wang, Z.Y., Extensions of Kantorovich theorem to variational inequality problems, *Z. Angew. Math. Mech.*, **88** (2008), 179–190.

[726] Wang, Z.Y., Extensions of the Newton–Kantorovich theorem to variational inequality problems, preprint 2009,
http://math.nju.edu.cn/~zywang/paper/Kantorovich_VIP.pdf.

[727] Wang, Z., Shen, Z., Kantorovich theorem for variational inequalities, *Applied Mathematics and Mechanics*, **25**, (2004), 1291–1297.

[728] Wedin, P.A., Perturbation theory for pseudo–inverse, *BIT*, **13** (1973), 217–232.

[729] Werner, W., Über ein Verfahren der Ordung $1 + \sqrt{2}$ zur Nullstellenbestmmung, Numer. Math., 32, (1979), 333–342.

[730] Werner, W., Newton–like method for the computation of fixed points, *Comput. Math. Appl.*, **10**, 1, (1984), 77–86.

[731] Wolfe, M.A., A quasi–Newton method with memory for unconstrained function minimization, *J. Inst. Math. Appl.*, **15**, (1975), 85–94.

[732] Wolfe, M.A., Extended iterative methods for the solution of operator equations, *Numer. Math.*, **31**, (1978), 153–174.

[733] Wolfe, M.A., An existence–convergence theorem for a class of iterative methods, *J. Optim. Theory Appl.*, **31**, (1), (1980), 125–129.

[734] Wolfe, M.A., A modification of Krawczyk's algorithm, *SIAM J. Numer. Anal.*, **17**, (3), (1980), 376–379.

[735] Wolfe, M.A., On the convergence of some methods for determining zeros of order–convex operators, *Computing*, **26**, (1), (1981), 45–56.

[736] Wright, S.J., Primal–Dual Interior–Point Methods, *SIAM*, Philadelphia, 1997.

[737] Wu, J.W., Brown, D.P., Global asymptotic stability in discrete systems, *J. Math. Anal. Appl.*, **140**, 1, (1989), 224–227.

[738] Xiao, B., Harker, P.T., A nonsmooth Newton method for variational inequalities, I: theory, *Math. Programming*, **65**, (1994), 151–194.

[739] Xiao, B., Harker, P.T., A nonsmooth Newton method for variational inequalities: II: numerical results, *Math. Programming*, **65**, (1994), 195–216.

[740] Xu, X., Li, C., Convergence of Newton's method for systems of equations with constant rank derivatives, *J. Comput. Math.*, **25** (2007), 705–718.

[741] Xue, Z.Q., Zhou, H.Y., Cho, Y.J., Iterative solution of nonlinear equations for m–accretive operators in Banach spaces, *J. Non. Con. Anal.*, **1**, 3, (2000), 313–320.

[742] Yakoubsohn, J.C., Finding zeros of analytic functions : α–theory for Secant type method, *J. of Complexity,* **15**, (1999), 239–281.

[743] Yamamoto, T., Error bounds for Newton's process derived from the Kantorovich conditions, *Japan J. Appl. Math.*, **2**, (1985), 285–292.

[744] Yamamoto, T., Newton's method and its applications, (Japanese) *Sugaku Expositions*, **1**, (1988), no. 2, 219–238, Sūgaku, 37, (1), (1985), 1–15.

[745] Yamamoto, T., A unified derivation of several error bounds for Newton's process, Proceedings of the international conference on computational and applied mathematics, (Leuven, 1984), *J. Comput. Appl. Math.*, **12/13**, (1985), 179–191.

[746] Yamamoto, T., Error bounds for Newton–like methods under Kantorovich type assumptions, *Japan J. Appl. Math.*, **3**, (2), (1986), 295–313.

[747] Yamamoto, T., A convergence theorem for Newton's method in Banach spaces, *Japan J. Appl. Math.*, **3**, (1), (1986), 37–52.

[748] Yamamoto, T., *Error bounds for Newton's method under the Kantorovich assumptions,* The merging of disciplines: new directions in pure, applied, and computational mathematics (Laramie, Wyo., 1985), Springer, New York, (1986), 197–208.

[749] Yamamoto, T., A convergence theorem for Newton–like methods in Banach spaces, *Numer. Math.*, **51**, (1987), 545–557.

[750] Yamamoto, T., Uniqueness of the solution in a Kantorovich–type theorem of Häussler for the Gauss–Newton method, *Japan J. Appl. Math.*, 6, (1), (1989), 77–81.

[751] Yamamoto, T., Historical developments in convergence analysis for Newton's and Newton–like methods, Numerical analysis 2000, Vol. IV, Optimization and nonlinear equations, *J. Comput. Appl. Math.*, **124**, (1–2), (2000), 1–23.

[752] Ye, Y., *Interior Point Algorithms: Theory and Analysis,* Wiley–Interscience Series in Discrete Mathematics and Optimization, John Wiley and Sons, 1997.

[753] Ypma, T.J., Numerical solution of systems of nonlinear algebraic equations, Ph. D. thesis, Oxford, 1982.

[754] Ypma, T.J., Affine invariant convergence results for Newton's methods, *BIT*, **22**, (1982), 108–118.

[755] Ypma, T.J., The effect of rounding error on Newton–like methods, *IMA J. Numer. Anal.*, **3**, (1983), 109–118.

[756] Ypma, T.J., Convergence of Newton–like iterative methods, *Numer. Math.*, **45**, (1984), 241–251.

References 339

[757] Ypma, T.J., Local convergence of inexact Newton methods, *SIAM J. Numer. Anal.*, **21**, 3, (1984), 583–590.

[758] Yau, L., Ben–Israel, A., The Newton and Halley methods for complex roots, *Amer. Math. Monthly,* **105**, 9, (1998), 806–818.

[759] Yohai, V.J., High breakdown-point and high efficiency robust estimates for regression. *Ann. Statist.,* **15**, (1987), 642-656.

[760] Yohai, V.J., Zamar, R., High Breakdown–Point estimates of Regression by means of the minimization of an efficient scale, *J. Amer. Stat. Assoc.,* **83**, 402, (1988), 406–413.

[761] Zaanen, A.C., *Linear analysis,* North–Holland Publ., Amsterdam, 1953.

[762] Zabrejko, P.P., K–metric and K–normed linear spaces: A survey, *Collect. Math.,* **48**, (4–6), (1997), 825–859.

[763] Zabrejko, P.P., Majorova, N.L., On the solvability of nonlinear Uryson integral equations (Russian), *Kach. Pribl. Metody Issled. Oper. Uravn.,* **3**, (1978), 61–73.

[764] Zabrejko, P.P., Nguen, D.F., The majorant method in the theory of Newton–Kantorovich approximations and the Pták error estimates, *Numer. Funct. Anal. Optim.,* **9**, (1987), 671–684.

[765] Zabrejko, P.P., Zlepko, P.P., *On majorants of Uryson integral operators (Russian),* Kach. Pribl. Metody Issled. Oper. Uravn., 8, (1983), 67–76.

[766] Zehnder, E.J., A remark about Newton's method, *Comm. Pure Appl. Math.,* **27**, (1974), 361–366.

[767] Zeidler, E., *Nonlinear Functional Analysis and Its Applications,* I: Fixed Point Theorems, Springer–Verlag, New–York, 1986.

[768] Zhang, C., Mapes, B.E., Soden, B.J., Bimodality in Tropical Water Vapour, *Quarterly Journal of the Royal Meteorological Society,* **1**(29), (2003), 2847–2866.

[769] Zhao, F., Wang, D., The theory of Smale's point estimation, and its applications, *J. Comput. Appl. Math.,* **60**, (1995), 253–259.

[770] Zinčenko, A.I., *Some approximate methods for solving operation equations with non–differentiable operators,* Dopovidi Akad. Nauk Ukrain. RSR, (1963), 156–161.

[771] Zlepko, P.P., Migovich, F.M., an application of a modification of the Newton–Kantorovich method to the approximate construction of implicit functions (Ukrainian), *Ukrainskii Mathematischeskii Zhürnal,* **30**, 2, (1978), 222–226.

[772] Zuhe, S., Wolfe, M.A., A note on the comparison of the Kantorovich and Moore theorems, *Nonlinear Anal.,* **15**, 3, (1990), 229–232.

Glossary of Symbols

$\mathbb{R}^n$	real n–dimensional space
$\mathbb{C}^n$	complex n–dimensional space
$X \times Y, X \times X = X^2$	Cartesian product space of X and Y
$e^1, \ldots, e^n$	the coordinate vectors of $\mathbb{R}^n$
$x = (x_1, \ldots, x_n)^T$	column vector with component x_i
x^T	the transpose of x
$\{x_n\}_{n \geq 0}$	sequence of points from X
$\|\cdot\|$	norm on X
$\|\cdot\|_p$	L_p norm
$|\cdot|$	absolute value symbol
$\angle(u, v)$	the angle between two vectors u and v
$x \longrightarrow x^\star$	x converges to $x^\star$ strongly
$\liminf$	lower limit for real sequence
$\limsup$	upper limit for real sequence
$\prec$	preference relation
$\dim$	dimension
(x, y)	set $\{z \in X \mid z = tx + (1-t)y,\ t \in [0,1]\}$
$U(x_0, R)$ or $\mathbb{B}_R(x_0)$	open ball $\{z \in X \mid \|x_0 - z\| < R\}$
$\overline{U}(x_0, R)$ or $\overline{\mathbb{B}}_R(x_0)$	closed ball $\{z \in X \mid \|x_0 - z\| \leq R\}$
$U(R) = U(0, R)$	ball centered at the zero element in X and of radius R
$U, \overline{U}$	open, closed balls, respectively no particular reference to X, x_0 or R
$M = \{m_{ij}\}$	matrix $1 \leq i, j \leq n$
M^{-1}	inverse of M
M^+	generalized inverse of M

Glossary of Symbols

$\det M$ or $|M|$	determinant of M
M^k	the kth power of M
$\operatorname{rank} M$	rank of M
I or $\mathbf{I}$	identity matrix (operator)
L	linear operator
L^{-1}	inverse
$f : D \subseteq X \to Y$	an operator with domain D included in X, and values in Y
$\partial f(x)$	the subdifferential of f at x
$f'(x;d)$	the generalized directional derivative of f at x in the direction d
$f'(x)$ or $\nabla f(x)$, $f''(x)$ or $\nabla^2 f(x)$	first, second Fréchet–derivatives of f evaluated at x
$F : D \subseteq X \rightrightarrows Y$	a set–valued mapping with domain D included in X, and values in the subsets of Y
$\operatorname{gph} F$	graph of F
$\operatorname{dist}(x, A)$	a distance of x to A
$F^{-1} : Y \rightrightarrows X$	a inverse mapping to $F : X \rightrightarrows Y$

Index

τ–estimation, 265, 268, 270, 272, 275

(CNTM) method, *see* Chebyshev–Newton–type method

(CSTM) method, *see* Chebyshev–Secant–type method

(DMCMSM) method, *see* Directional maximal–component–modulus Secant method

(DNGSM) method, *see* Directional near–gradient Secant method

(DNLM) method, *see* Directional Newton–like method

(DNM) method, *see* Directional Newton method

(DRNLM) method, *see* Directional residual Newton–like method

(DSM) method, *see* Directional Secant method

(NGM) method, *see* Newton–Gauss method

(NJM) method, *see* Newton–Josephy method

(NM) method, *see* Newton's method

(NTM) method, *see* Newton–type method

(SM) method, *see* Secant method

(TSDNM) method, *see* Two–step directional Newton method

(TSNM) method, 4

(TSNTM) method, *see* Two–step Newton–type method

Abelian group, 172, 174, 175, 177, 179, 180

Baker–Campbell–Hausdorff formula, 172

Banach Lemma, 14, 23, 53, 54, 67, 69, 71, 95, 105, 117, 141, 174, 184, 190, 192, 197, 202, 274

Banach space, 3, 13, 16, 23, 37, 39, 49, 54, 61, 70, 71, 73, 96, 99, 105, 123, 124, 137–139, 149, 151, 159, 160, 165, 171, 195, 247, 272, 275

Banach space with convergence structure, 137

biomedical growth, 286

biomedical problem, 283

center–Lipschitz condition, *see* Lipschitz's condition

Chandrasekhar equation, 3

Chebyshev–Newton–type method, 61

Chebyshev–Secant–type method, 61

Clarke's Jacobian, 41

coderivative, 39

constant rank derivatives, 109

Directional maximal–component–modulus Secant method, 222

Directional near–gradient Secant method, 222

Directional Newton method, 222, 233, 245

Directional Newton–like method, 233

Directional Newton–type method, 233

Directional residual Newton–like method, 234

Directional Secant method, 222

Directional Secant–type Method, 221

Directional two–step method, 245

directionnally differentiable, 41

divided difference, 38, 61, 62, 65, 72, 99

Index

eff ciency index, 3, 4
excess, 39, 44

f nite element, 159, 163, 164
f xed point, 39, 43, 45
f xed points theorem, 38, 39
Fréchet differentiable, 3, 13, 21, 29, 38, 49, 50, 55, 59, 61, 65, 70, 72, 73, 84, 94, 99, 109, 116, 120, 159, 205, 209, 213, 214, 217, 218

generalized equation, 37, 38
geodesic, 166
Green function, 28
Green's kernel, 3

Hölder's condition, 123, 180
Hausdorff–Pompeiu excess, 39
Hilbert space, 195
Hyperbolastic differential equation, 285
Hyperbolastic equation, 288
Hyperbolastic growth, 283–289

indicator function, 37
interior point methods, 155–157

Jacobian, 41

LCP problem, 155
Lie group, 171, 172
Lipschitz's condition, 49, 159, 160, 164, 173, 174, 177
Lipschitz–like property, *see* pseudo–Lipschitz
LP problem, 155

mesh independence principle, 195
metric regularity, 39

Newton's method, 38, 155, 159, 171, 173, 174, 177, 195
Newton–Gauss method, 109
Newton–Josephy method, 205
Newton–Kantorovich condition, 26–28, 34, 58, 80, 82, 90, 109, 121, 123, 135, 151, 155, 159, 163, 177, 193, 199, 215, 216, 222, 245

Newton–Kantorovich Theorem, 49, 59, 81, 82, 90, 155, 156, 159, 162, 168, 181, 182, 187, 198
Newton–type method, 4, 99
nonlinear integral equation, *see* Chandrasekhar equation

pseudo–Lipschitz, 39, 42–46

QP problem, 155

recurrent functions, 3, 4, 73, 99, 107, 109, 110, 123, 137, 146, 205, 206, 233, 245, 247, 260
recurrent sequences, 245, 247, 248, 251, 254
regression model, 265
Riemannian manifold, 165, 168, 171
Riemannian metric, 172
Riemannian–Newton method, 165

Secant method, 38, 62
semidef nite program, 187
semilocal convergence, 3, 4, 13, 21, 24, 49, 50, 55, 57, 61, 62, 65, 66, 70, 73, 76, 84, 85, 94, 99–101, 104, 109, 110, 116, 123, 127, 134, 137, 139, 149, 157, 165, 168, 176, 177, 180, 181, 190, 191, 205, 206, 209, 221–223, 229, 233–235, 239, 245–247, 254, 260
semismooth operator, 40, 47
semismoothing Newton method, 47
set–valued mapping, 37–39, 42, 43, 45, 46
shape optimization, 195
slant consistency property, 46
slant differentiability, 37–42, 45, 46
slanting function, 40–42, 46
smoothing Newton method, 46
Steffensen's method, 38
subdifferential operator, 37

Tabaistic regression, 277
tangent space, 171
The shadowing lemma, 181, 183

Two–step directional Newton method, 247
Two–step Newton–type method, 3, 4

variational inclusion, 37
variational inequality, 37, 38, 205, 206
vector f eld, 165, 172

Zabrejko–Zincenko conditions, 99